Lecture Notes in Bioinformat

Edited by S. Istrail, P. Pevzner, and M. W

Subseries of Lecture Notes in Computer Science

Vincent Danos Vincent Schachter (Eds.)

Computational Methods in Systems Biology

International Conference CMSB 2004
Paris, France, May 26 - 28, 2004
Revised Selected Papers

Springer

Volume Editors

Vincent Danos
Université Paris 7
Equipe PPS Case 7014
2 place Jussieu, 75251 Paris Cedex 05, France
E-mail: Vincent.Danos@pps.jussieu.fr

Vincent Schachter
CNRG Genoscope
2 rue Gaston Cremieux, 91000 Evry, France
E-mail: vs@genoscope.cns.fr

Library of Congress Control Number: 2005922242

CR Subject Classification (1998): I.6, D.2.4, J.3, H.2.8, F.1.1

ISSN 0302-9743
ISBN 3-540-25375-0 Springer Berlin Heidelberg New York

Springer is a part of Springer Science+Business Media

springeronline.com

Printed in Germany

Typesetting: Camera-ready by author, data conversion by Scientific Publishing Services, Chennai, India
Printed on acid-free paper SPIN: 11409083 06/3142 5 4 3 2 1 0

Preface

The Computational Methods in Systems Biology (CMSB) workshop series was established in 2003 by Corrado Priami. The purpose of the workshop series is to help catalyze the convergence between computer scientists interested in language design, concurrency theory, software engineering or program verification, and physicists, mathematicians and biologists interested in the systems-level understanding of cellular processes. Systems biology was perceived as being increasingly in search of sophisticated modeling frameworks whether for representing and processing system-level dynamics or for model analysis, comparison and refinement. One has here a clear-cut case of a must-explore field of application for the formal methods developed in computer science in the last decade.

This proceedings consists of papers from the CMSB 2003 workshop. A good third of the 24 papers published here have a distinct formal methods origin; we take this as a confirmation that a synergy is building that will help solidify CMSB as a forum for cross-community exchange, thereby opening new theoretical avenues and making the field less of a potential application and more of a real one. Publication in Springer's new Lecture Notes in Bioinformatics (LNBI) offers particular visibility and impact, which we gratefully acknowledge.

Our keynote speakers, Alfonso Valencia and Trey Ideker, gave challenging and somewhat humbling lectures: they made it clear that strong applications to systems biology are still some way ahead. We thank them all the more for accepting the invitation to speak and for the clarity and excitement they brought to the conference. We also wish to thank René Thomas for his keynote lecture on recent mathematical advances in the qualitative analysis of genetic regulation networks. As one can tell from the proceedings, his work has inspired many recent applications of formal methods to the engineering of biological models.

We are glad to take here the opportunity to express our gratitude to the members of the program committee and to the referees for their effort in the paper selection process and for their willingness to participate in the open-minded debate needed given the interdisciplinary nature of the area of computational systems biology. We would also like to thank the authors for their interest in the workshop and for their high-quality submissions and communications.

Finally, we wish to extend our warmest thanks to Monique Meugnier, Catherine Sarlande and Serge Smidtas for their invaluable help in organizing the workshop, and to the participating institutions, Genoscope, Genopole, CNRS, University of Paris 7, and the BioPathways Consortium, which provided financial support.

Conference web-site: `http://www.biopathways.org/CMSB04/`

Vincent Danos
Vincent Schachter

Table of Contents

Long Papers

Short Papers

Invited Contributions

An Explicit Upper Bound for the Approximation Ratio of the Maximum Gene Regulatory Network Problem

Sergio Pozzi[1], Gianluca Della Vedova[2], and Giancarlo Mauri[1]

[1] DISCo, Univ. Milano-Bicocca
[2] Dip. Statistica, Univ. Milano-Bicocca
sergio.pozzi@disco.unimib.it
{giancarlo.mauri, gianluca.dellavedova}@unimib.it

Abstract. One of the combinatorial models for the biological problem of inferring gene regulation networks is the MAXIMUM GENE REGULATORY NETWORK PROBLEM, shortly MGRN, proposed in [2]. The problem is **NP**-hard [2], consequently the attention has shifted towards approximation algorithms, leading to a polynomial-time 1/2-approximation algorithm [2], while no upper bound on the possible approximation ratio was previously known.

In this paper we make a first step towards closing the gap between the best known and the best possible approximation factors, by showing that no polynomial-time approximation algorithm can have a factor better than $1 - \frac{1/8}{1+e^2}$ unless **RP**=**NP**.

1 Introduction

The completion of the Human Genome project [9, 3] has only given more importance to the problem of determining the processes regulating the methabolism of living beings. The knowledge of all genetic sequences of an organism is just the first necessary step in understanding which of these sequences determine how those sequences are actually related to the phenotypes, as it is commonly believed that the dynamics of a living organism is determined throught some complicated and orchestrated interactions between thousands of genes and their products.

A Gene Network can be thought of as a set of molecular components such as genes, proteins and other molecules, interacting to collectively carry out some cellular functions. The advent of DNA microarray technology has led to easily obtaining huge amount of data regarding various aspects of cellular behavior, making possible to identify the interactions occuring among the various elements of a genetic system. Anyway the amount of data does not imply that the overall quality of data is sufficient to understand the various interaction, in fact these data are actually insufficient in granularity to uniquely determine the underlying network of interactions. Building the complex causal gene network of a genetic system on the basis of these sampled data is then a tipical inference and reverse engineering task.

V. Danos and V. Schachter (Eds.): CMSB 2004, LNBI 3082, pp. 1–8, 2005.

A number of different gene network models have been proposed in literature, each of them resorting to some simplifying assumptions either of biological or computational nature. In this paper we will study the boolean network models, where the state of each gene can be only dicotomic, that is active or not active. This model was already recognized to give a valid description of a genetic system in [7]. Boolean models are rich enough to represent interesting interactions among elements and, even if they are sometimes too simplistic [6], they allow to analyze briefly more complex systems. Actually this fact does not detract to the result of our paper, as we will show that a certain formulation of the gene network inference problem cannot be approximated efficiently, and this inapproximability result is very likely to be extended to more refined models.

In [1] modeling genes as boolean switches has allowed to study the problem of reverse engineering the gene networks by devising experiments in which, these switches are strategically manipulated (turned on and off) and then observing the behavior of the whole system. The main limit of this model is that the number of experiments that have to be performed in order to reconstruct a gene network of bounded in-degree D over n genes is $\Omega\left(n^D\right)$. In an other boolean GN model [5], the causal relations among network elements is derived on the basis of the mutual information among them. In our paper we will analyze a particular boolean model introduced in [2]. As will be explained in Sect. 2, this model is based on a simple combinatorial description with some biological evidence.

In [2], the problem of determining the causal relations among network elements has been proved to be **NP**-hard, consequently there has been much attention to designing approximation algorithms for the problem. The best known result in such direction is the 1/2-approximation algorithm of [2] (in this paper the approximation ratio of an algorithm is an upper bound of the ratio between the value of the approximate solution and the value of an optimal solution).

In this paper a first inapproximability result for the gene network inference problem based on this model is derived, by showing that it is unlikely that there exists an efficient approximation algorithm that can guarantee to obtain a $1-\frac{(1/8)}{1+e^2}$ ratio.

Our paper is organized as follows: initially we will present formally the MGRN problem, together with some known approximability results.

Successively we will give a probabilistic reduction from instances of MaxE3Sat to MGRN ones. This reduction uses a previuolsy known reduction from instances of MaxE3Sat to instances of MaxE3Sat-B, originally proposed by Trevisan [8]. Our reduction extends the Trevisan reduction to the MGRN problem. We will conclude the paper by showing that a consequence of our reduction is that no polynomial-time approximation algorithm for the MGRN problem with approximation ratio $1-\frac{1/8}{1+e^2}$ can exist, unless **RP**=**NP**.

2 The Maximum Gene Regulatory Problem

A Genetic Network in which an element can only activate or inhibit other elements, can be viewed as a directed graph in which the nodes represent the genes and arcs represent the interactions between genes. Moreover each arc (v, w) is labeled by A or I, according to the fact the gene represented by v *activates* or *inhibits* the activity of the gene represented by w. Such graphs can be built using experiments data relative to gene expression dynamics [2]. In order to suggest the causal genetic network on the basis of the edge labeled directed graph, activating/inhibiting edges representing spurious interactions must be deleted. The task of deleting spurious interactions has to be done with the following constraints:

- A gene (a node on the graph) cannot be both of activating and inhibiting type.
- The number of genes that are *controlled* (that is vertices that have both A-labeled and I-labeled incoming arcs) must be maximized.

Both kinds of constraints find their justification in biological evidence and consistency with the parsimonious principle. As a final result of these two guiding assumptions a combinatorial optimization problem on graphs has been defined in [2]

Problem 1. MAXIMUM GENE REGULATION PROBLEM (shortly MGRN). The instance is a directed graph $G = (V, E)$, where each arc is labeled by either A or I. The goal is assigning to each vertex a label that is either A or I, so that, after deleting all arcs (v, w) with label different from that of v, the number of *controlled* vertices is maximized.

It is hopeless to devise efficient exact algorithm for the MGRN problem, since the problem is **NP**-hard, even for directed acyclic graphs of constant in/out-degree [2]. For this reason in the last few years the attention has been turned into finding efficient approximate solutions, showing that the solution having at least one half of the optimal number of controlled vertices, can be found in polynomial time [2], but it was not previously known if a polynomial-time approximation scheme (PTAS) was possible for such problem.

In our paper we will settle the question, by proving that it is not possible to describe a polynomial-time approximation algorithm with guaranteed ratio strictly better than $1 - \frac{1/8}{1+e^2}$, unless **RP=NP**.

3 A Better Reduction

In [2] it has been proved that MGRN is **NP**-hard. Here we will show a new reduction from MAXE3SAT-B to MGRN; our reduction is stronger, since it allows to prove a better inapproximation results.

The reduction associates to an instance of MAXE3SAT-B an instance of MGRN as follows: for each clause C_i we have a *clause gadget* consisting of two vertices C_i^1, C_i^2 and the A-labeled arc from C_i^1 to C_i^2. For each variable x_i we have the *variable gadget* consisting of two vertices x_i^T, x_i^F and no arc. If the total number of variables is n, then we have also $\lfloor n/2 \rfloor$ *assignment gadgets*, each gadget is made of $2(B + 1)$ vertices, half of which are labeled red and half are labeled blue. For $1 \leq i \leq \lfloor n/2 \rfloor$ all vertices x_{2i-1}^T, x_{2i-1}^F, x_{2i}^T, x_{2i}^F have an outgoing arc to each of the vertices of the i-th assignment gadgets. More precisely all red vertices have A-labeled arcs incoming from x_{2i-1}^T and x_{2i-1}^F and I-labeled arcs incoming from x_{2i}^T and x_{2i}^F, while all blue vertices have I-labeled arcs incoming from x_{2i-1}^T and x_{2i-1}^F and A-labeled arcs incoming from x_{2i}^T and x_{2i}^F. An assignment gadget and the two corresponding vertex gadgets are represented in Fig. 1.

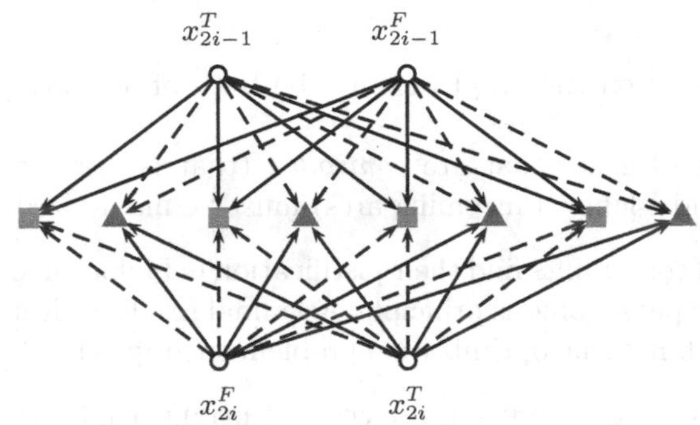

Fig. 1. Example of vertex and assignment gadget, red vertices are represented by squares and blue vertices by triangles. A-labeled edges in solid lines, I-labeled edges in dashed lines

Actually there is a minor problem if the number of variables is odd. In this case the last assignment gadget is different, as three variables are connected to it, as shown in Fig. 2.

The reduction can now be completed with the encoding of each clause $C_i = x_{i_1}^{\alpha_{i_1}} \vee x_{i_2}^{\alpha_{i_2}} \vee x_{i_3}^{\alpha_{i_3}}$, where each exponent α_{i_j} is equal to T or F, according to the fact that the corresponding variable is or is not negated in the clause. For each clause C_i there are three I-labeled arcs incoming in C_i^2 and outgoing from vertex gadgets associated to the vertices appearing in the clause, more precisely the arcs are outgoing from the actual vertices encoding the variable and the fact that the variable is or is not negated in the formula. Formally for each clause $C_i = x_{i_1}^{\alpha_{i_1}} \vee x_{i_2}^{\alpha_{i_2}} \vee x_{i_3}^{\alpha_{i_3}}$ there are the three arcs $(x_{i_1}^{\alpha_{i_1}}, C_i^2)$, $(x_{i_2}^{\alpha_{i_2}}, C_i^2)$, $(x_{i_3}^{\alpha_{i_3}}, C_i^2)$. In Fig.3 is represented an example of encoding.

In the following of the paper we will denote with F an instance of MaxE3Sat-B and with G the instance of MGRN that is associated to F with the reduction we have just described. Moreover we will denote with $opt(F)$ and $opt(G)$ respectively the maximum number of clauses of F that are satisfiable by a single assignment and the optimum of the instance G.

The following lemma is the foundation of our inapproximability result.

Lemma 3.1. *Let F be an instance of the* MaxE3Sat-B *problem with n_B boolean variables and m_B clauses, and let G be the instance of* MGRN *that is associated to F. Then it is possible to associate to any solution of F with value x a solution of G of value $n_B(B+1)+x$. Vice versa it is possible to associate to any solution of G of value $n_B(B+1)+x$ a solution of F of value at least x.*

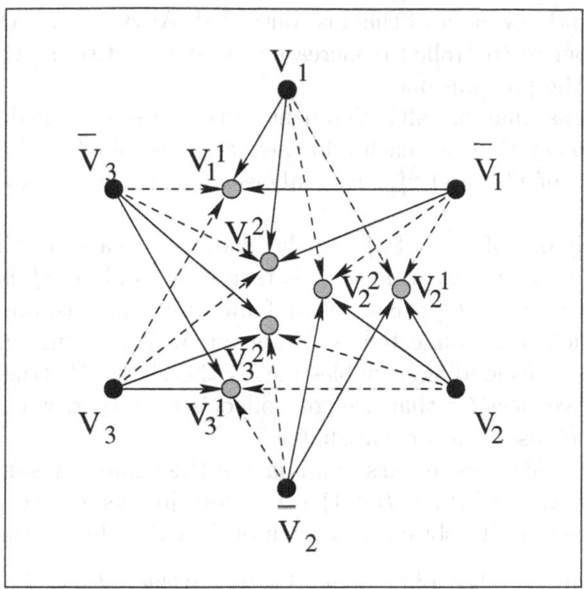

Fig. 2. Three variables connected to an assignment gadget with $B = 2$. A-labeled edges in solid lines, I-labeled edges in dashed lines

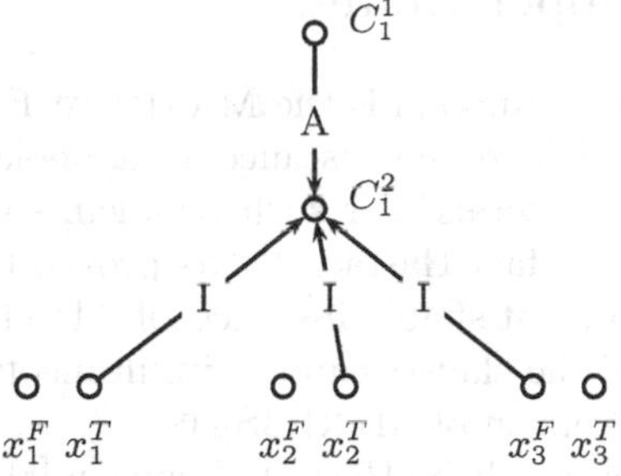

Fig. 3. Encoding of the clause $C_1 = x_1 \vee x_2 \vee \neg x_3$

Proof. Initially assume we have a solution of F. Please notice that the vertices that are relevant in computing the value of the solution are the vertices which have both A-labeled and I-labeled arcs and, by construction, the only such vertices are the assignment vertices and the vertices C_j^2.

Without loss of generality we can assume that all assignment vertices must be controlled. In fact if all assignment vertices of two certain variable x_i are not controlled and i is odd, then we obtain a better solution by A-labeling x_i^T, x_{i+1}^T and I-labeling x_i^F, x_{i+1}^F (if i is even, then we have to A-label x_i^T, x_{i-1}^T and I-label x_i^F, x_{i-1}^F). Now all $2B+2$ assignment vertices of x_i are now controlled, but we do not know how many of the $2B$ clause gadgets to which x_i and x_{i+1} are connected are controlled. In the worst case they all were controlled before the modification, and now none of them is controlled. Anyway after the modification the total number of controlled is increased by at least two, so the new solution is better than the previous one.

Now we can assume that all assignment vertices are controlled. This condition is equivalent to say that, for each odd i, exactly one of x_i^T and x_i^F is A-labeled and exactly one of x_{i+1}^T and x_{i+1}^F is A-labeled, which can be shown by trying all possibilities.

Since exactly one of x_i^T and x_i^F is A-labeled, we can assume that the labeling encodes a truth assignment, that is x_i is true if and only if x_i^T is A-labeled. By construction each vertex C_j^2 is controlled if and only if at least one of the variable vertices to which it is connected is A-labeled, which in turn means that the assignment of corresponding variables makes the clause C_j true. Consequently the number of vertices C_j^2 that are controlled is exactly x, where x is equal to the number of clauses that are satisfiable.

Now we are able to prove the second part of the lemma. Assume that we have a solution of G with value $n_B(B+1)+x$. Then, just as for the first part of the proof, it is immediate to obtain a solution of F with value at least x.

An immediate corollary of Lemma 3.1 is that if the instance F of MaxE3Sat-B is satisfiable, then the instance G of MGRN has optimum $n_B(B+1)+m_B$.

4 An Explicit Upper Bound

The starting point of our reduction is the Maximum Exact 3-Satisfiability (MaxE3Sat) problem, where the instance is a boolean formula where each clause contains exactly 3 literals. For such problem some strong inapproximability results are known; in fact Håstad [4] has proved that for every $\delta > 0$, it is **NP**-hard to distinguish a satisfiable instance of MaxE3Sat from an instance where at most $7/8+\delta$ of the clauses can be simultaneously satisfied; we will call such problem a *gapped* version of MaxE3Sat.

Bulding upon the last result by Håstad, Trevisan [8] has devised a stochastic reduction from an instance I of MaxE3Sat to an instance F of MaxE3Sat-B, that is in F each literal appears in at most B clasuses. In the following we will denote by I, F and G respectively an instance of MaxE3Sat, MaxE3Sat-B and MGRN. Morever we will denote by n, m respectively the number of variables and of clauses of I, by n_B, m_B respectively the number of variables and of clauses of F. Consequently the maximum number of vertices of G that might be controlled is $(B+1)n_B+m_B$. The probabilistic reduction of [8] has the following properties:

1. if I is satisfiable then F is satisfiable;
2. for any sufficiently large B, then with probability at least $3/4-o(1)$ over the random choices made in the construction of F, if there is an assignment that satisfies at least a fraction $7/8+5/\sqrt{B}$ of the clauses of F, then there is an assignment that satisfies at least a fraction $7/8+1/\sqrt{B}$ of the clauses of I; furthermore $m_B > \left(\frac{B}{e^2}-4\right)n_B$.

Following the same ideas of [8], we will give a probabilistic reduction from instances of MAXE3SAT to instances of MGRN. Our reduction is actually a composition of the reduction in [8] (i.e. a reduction from MAXE3SAT to MAXE3SAT-B) and the reduction proposed in Sect. 4 (i.e. a reduction from MAXE3SAT-B to MGRN). Now we are ready to prove the fundamental feature of the our probabilistic reduction from MAXE3SAT to MGRN.

Lemma 4.1. *Let I be an instance of* MAXE3SAT *with n variables and m clauses, and let G be instance of* MGRN *associated to I by our reduction. Then for sufficiently large B and with probability at least $3/4-o(1)$, if there is a label assignment to the vertices of G such that at least $\left(1-\frac{\frac{1}{8}-\frac{5}{\sqrt{B}}}{1+\frac{(B+1)e^2}{(B-4e^2)}}\right)((B+1)n_B+m_B)$ vertices are actually controlled, then at least $\left(7/8+1/\sqrt{B}\right)m$ clauses of I can be satisfied.*

Proof. Let I be an instance of MAXE3SAT and let us suppose there exists a label assignment a to vertices of G such that its measure $m(a)$ is at least $\left(1-\frac{\frac{1}{8}-\frac{5}{\sqrt{B}}}{1+\frac{(B+1)e^2}{(B-4e^2)}}\right)((B+1)n_B+m_B)$, then $m_B > \left(\frac{B}{e^2}-4\right)n_B$ with probability at least $3/4-o(1)$. Consequently

$$\begin{aligned} m(a) &\geq \left(1-\frac{\frac{1}{8}-\frac{5}{\sqrt{B}}}{1+\frac{e^2(B+1)}{B-4e^2}}\right)((B+1)\,n_B+m_B) > \\ &> \left(1-\frac{\frac{1}{8}-\frac{5}{\sqrt{B}}}{1+(B+1)\,\frac{n_B}{m_B}}\right)((B+1)\,n_B+m_B) = \\ &= \left(1-\frac{\left(\frac{1}{8}-\frac{5}{\sqrt{B}}\right)m_B}{m_B+(B+1)\,n_B}\right)((B+1)\,n_B+m_B) = \\ &= m_b+(B+1)n_B-\frac{1}{8}m_B+\frac{5}{\sqrt{B}}m_B = (B+1)\,n_B+\left(\frac{7}{8}+\frac{5}{\sqrt{B}}\right)m_B \end{aligned}$$

By Lemma 3.1 there exists a solution of F satisfying at least $(7/8+5/\sqrt{B})m_B$ clauses. Applying the second property of the reduction in [8], there exists (with probability at least $3/4-o(1)$) a solution of I satisfying at least $(7/8+1/\sqrt{B})m$ clauses.

The following corollary is our main contribution.

Corollary 4.2. *For any $\delta > 0$, it is not possibile to approximate the* MGRN *problem within a factor $1-\frac{1/8}{1+e^2}+\delta$, unless* ***NP=RP****.*

Proof. First notice that it is not possible to approximate the MGRN problem within a factor $1-\frac{\frac{1}{8}-\frac{5}{\sqrt{B}}}{1+\frac{(B+1)e^2}{(B-4e^2)}}$ unless **NP=RP**. Otherwise we could solve the gapped version MaxE3Sat with a polynomial-time probabilistic algorithm. In fact let I be an instance of such gapped version, and let G be the instance of MGRN associated to I. If the solution returned by the approximation algorithm has value more than $\left(1-\frac{\frac{1}{8}-\frac{5}{\sqrt{B}}}{1+\frac{(B+1)e^2}{(B-4e^2)}}\right)((B+1)n_B+m_B)$ then, by Lemma 4.1, with probability $3/4-o(1)$, I is a satisfiable instance of gapped MaxE3Sat. Otherwise the solution returned by the algorithm has value at most $\left(1-\frac{\frac{1}{8}-\frac{5}{\sqrt{B}}}{1+\frac{(B+1)e^2}{(B-4e^2)}}\right)((B+1)n_B+m_B)$ consequently, with probability 1, at most a fraction $7/8+\delta$ of the clauses are satisfied, hence we would have an algorithm in **RP**for the gapped version of MaxE3Sat. This would imply that **NP**=**RP**.

Without loss of generality we can restrict our interest only to large values of B. Since $\lim_{B\to\infty} 1-\frac{\frac{1}{8}-\frac{5}{\sqrt{B}}}{1+\frac{(B+1)e^2}{(B-4e^2)}} = 1-\frac{1/8}{1+e^2}$, for any $\delta>0$ taking a sufficiently large B completes the proof.

Acknowledgments

This work has been partially supported by FIRB project "Bioinformatica per la Genomica e la Proteomica".

References

1. T. Akutsu, S. Kuhara, O. Maruyama, and S. Miyano. Identification of gene regulatory networks by strategic gene disruptions and gene overexpressions. *Proc. 9th Symp. on Discrete Algorithms (SODA)*, pages 695–702, 1998.
2. T. Chen, V. Filkov, and S. S. Skiena. Identifying gene regulatory networks from experimental data. *Parallel Computing*, 27:317–330, 1999.
3. I. H. G. S. Consortium. Initial sequencing and analysis of the human genome. *Nature*, 409:860–921, February 2001.
4. J. Håstad. Some optimal inapproximability results. *Journal of the ACM*, 48:798–859, 2001.
5. S. Liang, S. Fuhrman, and R. Somogyi. Reveal, a general reverse engineering algorithm for inference of genetic network architectures. *Proc. 5th Pacific Symposium on Biocomputing (PSB)*, pages 18–29, 1998.
6. C. Soulé. Graphic requirements for multistationarity. *ComPlexUs*, 1:123–133, 2003.
7. R. Thomas, A. Gathoye, and L. A. Lambert. A complex control circuit. regulation of immunity in temperate bacteriophage. *European Journal of Biochemistry*, 71:211–227, 1976.
8. L. Trevisan. Non-approximability results for optimization problems on bounded degree instances. *Proc. 33rd Symp. Theory of Computing (STOC)*, pages 453–461, 2001.
9. J. C. Venter, M. D. Adams, E. W. Myers, and et. al. The sequence of the human genome. *Science*, 291:1304–1351, 2001.

Autonomous Mobile Robot Control Based on White Blood Cell Chemotaxis

Matthew D. Onsum and Adam P. Arkin

[1] Department of Mechanical Engineering, University of California, Berkeley,
[2] Department of Bioengineering, University of California, and Physical Biosciences Division, Lawrence Berkeley National Laboratory, Howard Hughes Medical Institute, 1 Cyclotron Road, MS 3-144, Berkeley, CA 94720-1770

Abstract. This paper presents a biologically inspired algorithm to control an autonomous robot tracking a target. The algorithm is designed to mimic the behavior of a human neutrophil, a type of white blood cell that travels to sites of infection and digests bacterial antagonists. Neutrophils are known to be highly sensitive to low levels of chemical stimuli, robust to noise, and are capable navigating unknown terrain, all qualities that would be desired in an autonomous robot. In this paper we model a neutrophil as a collaborative control system, demonstrate the robustness of this algorithm, and suggest a computationally cheap method of implementation. Our simulations show that the performance of the robot is unaffected by constant disturbances and it is robust to random noise levels up to 5 times the tracking signal. Additionally, we demonstrate that this algorithm, as well the current models of neutrophil chemotaxis, are equivalent to a sensor fusion problem that optimizes directional sensing in the presence of noise.

1 Introduction

This paper presents a collaborative control algorithm based on the behavior of a neutrophil, a class of white blood cell. Created in the bone marrow, these cells passively travel through the blood stream, until they sense the chemical traces of an invading bacteria. At this point, they leave the blood stream, crawl through the endothelial cells to the site of infection and digest the intruder. Figure 1 shows a neutrophil about to digest a bacterium [11]. The ability of a neutrophil to move up a chemical gradient is referred to as *chemotaxis*. While this behavior is remarkable it is believed that the underlying mechanisms are simple, that is, there is not a high level algorithm or form of intelligence within the cell. It is more likely that there is a combination of simple controls [5][9]. Despite this assumed simplicity, it is believed that through the evolutionary process these cells have been optimized for their task [9]. This optimal design is what we attempt to reproduce here. The main contributions of this paper are 1.) we extend the work of Goldberg and Chen [2] by deriving a relationship for the performance/robustness tradeoff that was hinted at in their analysis

V. Danos and V. Schachter (Eds.): CMSB 2004, LNBI 3082, pp. 9–19, 2005.

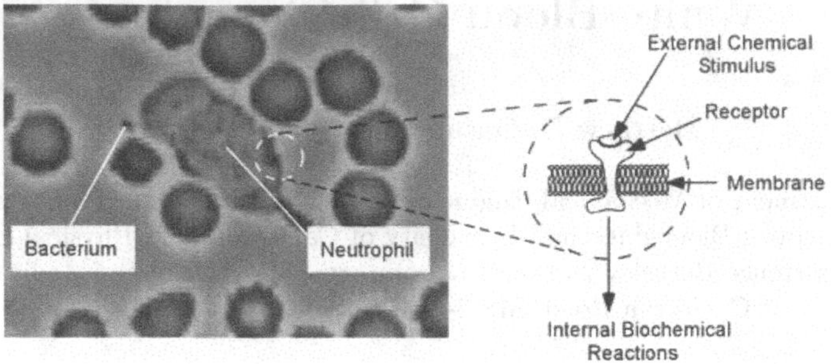

Fig. 1. A Neutrophil tracking a bacterium [11], and a schematic diagram of a trans-membrane receptor

of collaborative control systems 2.) we show that current neutrophil sensing models are a special case of an optimal sensor fusion problem 3.) we suggest a computationally fast method for implementing this algorithm. Before going into our analysis and simulations, we review some of the observed behavior in neutrophils and collaborative control and elucidate how the two relate.

1.1 Neutrophil Review

This section presents a simplified view of the neutrophil sensory and actuation system. Trans-membrane receptors are evenly distributed around the periphery of the cell: their function is to transmit information from the environment to within the cell [8]. When a receptor binds to a signalling chemical it will initiate a series of chemical reactions that instruct the cell to move in the direction of the activated receptor. Figure 1 shows a schematic diagram of a transmembrane receptor with a bound chemical stimulus. Since there are many receptors (around 10,000) there is a competition as to which direction to move. This competition results in a net motion towards the target since the receptors reading the strongest signal (hence closer to the target) will "pull" harder than receptors reading a weaker signal. This sort of sensing scheme can be thought of as a vector sum: the true signal is proportional to the sum of the sensed signal at each receptor times its normal vector.

We hypothesize that this vector sum approach to signal evaluation makes the cell robust to noise. For example, if the chemical concentration is constant than each receptor detects the same signal, and each pulls with same force; therefore the cell will have no net motion. This is consistent with what is observed in the cell– it will only respond to a chemical gradient [13] [14]. Therefore a cell can adapt to any constant level of chemical stimulus and similarly, it is able to reject any constant noise or bias in its environment or sensing pathways. In the case of non-deterministic noise, there will not be perfect noise rejection, but one would expect some filtering or partial noise subtraction. This point was explored with our model.

In addition to noise-robustness, neutrophils have been observed to be very sensitive to changes in chemical stimulus. In fact, it has been shown that they respond to chemical gradients as low as a 2% difference across their length [3]

[14]. This has led many investigators to conclude that there must be some internal amplification within the cell. One hypothesis in that there is receptor coactivation, which means that when one receptor becomes active it sequesters important signalling chemicals towards its neighbors, thus making them more sensitive. The net effect of receptor coactivation is to make the up-gradient receptors more sensitive then the down-gradient receptors, which some authors refer to as "frontness" and "backness", respectively.

1.2 Collaborative Control Review

In collaborative control systems *multiple* sources share control of a single robot [2]. A source is an element that relates information about the environment and current state of the robot with the robot's objective and produces a control output. Sources can take many forms. The robot can be controlled by multiple sensors (*sensor fusion*), control processes (*subsumption*) or human operators. Essentially, collaborative control is an average control based on many sources trying to accomplish the same task. In fact, this is where we see the principle advantage: in a noisy environment the average control will be better than a single control. This advantage arises from the Central Limit Theorem and good engineering sense: multiple measurements will give you a better approximation of the "true signal" [2].

1.3 The Connection

We assume that the cell's ability to detect the direction of a chemical gradient arises from the collaboration of its receptors. Each receptor measures the signal and tries to move the cell in its normal direction. The sum of these controls results in the motion of the cell. As discussed above this "vector sum" methodology will be immune to constant noise. Furthermore, in our simulations we show that while it cannot completely reject random noise, it is able to track and pursue a target in the presence of high noise to signal ratios. We also demonstrate the advantages of source coactivation, which we formulate as an optimal direction sensing problem.

2 Related Work

The study of neutrophil chemotaxis is an active field, yet much about this system remains unknown. Current work focuses on understanding the chemical reactions that relate external stimulus to motion. This is important in understanding and preventing cancer metastasis as well as for engineering drugs that will cause an optimal immune response. The fundamental work on chemotaxis was done by Sally Zigmond in the 1970's [13] and an excellent review of recent work can be found in [14]. Current models of receptor coactivation and its relationship to signal amplification can be found in [3] [4]. The connection between neutrophil chemotaxis and cancer is discussed in [8] and good introductory papers on using engineering approaches in biology are [5] [9].

Much of the inspiration for this work came from Goldberg and Chen's analysis of collaborative control systems [2]. This paper is an excellent introduction to the field and formalizes a measure of robustness due to failing sources. Gerkey [1] attempted to refute the results of Goldberg and Chen, but their results were inconclusive. After careful examination of both papers, it is apparent that their disagreement stems from a misunderstanding of Goldberg and Chen's performance metric. Specifically, Goldberg's metric did not take into account the speed of the robot, where the robot in [1] traveled at a constant velocity.

3 Results

This section analyzes the robustness and performance properties of collaborative control systems. We first consider a system similar to that of Goldberg and Chen [2] and extend their work by formulating the robustness property that was hinted at in their paper. Next we derive a robustness/performance relationship for a collaborative control system by recasting the problem as a constrained optimization. Finally we show that our formulation of the collaborative control problem can be related to current neutrophil models. This relationship is important for two reasons: our robustness/performance analysis applies to neutrophils (this has not been shown in the literature) and second, the neutrophil models provide us with a computationally cheaper algorithm then our optimization.

3.1 Noise Robustness in Distributed Sensing Systems

Derivation. Figure 2 shows a schematic diagram of our proposed system. A robot of unit diameter has N sources distributed around its periphery (dark circles in the figure). Each source senses the local concentration of the surrounding chemical field. The outputs of the individual sensors are then combined (eg, averaged) and sent to the robot's actuators.

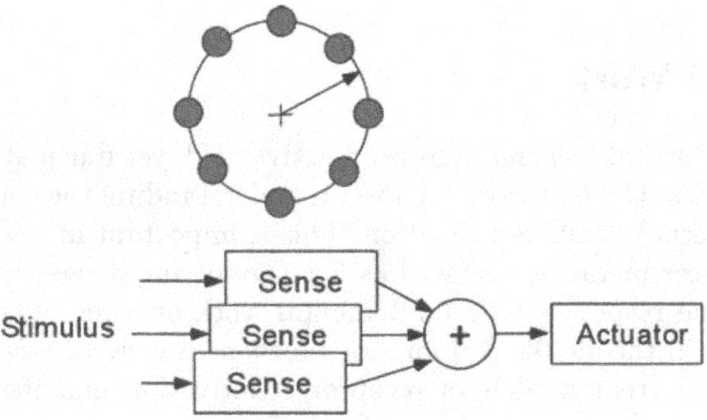

Fig. 2. Schematic diagram of simulation: A circular robot combines the input from its N sensor into a single signal that is then sent an actuator

We start by defining the output of each individual sensor, s_i:

$$s_i = (g_i + \nu_i)\overrightarrow{n_i} \tag{1}$$

$$\nu_i \sim \mathbb{N}(m, \sigma) \tag{2}$$

where g_i is the chemoattractant concentration, ν_i is sensor noise, $\overrightarrow{n_i}$ is the outward pointing normal vector at the i^{th}sensor. We assume that the noise on each sensor, ν_i is an independent normally distributed random variable with mean m and variance σ (equation 2). Our problem to combine the sensors readings so that that our robot moves in the correct direction. A simple and obvious scheme is to add the sensor readings; the resulting vector, $\overrightarrow{s}$, gives the direction and magnitude of the sensed signal. That is:

$$\overrightarrow{s} = \sum_{i=1}^{N} (g_i + \nu_i)\overrightarrow{n_i} \tag{3}$$

$$= \sum_{i=1}^{N} g_i\overrightarrow{n_i} + \sum_{i=1}^{\frac{N}{2}} \xi_i\overrightarrow{n_i} \tag{4}$$

$$\xi_i \sim \mathbb{N}(0, 2\sigma) \tag{5}$$

For large N equations 4 and 5 become:

$$\lim_{N\to\infty} \overrightarrow{s} = \nabla g + \mu \tag{6}$$

$$\mu \sim \mathbb{N}\left(\begin{bmatrix}0\\0\end{bmatrix}, \pi\sigma \mathrm{I}_2\right) \tag{7}$$

where I_2 is the two dimensional identity matrix. Equations 6 and 7 show that for many sensors, $\overrightarrow{s}$ approximates the gradient of the external field plus a zero mean gaussian noise process. This is significant because it shows that the vector sum of the sensors will cancel out any constant disturbances. In this sense, our collaborative control strategy acts as a high-pass filter. We now show the effectiveness of this control strategy with simulations.

Simulation. Our simulation consists of a circular robot with unit radius moving in a linear chemical gradient. The task of the robot is to move 10 units up-gradient (the positive y-direction) of its starting point, and its performance P is measured by the ratio of the minimum path length, L_{min}, to its total path length, L. That is,

$$P = \frac{L_{min}}{L} \tag{8}$$

The signal the robots senses is described by equation 3, where g linearly increases in the y-direction with unity slope. At each time step, the robot calculates $\overrightarrow{s}$ and then moves in that direction 1 unit. Clearly, any deviation from a straight path will decrease the value of P. Figure 3 shows the performance of the algorithm described by equation 3 for various levels sensor noise. The noise added to each

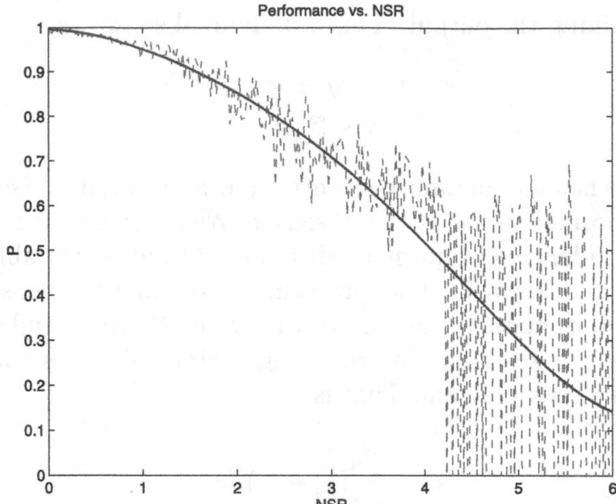

Fig. 3. Performance degradation as a function of random noise-to-signal ratio (NSR) for the algorithm described in equation 3. The red dashed line is taken from 300 experiments and the solid blue line a polynomial fit to the data

sensor is $\nu_i \sim \mathbb{N}\left(0, NSR^2\right)$. Remarkably, the robot is able to reliably accomplish its task (although with diminishing performance)until the $NSR \approx 4$. At this point, the robot is no longer able to distinguish between sensor noise and the true signal and therefore its motion appears brownian.

3.2 Improving Performance

While the algorithm presented above proved to be robust, it is also very conservative– it equally weights the signal from each sensor even though some will have a better NSR then others. In particular, the up-gradient sensors will have have a lower NSR then the down-gradient sensors and therefore they should contribute more to the estimated signal. So instead of taking a vector-sum of the sensor readings, we propose taking a weighted average, where the weighting will be a function of the NSR.

Derivation. We begin by defining our signal s and weighting vector $w \in \mathbb{R}^N$

$$s = w^T(g + \nu) \tag{9}$$

$$\sum_{i=1}^{N} w_i = 1 \tag{10}$$

where $g, \nu \in \mathbb{R}^N$ are the detected signal and noise vectors, respectively. (We have dropped the vector notation for s and will now assume that we are in polar coordinates with the origin at the cell's center.) Our new problem is to choose w

such that we maximize the detected signal in the correct direction. We formulate this optimization as

$$\hat{w} = \arg\max_{w} \left(\mathbf{E}\left[s\right]^2 - var(s)\right) \quad (11)$$

The first term on the right in equation 11 causes w to be high near sensors receiving a high signal, and the second term penalizes the objective function when the variance in the noise is high. Therefore, if there is high uncertainty in the sensor reading then second term in equation 11 prevents the weighting of noisy sensors.

After some manipulation, equation 11 can be put in the following quadratic form:

$$\hat{w} = \arg\max_{w} (w^T A w) \quad (12)$$

$$A = gg^T - \sigma \mathrm{I_N} \quad (13)$$

$$\sigma = NSR^2 \quad (14)$$

This can be solved analytically with the Lagrange Dual Function.

$$\hat{w} = A^{-1}\lambda e \quad (15)$$

$$\lambda = \frac{1}{e^T R^{-1} e} \quad (16)$$

where $e \in \mathbb{R}^N$ is a vector of ones. However this formulation allows for negative values of w_i because A is sign indefinite. This is not useful for our application so we solve equation 12 numerically with the added constraint $w_i \geq 0$.

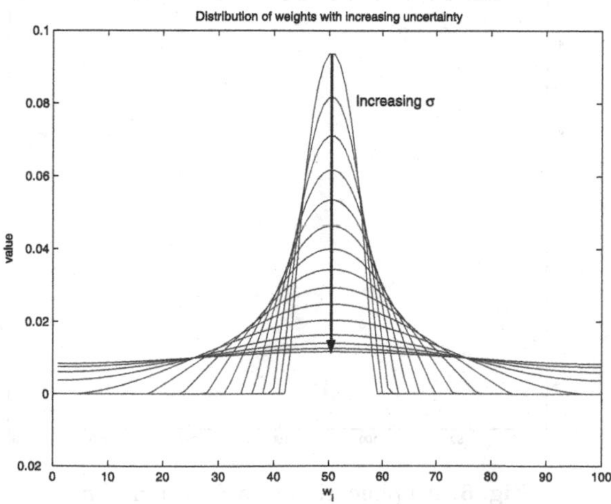

Fig. 4. Polarity vector w as noise variance, σ, increased from $1 \rightarrow 300$

Simulation Figure 4 shows how the weighting vector w changes with increasing noise level, $\sigma = 1 \rightarrow 300$ (the values are evenly spaced on a log scale), and g is a linear gradient with unity slope. For small σ, w heavily weights the sensors near the highest value of g and as σ increases, w weights the sensors evenly.

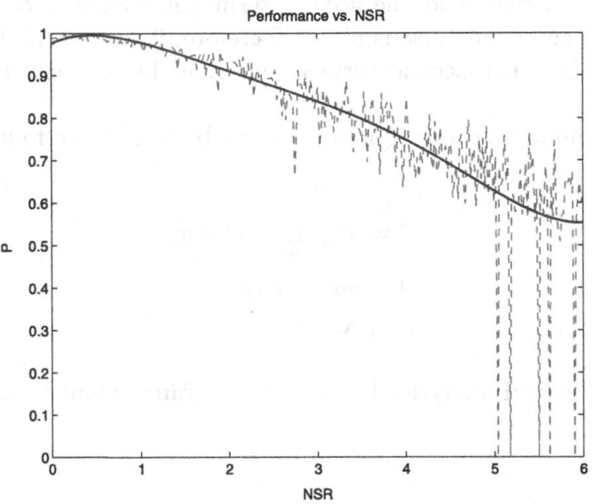

Fig. 5. Performance degradation as function of random noise-to-signal ratio (NSR) for the algorithm described in equations 10 and 12. The red dashed line is taken from 300 experiments and the solid blue line a polynomial fit to the data

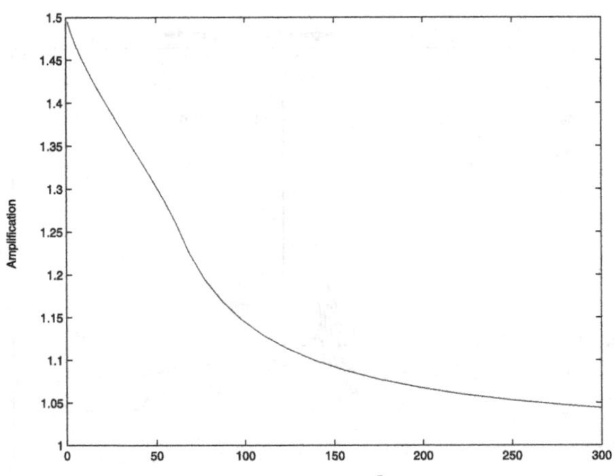

Fig. 6. Amplification as a function of σ

We tested this new algorithm using the same simulation strategy as in section 3.1, except that at each time step w is recalculated. Figure 5 shows the performance of this algorithm with increasing NSR. In this case w was calculated assuming $\sigma = 16$ so that the algorithm would be robust to $NSR = 4$. Comparing figures 3 and 5 shows that our new algorithm improved performance for $NSR = 1 \rightarrow 4$ and that it did not fail until after $NSR \approx 5$. When we optimized for $NSR \geq 10$, which results in a "flat" w, the performance of the system was similar to 3, and when we optimized for $NSR < 10$ all of our results looked very similar to figure 5.

In addition to improving noise robustness, the asymmetric weighting of the sensors amplifies the detected signal. We define amplification as the ratio of s from equation 10 to the slope of the gradient (in this case the slope is one). Figure 6 quantifies the signal amplification for different w vectors, where w is parameterized by the noise variance σ. As would be expected from equation 10, a highly asymmetric w will give a higher amplification since up-gradient sensors make a larger contribution to s, and as $\sigma \rightarrow \infty$ the amplification will go to one. Our performance simulations did not include the effect of the amplification since the speed of the cell was held constant. This was done so that we could directly assess the effect of w on direction sensing and not on the speed of the robot. However, we include the results of figure 6 to demonstrate that the signal amplification discussed in [13] and [14] could come from the weighted collaboration of the cell's sensors.

3.3 Connection to Neutrophil Models

In the formulation above we showed that the preferential weighting of sensors leads to increased performance and noise robustness of a gradient sensing robot. The models of [3], [10], and [7] describe the dynamics and asymmetric distribution of PH (Pleckstrin Homology) proteins in chemotaxing Eukaryotes (specifically neutrophils and *Dictyostelium discoideum*). The PH proteins accumulate at the front or up-gradient region of the cell membrane and it is thought that this makes the front more sensitive to chemical stimuli then the back. Their models differ in how this asymmetry of proteins evolve and how the dynamics relate to the known biochemistry. However, their models are similar in that when presented with a temporarily stable gradient the distribution of PH proteins is similar to our calculated w vector (with slight modification to their parameters we can make them equivalent). Therefore the performance and robustness results of our algorithm can be directly applied to their models. In this sense, their models are a special case of our optimal sensing formulation.

Additionally, this similarity leads to a novel way of implementing our algorithm on an autonomous robot. Instead of computing the optimal w vector at each time point, we can design a circuit for each sensor that mimics these neutrophil models. For example, each sensor could be given the following dynamics:

$$\frac{da_i}{dt} = g_i - k_a a \tag{17}$$

$$\frac{db_i}{dt} = k_s g_i - k_b b \tag{18}$$

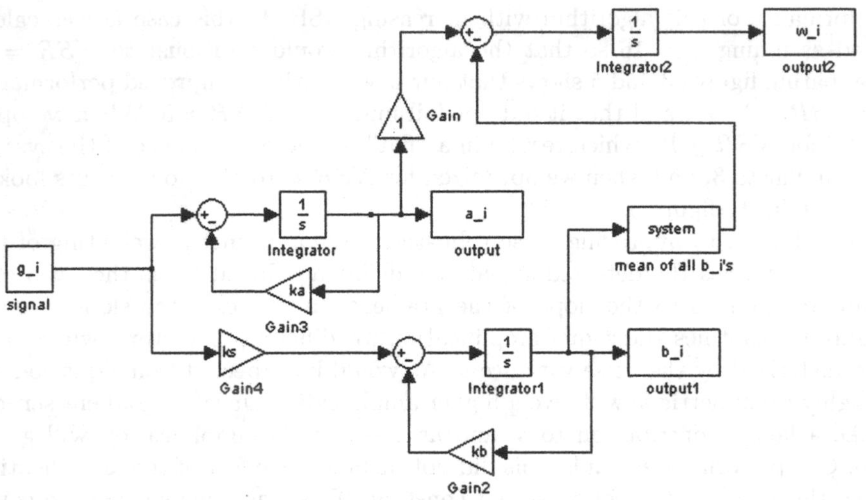

Fig. 7. Schematic diagram of proposed circuit for computing w_i

$$\frac{dw_i}{dt} = a_i - \frac{1}{N}\sum_{i=1}^{N} b_i - k_w w_i \tag{19}$$

Equation 19 describes how the the weight, w_i, on each sensor changes as a function of a and b. We refer to a as the "activator" dynamics– a_i increases with increasing g_i which in turn increases w_i. We refer to b as the "global inhibitor"– it subtracts off the mean value of the received signal for all w_i. The interplay of the activator and global inhibitor leads to an asymmetric distribution of w_i's. The constants, k_x, are design variables that will affect the dynamics and distribution of w. This method of sensor weighting can be implemented in hardware with 7 op-amps (3 integrators and 4 gains) per sensor and one summing circuit (for computing the mean). Figure 7 shows the configuration of this circuit. Clearly this will be a much faster method of computing w then by solving the quadratic optimization problem.

4 Conclusions and Future Work

This paper has presented and analyzed a biologically inspired algorithm for collaborative control. We have evaluated the algorithm's performance and noise robustness and have suggested how to implement the algorithm with hardware. We have also shown how current neutrophil models are similar to our formulation, which suggests that neutrophils are robust to noise in their signaling pathway. Our next task will be to evaluate how the dynamics of w and the resulting amplification effect the performance of a robot in an obstacle field. Preliminary results show that if w has fast dynamics (that is, can quickly redistribute) the robot is better at avoiding obstacles but it is more sensitive to noise.

References

1. B. Gerkey, M. Matari, G. Sukhatme, Exploiting Physical Dynamics for Concurrent Control of a Mobile Robot, *IEEE International Conference on Robotics and Automation*, 2002
2. K. Goldberg, B. Chen, Collaborative Control of Robot Motion: Robustness to Error, *IEEE Internaltional Conference on Robots and Systems*, 2001
3. A. Levchenko, P. Iglesias, Models of Eukaryotic Gradient Sensing: Application to Chemotaxis of Amoebae and Neutrophils, *Biophysical Journal* 82:50-63, 2002
4. P. Iglesias, A. Levchenko, Modeling the Cell's Guidance System, *Science STKE* 148:1-12, 2002
5. H. Kitano, Systems Biology: A Brief Overview, *Science*, 295:1662-1664, 2002.
6. P. Maes, R. Brooks, Learning to Coordinate Behaviors, *National Conference on Artificial Intelligence*, 1990
7. H. Meinhardt, Orientation of Chemotactic Cells and Growth Cones: Models and Mechanisms *Journal of Cell Science*, 112:2867-2874, 1999
8. A. Muller et.al., Involvement of Chemokine Receptors in Breast Cancer Metastasis, *Nature*, 410:50-56, 2001
9. C. Rao, A. Arkin, Control Motifs for Intracellular Regulatory Networks,*Ann. Rev. Biomed. Eng.* 3:91-419, 2001
10. W. Rappel, et.al., Establishing Direction during Chemotaxis in Eukaryotic Cells *Biophysical Journal*, 83:1361-1367, 2002
11. This picture was taken from a film done by David Rogers at Vanderbilt University circa 1950. The movie can be found at http://expmed.bwh.harvard.edu/projects/motility.html
12. G. Servant et.al., Dynamics of a Chemoattractant Receptor in Living Neutrophils during Chemotaxis, *Mol.Bio.Cell*, 10:1163-1178, 1999
13. S. Zigmond, Mechanisms of Sensing Chemical Gradients by Polymorphonuclear Leukocytes, *Nature*, 249: 450-52, 1974
14. S. Zigmond, P. Deverotcs, Chemotaxis in Eukaryotic Cells: A Focus on Leukocytes and Dictyostelium, *Ann. Rev. Biol*, 4:649-86, 1988

Beta Binders for Biological Interactions*

Corrado Priami and Paola Quaglia

Dipartimento di Informatica e Telecomunicazioni,
Università di Trento, Italy

Abstract. This paper presents binders and operators, in the process calculi tradition, to reason about biological interactions.

Special binders are added to wrap a process just as membranes enclose some living matter and hence to mimick biological interfaces. A few operators are then added to the pi-calculus kernel to describe the dynamics of those interfaces.

1 Introduction

The literature on the formal specifications of living entities and their behaviour offers a number of descriptive examples in terms of either 'general purpose' process calculi (biochemical stochastic π-calculus [11, 9], CCS-R [4]) or *ad hoc* calculi whose definitions were intentionally and directly driven by some phenomena observed in life science (BioAmbients [10], Brane Calculi [3], Formal Molecular Biology [5, 6]).

Both approaches have their advantages and drawbacks. CCS [7], the π-calculus [8, 12], and all the calculi derived from them have a usually well-understood mathematical theory, and a number of associated tools for verification and analysis. On the other hand, w.r.t. the final goal of giving compact (if not graphical) representations of biological interactions, all these calculi suffer from being too much low-level. We mean by low-level that it is not possible to provide explicit representations or specifications of wrappers, membranes, enclosing surfaces, shapes, etc., while each of these elements is a main ingredient of the living matter (e.g., the nucleus of a cell is a distinct portion of it, cells are distinct and well-defined parts of a tissue, etc.). Modulo restriction or hiding operators, both CCS and π-calculus processes are just flat compositions of parallel subcomponents, and when analizying the evolution of any process it can become extremely hard to identify the derivatives of sub-parts of it.

The calculi directly born for formal biological reasoning take a quite different point of view. Their formal theory is less investigated than those for CCS or the π-calculus (at least for chronological reasons). They provide means to model 'borders' and hierarchies of entities, and usually come equipped with appealing graphical representations to identify elements at a simple glance.

* This work has been partially supported by the FIRB project "Modelli formali per Sistemi Biochimici".

V. Danos and V. Schachter (Eds.): CMSB 2004, LNBI 3082, pp. 20–33, 2005.

In this work we follow yet another approach. Our principal aim is to break down the user-driven coordination of interaction, and to enable promiscuity of communication. We believe this step be due to eventually provide predictive modeling, namely the possibility of describing either, e.g., the evolutionary behaviour of biological entities, or just inferring systems models from incomplete information. The final goal of this research direction is to set up a probabilistic/stochastic model for reasoning about the behaviour of biological entities. The binders and operators we introduce in this paper represent the possible non-deterministic kernel of such a model. Here we describe some primitives to be added to a π-calculus-like process calculus to specify biological phenomena with (as much as possible) ease of notation, while retaining (most of) the expressiveness of general-purpose calculi for concurrency, and giving non-determinism a first-citizen role throughout.

A special class of binders, called beta binders, are introduced and used to model processes encapsulated into *boxes* with interaction capabilities. To keep the formalism as simple as possible, we forbid nesting of boxes. Processes inside boxes (at least as a first approximation) can be thought of as standard π-calculus processes. Boxes (π-processes prefixed by beta binders) are called bio-processes. The following is a graphical representation of a system composed by two parallel bio-processes, each with one single site.

$$\boxed{x(y).\,P_1 \mid \overline{x}z.\,P_1'}^{\;x:\Gamma} \qquad \boxed{\overline{u}w.\,P_2 \mid P_2'}^{\;u:\Delta} \tag{1}$$

The pairs $x : \Gamma$ and $u : \Delta$ indicate the sites through which each box may interact with the external world. The components Γ and Δ, to be thought of as the types of the two sites, denote the interaction capabilities at x and u, respectively. More specifically, Γ and Δ are sets of names giving a compact and parametric view of the corresponding interfaces.

The pi-processes enclosed into boxes can evolve independently from each other and from the external world. For example the system in (1) may move to the following system after the occurrence of an internal communication on x in the leftmost box.

$$\boxed{P_1\{z/y\} \mid P_1'}^{\;x:\Gamma} \qquad \boxed{\overline{u}w.\,P_2 \mid P_2'}^{\;u:\Delta} \tag{2}$$

Moreover, the boxes in (1) may interact through the sites x and u under the proviso that Γ and Δ are not disjoint. Taking, e.g., $\Gamma = \{a, b\}$, the intuition behind the typing $x : \Gamma$ is that the process $x(y).\,P_1 \mid \overline{x}z.\,P_1'$ enclosed in the leftmost box may interact with either a box which exposes an a-typed site or a box offering a b-typed site. So, for instance, if Γ and Δ share a common element, the system drawn in (1) can evolve into the following one.

$$\begin{array}{cc} x : \Gamma & u : \Delta \\ \boxed{P_1\{w/y\} \mid \overline{x}z.\,P_1'} & \boxed{P_2 \mid P_2'} \end{array} \tag{3}$$

The above example shows how data (the name w) may flow from one biological entity (the rightmost box) to another through the appropriate sites (x and u).

Besides using intra- and inter-actions (as in (2) and (3), respectively), the evolution of boxes is described by a limited number of macro-operations: add a site to an interface; hide one of the sites of the interface; unhide a site that has been previously hidden; join two boxes together; and split a box in two.

Adding, hiding and unhiding sites have a fundamental role in modelling the dynamics of box interfaces and hence, e.g., the functional dependency of the interaction capabilities of biological components on their particular shape or folding.

The join and split operations, that actually have to do with the evolution of box structure rather than with the dynamics of interfaces, are described in a parametric way. In this case, as it was done for other modeling choices we made, we prefer to leave the description as undetermined as possible, to accomodate possible distinct instances of the same macro-behaviour.

Synopsis. In the next section we introduce the syntax and the semantics of the formal model. In Section 3 we run the model over the abstract specification of a virus attack. The following section collects considerations about the way the presented formalism could be used to model some interesting issues for the research community in systems biology. Among them: bi-directional communication; explicit handling of the quantity of energy possessed by biological entities; affinity; explicit reasoning about interactions of unknown type. Section 5 concludes the presentation with some final remarks.

2 Beta Binders and Bio-processes

This section presents the syntax and the semantics of the binders and operators introduced to model biological interactions.

2.1 Syntax

The formalism we use is much inspired by the π-calculus. For full details about it the interested reader is referred to [8, 12]. We just recall that the π-calculus is a calculus of names: names are the media and the values of communication. Here we take the same point of view, and, as in the π-calculus, assume the existence of a countably infinite set N of names (ranged over by lower-case letters).

Definition 1. *An* elementary beta binder *has either the form* $\beta(x : \Gamma)$ *or the form* $\beta^h(x : \Gamma)$*, where*

1. *the name* x *is the* subject *of the beta binder, and*
2. Γ *is the* type *of* x*. It is a non-empty set of names such that* $x \notin \Gamma$*.*

Intuitively, the elementary beta binder $\beta(x : \Gamma)$ is used to denote an active (potentially interacting) site of the box. Binders like $\beta^h(x : \Gamma)$ denote sites which have been hidden to forbid further interactions through them.

Definition 2. Composite beta binders *are generated by the following grammar:*

$$\boldsymbol{B} ::= \beta(x : \Gamma) \;\big|\; \beta^h(x : \Gamma) \;\big|\; \beta(x : \Gamma)\,\boldsymbol{B} \;\big|\; \beta^h(x : \Gamma)\,\boldsymbol{B}$$

A composite beta binder is said to be well-formed *when the subjects of its elementary components are all distinct. We let well-formed beta binders be ranged over by* $\boldsymbol{B}, \boldsymbol{B}_1, \boldsymbol{B}_2, \ldots, \boldsymbol{B}', \ldots$*.*

The set of the subjects of all the elementary beta binders in $\boldsymbol{B}$ *is denoted by* sub($\boldsymbol{B}$)*, and we write* $\boldsymbol{B} = \boldsymbol{B}_1\boldsymbol{B}_2$ *to mean that* $\boldsymbol{B}$ *is the beta binder given by the juxtaposition of* $\boldsymbol{B}_1$ *and* $\boldsymbol{B}_2$*.*

Also, the metavariables $\boldsymbol{B}^*, \boldsymbol{B}_1^*, \boldsymbol{B}_2^*, \ldots$ *stay for either a well-formed beta binder or the empty string. The above notation for the subject function and for juxtaposition is extended to these metavariables in the natural way.*

Processes encapsulated into boxes (ranged over by capital letters distinct from B) are given by the following syntax.

$$P ::= \mathsf{nil} \;\big|\; x(w).\,P \;\big|\; \overline{x}y.\,P \;\big|\; P \mid P \;\big|\; \nu y\, P \;\big|\; !\,P \;\big|$$
$$\mathsf{expose}(x, \Gamma)\,.\,P \;\big|\; \mathsf{hide}(x)\,.\,P \;\big|\; \mathsf{unhide}(x)\,.\,P$$

For simplicity, despite the difference w.r.t. the usual π-calculus syntax, we refer to the processes generated by the above grammar as to pi-processes. The deadlocked process nil, input and output prefixes ($x(w).\,P$ and $\overline{x}y.\,P$, respectively), parallel composition ($P \mid P$), restriction ($\nu y\, P$), and the bang operator (!) have exactly the same meaning as in the π-calculus.

The expose, hide, and unhide prefixes are intended for changing the external interface of boxes and their intuitive meaning is best given after the formal definition of bio-processes.

Bio-processes (ranged over by $B, B_1, \ldots, B', \ldots$) are generated by the following grammar:

$$B ::= \mathsf{Nil} \;\big|\; \boldsymbol{B}[\,P\,] \;\big|\; B \parallel B$$

Nil denotes the deadlocked bio-process and is the neutral element of the parallel composition of bio-processes, written $B \parallel B$. But for Nil, the simplest form of bio-process is given by a pi-process encapsulated into a beta binder ($\boldsymbol{B}[\,P\,]$).

To any bio-process consisting of n parallel components corresponds a simple graphical notation, given by n distinct boxes, one per parallel component. Each box contains a pi-process and has as many sites (hidden or not) as the number of elementary beta binders in the composite binder. As an example, the box corresponding to the bio-process $\beta(x : \Gamma)\,\beta^h(y : \Delta)\,[\,P\,]$ is drawn below.

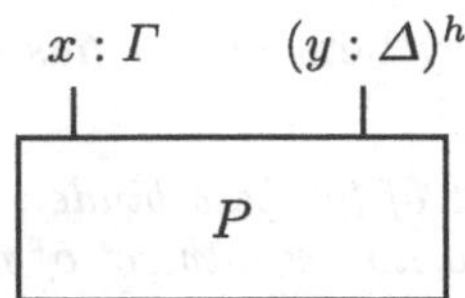

The position of the two sites along the perimeter of the box is irrelevant, just as the relative positions of $\beta(x : \Gamma)$ and $\beta^h(y : \Delta)$ in $\beta(x : \Gamma)\, \beta^h(y : \Delta) [\, P \,]$, which will be let to be structurally congruent to $\beta^h(y : \Delta)\, \beta(x : \Gamma) [\, P \,]$. For similar reasons, the relative position of the boxes corresponding to the parallel subcomponents of a bio-process is irrelavant.

We can now comment on the newly introduced operators for pi-processes. The prefix $\mathsf{expose}(x,\, \Gamma) \,.\, P$ calls for adding a new site of type Γ to the box containing the pi-process. To meet the requirement of beta binder well-formedness, the name of the added site has to be fresh w.r.t. the already existing sites. The prefix $\mathsf{hide}(x) \,.\, P$ ($\mathsf{unhide}(x) \,.\, P$, resp.) reads as a request to hide (unhide, resp.) the site x of the enclosing box.

The single other issue we want to mention here about the special prefixes $\mathsf{expose}(x,\, \Gamma)$, $\mathsf{hide}(x)$, and $\mathsf{unhide}(x)$ is relevant for name substitutions. Neither $\mathsf{hide}(x)$ nor $\mathsf{unhide}(x)$ acts as a binder for x, while the occurrence of the prefix $\mathsf{expose}(x,\, \Gamma)$ in the pi-process $\mathsf{expose}(x,\, \Gamma) \,.\, P$ is a binder for x in P. Substitution of names for names over the extended pi-processes considered above is then defined as a mere extension of the usual π-calculus substitution.

Notice that name substitution may affect the Γ type declared in expose prefixes. Hence name substitution is a simple way to change the specificity of a certain (still to be exposed) site. A more general way of modifying interface types will be commented later on.

2.2 Semantics

The semantics of bio-processes is given in the chemical style [2], in terms of a reduction relation which in turn uses a structural congruence relation.

The definition of the reduction relation uses both a structural congruence over bio-processes and a structural congruence over pi-processes. We overload the same symbol to denote both of them, and let the context disambiguate the intended congruence.

Definition 3. Structural congruence *over pi-processes, denoted* $\equiv$*, is the smallest relation which satisfies the laws in Table 1(a).* Structural congruence *over bio-processes, denoted* $\equiv$*, is the smallest relation which satisfies the laws in Table 1(b) where* $\hat{\beta}$ *ranges over* $\{\beta, \beta^h\}$.

The laws of structural congruence over pi-process are the typical π-calculus axioms. The following holds of the laws over bio-processes.

- The first axiom states that the structural congruence of pi-processes is reflected at the upper level as congruence of bio-processes.

Table 1. Laws for structural congruence

Pi-processes	$P_1 \equiv P_2$ provided P_1 is an α-converse of P_2 $P_1 \mid (P_2 \mid P_3) \equiv (P_1 \mid P_2) \mid P_3$ $P_1 \mid P_2 \equiv P_2 \mid P_1$ $P \mid \mathsf{nil} \equiv P$ $\nu z\, \nu w\, P \equiv \nu w\, \nu z\, P$ $\nu z\ \mathsf{nil} \equiv \mathsf{nil}$ $\nu z\, (P_1 \mid P_2) \equiv P_1 \mid \nu z\, P_2$ provided $z \notin \mathsf{fn}(P_1)$ $!\,P \equiv P \mid !\,P$	(a)
Bio-processes	$\boldsymbol{B}[\ P_1\ ] \equiv \boldsymbol{B}[\ P_2\ ]$ provided $P_1 \equiv P_2$ $B_1 \parallel (B_2 \parallel B_3) \equiv (B_1 \parallel B_2) \parallel B_3$ $B_1 \parallel B_2 \equiv B_2 \parallel B_1$ $B \parallel \mathsf{Nil} \equiv B$ $\boldsymbol{B}_1\boldsymbol{B}_2[\ P\ ] \equiv \boldsymbol{B}_2\boldsymbol{B}_1[\ P\ ]$ $\boldsymbol{B}^*\hat{\beta}(x : \Gamma)[\ P\ ] \equiv \boldsymbol{B}^*\hat{\beta}(y : \Gamma)[\ P\{y/x\}\ ]$ provided y fresh in P and $y \notin \mathsf{sub}(\boldsymbol{B}^*)$	(b)

- The next three laws are the monoidal axioms for the parallel composition of bio-processes.
- The fifth law declares, as we already mentioned, that the actual ordering of elementary beta binders within a composite binder is irrelevant.
- The last law states that the subject of elementary beta binders is a placeholder that can be changed at any time under the proviso that name clashes are avoided and well-formedness of beta binder is preserved.

The *reduction relation*, $\longrightarrow$, is the smallest relation over bio-processes obtained by applying the axioms and rules in Table 2.

The reduction relation describes the evolution within boxes (intra), as well as the interaction between boxes (inter), the dynamics of box interfaces (expose, hide, unhide), and the structural modification of boxes (join, split).

The rule intra lifts to the level of bio-processes any 'reduction' of the enclosed pi-process. Indeed no reduction relation is defined over pi-processes. Nonetheless the intra rule may be applied whenever the internal pi-process is structurally equivalent to a pi-process which could be reduced in the π-calculus sense.

The rule inter models interactions between boxes with complementary internal actions (input/output) over complementary sites (sites with non-disjoint types). Information flows from the box containing the pi-process which exhibits the output prefix to the box enclosing the pi-process which is ready to perform the input action.

The rules expose, hide, and unhide correspond to an unguarded occurence of the homonymous prefix in the internal pi-process and allow the dynamic modification of beta binders.

Table 2. Axioms and rules for the reduction relation

(intra)
$$\frac{P \equiv \nu\tilde{u}\,(x(w).\,P_1 \mid \overline{x}z.\,P_2 \mid P_3)}{\boldsymbol{B}[\; P \;] \longrightarrow \boldsymbol{B}[\; \nu\tilde{u}\,(P_1\{z/w\} \mid P_2 \mid P_3)\;]}$$

(inter)
$$\frac{P \equiv \nu\tilde{u}\,(x(w).\,P_1 \mid P_2) \qquad Q \equiv \nu\tilde{v}\,(\overline{y}z.\,Q_1 \mid Q_2)}{\beta(x:\Gamma)\,\boldsymbol{B}_1^*[\; P \;] \parallel \beta(y:\Delta)\,\boldsymbol{B}_2^*[\; Q \;] \longrightarrow \beta(x:\Gamma)\,\boldsymbol{B}_1^*[\; P' \;] \parallel \beta(y:\Delta)\,\boldsymbol{B}_2^*[\; Q' \;]}$$

where $P' = \nu\tilde{u}\,(P_1\{z/w\} \mid P_2)$ and $Q' = \nu\tilde{v}\,(Q_1 \mid Q_2)$

provided $\Gamma \cap \Delta \neq \emptyset$ and $x, z \notin \tilde{u}$ and $y, z \notin \tilde{v}$

(expose)
$$\frac{P \equiv \nu\tilde{u}\,(\mathsf{expose}(x,\,\Gamma)\,.\,P_1 \mid P_2)}{\boldsymbol{B}[\; P \;] \longrightarrow \boldsymbol{B}\,\beta(y:\Gamma)\,[\; \nu\tilde{u}\,(P_1\{y/x\} \mid P_2)\;]}$$

provided $y \notin \tilde{u}$, $y \notin \mathsf{sub}(\boldsymbol{B})$ and $y \notin \Gamma$

(hide)
$$\frac{P \equiv \nu\tilde{u}\,(\mathsf{hide}(x)\,.\,P_1 \mid P_2)}{\boldsymbol{B}^*\,\beta(x:\Gamma)\,[\; P \;] \longrightarrow \boldsymbol{B}^*\,\beta^h(x:\Gamma)\,[\; \nu\tilde{u}\,(P_1 \mid P_2)\;]}$$

provided $x \notin \tilde{u}$

(unhide)
$$\frac{P \equiv \nu\tilde{u}\,(\mathsf{unhide}(x)\,.\,P_1 \mid P_2)}{\boldsymbol{B}^*\,\beta^h(x:\Gamma)\,[\; P \;] \longrightarrow \boldsymbol{B}^*\,\beta(x:\Gamma)\,[\; \nu\tilde{u}\,(P_1 \mid P_2)\;]}$$

provided $x \notin \tilde{u}$

(redex)
$$\frac{B \longrightarrow B'}{B \parallel B'' \longrightarrow B' \parallel B''}$$

(struct)
$$\frac{B_1 \equiv B_1' \qquad B_1' \longrightarrow B_2}{B_1 \longrightarrow B_2}$$

(join) $\boldsymbol{B}_1[\; P_1 \;] \parallel \boldsymbol{B}_2[\; P_2 \;] \longrightarrow \boldsymbol{B}[\; P_1\sigma_1 \mid P_2\sigma_2 \;]$

provided that f_{join} is defined in $(\boldsymbol{B}_1, \boldsymbol{B}_2, P_1, P_2)$

and with $f_{join}(\boldsymbol{B}_1, \boldsymbol{B}_2, P_1, P_2) = (\boldsymbol{B}, \sigma_1, \sigma_2)$

(split) $\boldsymbol{B}[\; P_1 \mid P_2 \;] \longrightarrow \boldsymbol{B}_1[\; P_1\sigma_1 \;] \parallel \boldsymbol{B}_2[\; P_2\sigma_2 \;]$

provided that f_{split} is defined in $(\boldsymbol{B}, P_1, P_2)$

and with $f_{split}(\boldsymbol{B}, P_1, P_2) = (\boldsymbol{B}_1, \boldsymbol{B}_2, \sigma_1, \sigma_2)$

The rule expose causes the addition of an extra site with the declared type. The name x declared in the prefix expose(x, Γ) is a placeholder which can be renamed to meet the requirement of well-formedness of the enclosing beta binder.

The rules hide and unhide force the specified site to become hidden and unhidden, respectively. They cannot be applied if the site does not occur unhidden, respectively hidden, in the enclosing beta binder.

The rules redex and struct are typical rules of reduction semantics. They are meant, respectively, to interpret the reduction of a subcomponent as a reduction of the global system, and to infer a reduction after a proper structural shuffling of the process at hand.

The axiom join models the merge of boxes. The rule, being parametric w.r.t. the function f_{join}, is indeed an axiom schema. The function f_{join} determines the actual interface of the bio-process resulting from the aggregation of boxes, as well as possible renamings of the enclosed pi-processes. A simple instance of f_{join} is the following:

$$\begin{aligned} \lambda \boldsymbol{B}_1 \boldsymbol{B}_2 P_1 P_2.\ & \text{if } (\boldsymbol{B}_1 = \boldsymbol{B}_1^* \, \beta(x : \Gamma) \text{ and } \boldsymbol{B}_2 = \boldsymbol{B}_2^* \, \beta(y : \Delta) \text{ and} \\ & \quad \Gamma \cap \Delta \neq \emptyset) \\ & \text{then } (\boldsymbol{B}_1, \sigma_{id}, \{x/y\}) \\ & \text{else } \bot \end{aligned} \tag{4}$$

where σ_{id} stays for the identity substitution and $\bot$ for undefinedness. Assuming the above definition of the function f_{join}, the axiom join corresponds to the graphical reduction drawn below.

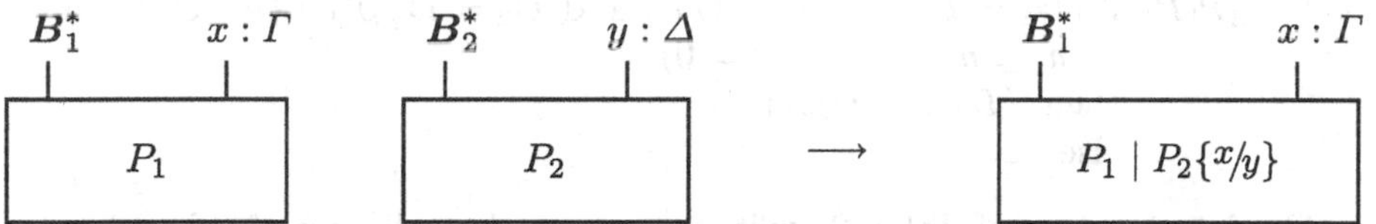

Such a reduction, in a formal context where nesting of boxes is intentionally avoided, can be used to render biological *endocytosis*, namely the absorbtion of substances from the external environment. Indeed, the pi-component of the absorbed box (P_2) is moved into the absorbing bio-process, and the engulfed material has no longer a proper beta binder to interact with the external world. The specific instance of f_{join} defined in (4) assumes that the endocytosis reduction can take place only if absorbing and absorbed bio-process have complementary sites (elementary beta binders with non disjoint types), and also states that the absorbed process can keep interacting with the external environment through the same site which has been used to absorb it ($\sigma_2 = \{x/y\}$).

Distinct definitions of f_{join} would correspond to different intuitions on the conditions leading to the merge of biological entities. For instance, as in the living matter the act of merging entities is time-consuming, one could imagine to hide the site used for absorbing external material ($x : \Gamma$ in the above example). This interpretation, in a formalism equipped with duration of actions, could be further refined to unhide the site after an appropriate quantity of time.

Orthogonally, and more interestingly, quantitative information about boxes and sites, or even just the types of the involved beta binders, could be used to break down the possible symmetry deriving from the interplay of struct and join. Think, e.g., that using the specific instance of join given by taking f_{join} as defined in (4), one could infer

$$\beta(x:\{a\})\,[\,P_1\,] \parallel \beta(y:\{a\})\,[\,P_2\,] \longrightarrow \beta(x:\{a\})\,[\,P_1 \mid P_2\{x/y\}\,]$$

as well as

$$\beta(x:\{a\})\,[\,P_1\,] \parallel \beta(y:\{a\})\,[\,P_2\,] \longrightarrow \beta(y:\{a\})\,[\,P_1\{y/x\} \mid P_2\,].$$

Biologically speaking, the above symmetry corresponds, e.g., to assessing that a bacterium can engulf a macrophage in the same way as a macrophage can engulf a bacterium. If the sites (or their types) are augumented with quantitative information, however, it becomes fairly easy to cut out one of the two reductions above. Imagine, for instance, to decorate sites with an integer number representing the 'endocytosis propension' of the site. Then the definition of f_{join} could reflect the intuition that in a join reduction the absorbing process is the one with bigger endocytosis propension.

An analogous effect could also be achieved by assuming a partial ordering $\sqsubseteq$ on a subset of names (say $n_1 \sqsubseteq n_2 \sqsubseteq \ldots$), then ensuring that the type of each site contains one of these special names, and finally discriminating the endocytosis potentials of bio-processes on the basis of the partial order. For example, a suitable definition of f_{join} could be:

$$\begin{aligned}\lambda \boldsymbol{B}_1 \boldsymbol{B}_2 P_1 P_2.\ \text{if } & (\boldsymbol{B}_1 = \boldsymbol{B}_1^*\,\beta(x:\{n_j\}\cup\Gamma) \text{ and } \boldsymbol{B}_2 = \boldsymbol{B}_2^*\,\beta(y:\{n_i\}\cup\Delta) \text{ and}\\ & \quad n_i \sqsubseteq n_j \text{ and } \Gamma\cap\Delta\neq\emptyset)\\ & \text{then } (\boldsymbol{B}_1, \sigma_{id}, \{x/y\})\\ & \text{else } \bot.\end{aligned}$$

The latest axiom of Table 2, split, formalizes the splitting of a box in two parts, each of them taking away a subcomponent of the content of the original box. Analogously to join, the rule split is an axiom schema depending on the specific definition of the function f_{split}. This function is meant to refine the conditions under which a bio-process can be split in two boxes.

From the biological point of view, this axiom corresponds to *meiosis*, which is typical of reproductive cells and consists in the separation of a cell and of the contained genetic material. Similarily to the case of join, the precise hypothesis leading to meiosis can be tuned by appropriately defining f_{split}.

We commented on the fact that our treatment of endocytosis is largely influenced by the choice of preventing nesting of boxes. Also, with the same care as in the case of endocytosis, split can be used to render *exocytosis*, i.e. the expulsion of biological sub-components. In particular, it is possible to encode it by exploiting the split rule and possibly appropriate occurrences of the expose prefix.

A derived operation which could reveal interesting for modeling biological phenomena is relative to changing the type of sites. Assume for instance that the bio-process $\boldsymbol{B}[\ P\ ]$ has a site $x : \Delta$, and suppose that, due for instance to bio-chemical modifications of the component, or even to its evolutionary behaviour, the type of the site has to be changed to Γ. This situation could be rendered by modifying P into the following:

$$\mathsf{hide}(x)\,.\,\mathsf{expose}(y,\,\Gamma)\,.\,P\{y/x\}.$$

A final remark on the approach is relative to the typing of sites. It allows an improved promiscuity of interaction and we believe that this can be a relevant point in modeling biological behaviours. Here types are assumed to be just sets of names, and boxes are allowed to communicate over sites with non-disjoint types. Given two sites $x : \Gamma$ and $y : \Delta$, the most generic constraint for interaction over x and y would probably look like a predicate in the form "if Γ agrees with Δ". Indeed, it might well be the case that types more structured than sets of names could reveal useful to primitively model the specificity of some biological phenomena.

3 An Abstract Virus Attack

This section illustrates the use of our formalism to specify a biological scenario representing an abstract view of the interactions between a virus and cells of our immune system.

The phenomena we are going to model has three main actors [1]:

- a virus;
- a cell of our immune system (macrophage) which can engulf the virus by endocytosis, elaborate it, distill the antigene molecule and display the antigen on its surface;
- a lymphocyte which can recognize the antigen and then activate mechanisms of the immune reply.

The formal representation of the above scenario is given by the specification below, where we shortly denote an output action and an input action on channel x by $\overline{x}$ and x, respectively, if the parameter of the communication is not relevant.

$$\beta(x : \{v_1, \ldots, v_n\})\,[\ C\ ] \parallel \beta(y : \{v_1, v_1'\})\,[\ V_1\ ] \parallel \beta(z : \{a_1, a_1'\})\,[\ L_1\ ]$$

where:

$$C = !\,x(w).\ \mathsf{expose}(x,\,\{w\})\,.\,\overline{x} \mid C_1$$

$$V_1 = \overline{y}a_1.\,V_1^{res}$$

$$L_1 = z.\,L_1^{act}$$

The bio-process $\beta(x : \{v_1, \ldots, v_n\})\,[\ C\ ]$ represents the macrophage cell. It can engulf, through the site x, n different kinds of viruses, represented by $v_1, \ldots, v_n$.

The parallel component $\beta(y : \{v_1, v'_1\}) \left[\, V_1 \,\right]$ plays a virus which can attack two kinds of cells, the first ones by v_1, and the second ones by v'_1.

The bio-process $\beta(z : \{a_1, a'_1\}) \left[\, L_1 \,\right]$ stays for a specialized lymphocyte which can recognize both the antigen a_1 associated with viruses of sort v_1 and the antigen a'_1 relative to v'_1-viruses.

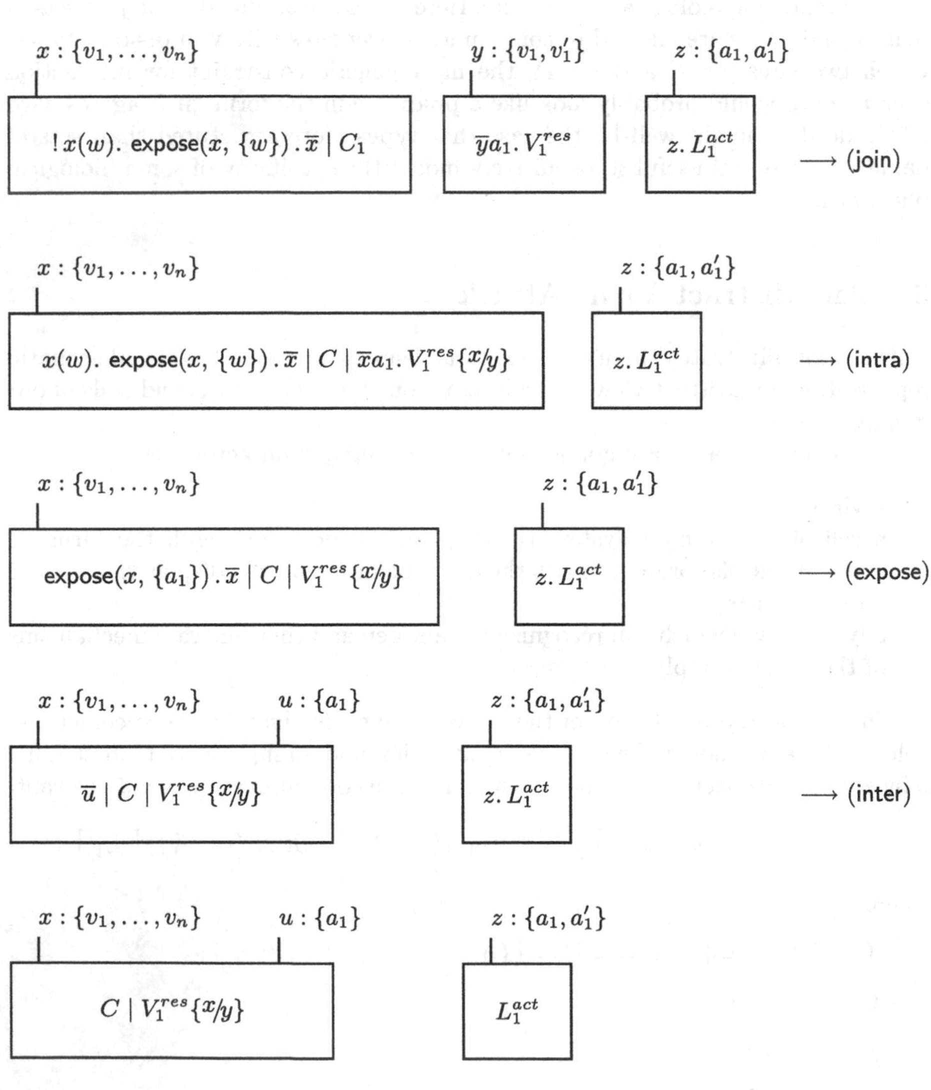

Fig. 1. Evolution of the virus attack

Fig. 1 shows one of the possible evolutions of the system. First the macrophage recognizes $\beta(y : \{v_1, v_1'\}) \left[V_1 \right]$ as a virus that it can elaborate, then it engulfs V_1 by an a join reduction where f_{join} is taken to be the function defined in (4). The next intra reduction, which involves the communication primitives $x(w)$ and $\overline{x}a_1$ shows how the macrophage degrades the virus by extracting from it the antigen molecule a_1 and leaving the residual V_1^{res}. After that, the antigen a_1 is displayed on the surface of the cell (expose reduction). The antigen is eventually recognized by the lymphocyte (inter reduction). This causes the activation of the lymphocyte which transforms into $\beta(z : \{a_1, a_1'\}) \left[L_1^{act} \right]$.

4 Further Considerations

In the following we comment on how the proposed formalism could handle a few issues which are of interest to the research community in systems biology, and are not yet handled in the process calculi approaches.

4.1 Two-Ways Communication

Biological interactions often involves a bi-directional flow of material (information). This is not properly rendered by our formalism, which sticks to the canonical π-calculus interpretation of communication as an agreement between a sender and a receiver.

Two-ways communication could be easily rendered in the formalism by substituting both the primitives for input and output prefix by a single communication prefix looking like the following:

$$x(?y, !w).\, P$$

and intuitively corresponding to the π-calculus sequence $x(y).\, \overline{x}w$ made atomic. With this communication primitive in the language, we would get, for example, the inter-reduction below.

$$\begin{aligned}\beta(x : \{a\}) \left[x(?u, !w).\, P \right] \parallel \beta(y : \{a\}) \left[y(?z, !v).\, Q \right] \longrightarrow \\ \beta(x : \{a\}) \left[P\{v/u\} \right] \parallel \beta(y : \{a\}) \left[Q\{w/z\} \right]\end{aligned}$$

4.2 Energy and Bound Replication

Another interesting topic is relative to the formal representation of the available quantity of energy. Energy can be either produced or consumed by living entities, but biochemical interactions are prevented to happen if the quantity of energy is not enough to go on. Our formalism does not tackle this issue at all. For instance, as it is usual in process calculi, we use $!\,P$ to mean "arbitrarily many copies of P are available" ($!\,P \equiv P \mid !\,P$). When modelling real living systems, this seems to be too strong an abstraction.

A special entity could be introduced to that purpose, say E^j with j positive real number, to represent that the available (quantity of) energy is equal to j.

Then each box could have its own energy, to be modified by computations and to monitor further interactions. For instance, letting $e(P)$ be some measure of the quantity of energy of P, the structural axiom for $!\,P$ could be replaced by the following.

$$!\,P \mid E^j \equiv P \mid !\,P \mid E^{j-e(P)} \text{ provided that } j - e(P) > 0.$$

4.3 Affinity

Another important point about general-purpose process calculi for biology which has not yet been mentioned is that they do not possess the machinery to deal with affinities. Some biological interactions take place on ports (elements, domains) which are not exactly the same but rather have a high level of affinity. On the other hand, by their own nature, process calculi impose that interactions take place on ports which are syntactically the same.

Assuming the existence of an affinity matrix 'names × names' which shows the affinity measure $\mathcal{P}(\_,\_)$ of pairs of names, we might represent interactions by affinity as, for instance, the following:

$$\beta(x:\{a\})\,[\; x(u).\,P \;] \parallel \beta(y:\{b\})\,[\; \overline{y}v.\,Q \;] \xrightarrow{\mathcal{P}(a,b)}$$
$$\beta(x:\{a\})\,[\; P\{v/u\} \;] \parallel \beta(y:\{b\})\,[\; Q \;]$$

with $\mathcal{P}(a,b) > 0$.

Namely, as it is shown above, instead of the classical unlabelled reduction relation, we could define a probabilistic relation. Indeed affinity is - or could be, by applying a suitable normalization - measured in the interval [0..1].

4.4 Reverse Engineering Unknown Activation Processes

Biology researchers are often puzzled by a number of events which they can observe to occur without having a clear understanding of the activation pathways leading to them. Our framework could be helpful in the investigation of those biological events. The formalism could be modified along the following main lines:

- allow a special *unknown* type for the sites of boxes, say '?';
- permit inter reductions between a ?-typed site and a Γ-typed site and impose that this kind of interaction modifies the ?-type into a new type which records its unknown origin, say $?\Gamma$;
- recursively allow inter reductions between a $?\Gamma$-typed site and a Δ-typed (or $?\Delta$-typed) site, and require that this interaction changes the type $?\Gamma$ into $?(\Gamma \cap \Delta)$.

In a setting like that drawn above, the study of 'unknown' biological activations could be carried out by specifying a system using boxes with appropriate ?-typed sites, then running it, and checking the biological evidence against the computations of the formal system. The unknown biological event could luckyly have some correspondence with the $?\Gamma$'s typing of the resulting formal process which most relates to the biological evidence.

5 Concluding Remarks

By their own nature, the communication paradigms of process calculi seem often to be too intentional for representing biological interactions.

The main aim of our work is to break down the strict complementarity which underlies the interaction mechanisms of process calculi, to enable promiscuity of communication, and to increment the global degree of non-determinism. This, we believe, could be an essential first step for predictive modeling of biological behaviour. The primitives described in this way could later on be refined and tuned by using probabilistic/stochastic measures.

Acknowledgements

We would like to thank anonymous referees for their useful comments and suggestions on the first draft of the paper, and Davide Prandi for fruitful discussions on the subject.

References

1. B. Alberts, A. Johnson, J. Lewis, M. Raff, K. Roberts, and P. Walter. *Molecular biology of the cell (IV ed.)*. Garland science, 2002.
2. G. Berry and G. Boudol. The chemical abstract machine. *Theoretical Computer Science*, 96(1):217–248, 1992.
3. L. Cardelli. Membrane interactions. In *BioConcur '03, Workshop on Concurrent Models in Molecular Biology*, 2003.
4. V. Danos and J. Krivine. Formal molecular biology done in CCS-R. In *BioConcur '03, Workshop on Concurrent Models in Molecular Biology*, 2003.
5. V. Danos and C. Laneve. Core formal molecular biology. In P. Degano, editor, *Proc. 12th European Symposium on Programming, ESOP 2003*, volume 2618 of *Lecture Notes in Comp. Sc.*, pages 302–318. Springer, 2003.
6. V. Danos and C. Laneve. Graphs for core molecular biology. In C. Priami, editor, *Proc. First Int. Workshop on Computational Methods in Systems Biology, CMSB 2003*, volume 2602 of *Lecture Notes in Comp. Sc.*, pages 34–46. Springer, 2003.
7. R. Milner. **Communication and Concurrency**. International Series in Computer Science. Prentice Hall, 1989.
8. R. Milner. **Communicating and mobile systems: the π-calculus**. Cambridge Universtity Press, 1999.
9. C. Priami, A. Regev, W. Silverman, and E. Shapiro. Application of a stochastic passing-name calculus to representation and simulation of molecular processes. *Information Processing Letters*, 80:25–31, 2001.
10. A. Regev, E.M. Panina, W. Silverman, L. Cardelli, and E. Shapiro. Bioambients: An abstraction for biological compartments. *Theoretical Computer Science*, 2003. to appear.
11. A. Regev, W. Silverman, and E. Shapiro. Representation and simulation of biochemical processes using the pi-calculus process algebra. In *Proceedings of the Pacific Symposium of Biocomputing 2001*, volume 6, pages 459–470, 2001.
12. D. Sangiorgi and D. Walker. **The π-calculus: a Theory of Mobile Processes**. Cambridge Universtity Press, 2001.

Biomimetic in Silico Devices

C. Anthony Hunt[1,3], Glen E.P. Ropella[1], Michael S. Roberts[2], and Li Yan[3]

[1] Dept. of Biopharmaceutical Scien es, Biosystems Group,
University of California, San Francisco, CA 94143-0446, USA
hunt@itsa.ucsf.edu; gepr@tempusdictum.com
http://biosystems.ucsf.edu

[2] Department of Medicine, University of Queensland, Princess Alexandra Hospital,
Woolloongabba, Q 4102 Australia
mroberts@soms.uq.edu.au
http://www.som.uq.edu.au/som/Research/therapeutics.shtml

[3] Joint UCSF/UC Berkeley Bioengineering Graduate Program
lyan@socrates.berkeley.edu
http://socrates.berkeley.edu/~lyan/

Abstract. We introduce biomimetic in silico devices, and means for validation along with methods for testing and refining them. The devices are *constructed* from adaptable software components designed to map logically to biological components at multiple levels of resolution. In this report we focus on the liver; the goal is to validate components that mimic features of the lobule (the hepatic primary functional unit) and dynamic aspects of liver behavior, structure, and function. An assembly of lobule-mimetic devices represents an in silico liver. We validate against outflow profiles for sucrose administered as a bolus to isolated, perfused rat livers. Acceptable in silico profiles are experimentally indistinguishable from those of the *in situ* referent. This new technology is intended to provide powerful new tools for challenging our understanding of how biological functional units function *in vivo*.

1 Introduction

Cells of the same type in the same tissue can experience different environments, and as a consequence exhibit quite different gene expression patterns [1]. This is just one of the problems faced when modeling biological systems at multiple levels of resolution. Network detail learned from experiments on isolated cells *in vitro* may not map directly to the tissue level. Reflecting on this reality, Noble rejects the reductionist bottom-up and the traditional top-down modeling and simulation approaches [2]. He makes the case for a "middle-out" strategy that focuses on the "functional level between genes and higher level function" [3], and calls for new ideas and new approaches to help move the field to the next level. The new class of biological analogue models presented here, which we refer to as biomimetic in silico devices (hereafter, devices), is an answer to that call. The devices are designed to generate biomimetic behaviors and are constructed from software components that map logically to biological components at multiple levels of resolution. The focus

V. Danos and V. Schachter (Eds.): CMSB 2004, LNBI 3082, pp. 34–42, 2005.

here is the liver. The data used for validation are outflow profiles from experiments on isolated, perfused rat livers (hereafter, perfused livers) given bolus doses of compounds of interest [4, 5]. Acceptable devices are hepato-mimetic in that they generate in silico outflow profiles that are experimentally indistinguishable from those of the *in situ* referent.

2 Device Design

2.1 Modeling Approach and Biological Data

Rather than the traditional inductive, analytic modeling approach, we used a constructive approach based in part on ideas and concepts from several sources, including compositional modeling [6]. We focus more on the aspects of structure and behavior that give rise to the data. We deconstruct the system into biologically recognizable components and processes that can be represented as software objects, agents, messages, and events. Next, we reconstruct using those objects within a software medium that handles probabilistic events, and can represent dynamic spatial heterogeneity. The process produces in silico *devices* [6] capable of biomimetic behaviors. Of course, these devices are also models. We use *device* to stress their modular, constructive nature, to emphasize the essential properties discussed below, and to distinguish them from traditional equational models.

The devices represent aspects of the anatomic structure and behavior of the functional unit of the liver that influences administered compounds. We conduct in silico experiments that follow protocols that mimic the original *in situ* experimental protocols. Because a device is not based on equations, we do not directly fit it to data. We use a Similarity Measure [7] to quantify the similarity between data generated by the device and data generated by the biological referent. Having completed that level of validation we run simulation experiments to address what-if questions and/or *grow* the device so that it accounts for additional, different data (e.g., perfused liver outflow profiles of additional compounds and/or hepatic imaging data).

2.2 Properties: Essential and Desired

We create a device from data with as few assumptions as possible, by first building and validating a simple, biomimetic device, and then iteratively improving it. Devices and their components must be reusable, revisable, and easily updatable when critical new data becomes available, without having to re-engineer the whole device. Device components, like their biological referents, need to be sufficiently flexible and adaptable to be useful in a variety of research contexts. They should be able to function at multiple levels of resolution, from molecule to organ. In addition, device and components must logically map to their biological counterparts. Spatial heterogeneity is a quintessential characteristic of organisms at each organizational level. So, it is essential that device components be capable of representing that heterogeneity at different levels of resolution as required by the problem. Finally, hepatic processes, drug disposition, and pharmacokinetic processes are characterized by probabilistic events. So, our biomimetic devices are exclusively event driven and most events can be probabilistic. These properties are deemed essential in part because they are expected to make this new technology easily accessible and useful to a majority of biomedical researchers.

2.3 Histological and Physiological Considerations

Changes in the architecture of hepatic fluid flow are associated with several disease states, and such alterations can influence a drug's disposition. These considerations suggest that a device must have a flexible means of representing that architecture at whatever level of detail is needed. We use directed graphs, with objects placed at graph nodes, to represent that architecture. Because the lobule is the primary structural and functional unit of the rat liver [8], the device must have a component that maps directly to the lobule. Hereafter refer to that component as a LOBULE[1].

Hepatocytes exhibit location-specific properties within lobules, including location-dependent expression of drug metabolizing enzymes [1]. Such intralobular heterogeneity requires that a LOBULE be capable of easily exhibiting heterogeneity and zonation when such properties are required. LOBULES must also be capable of representing specialized cell types, including endothelial, Kuppfer, and stellate cells, and their specialized behaviors in appropriate relative relationships, when needed, without forcing restructuring or redesign. The blood supply for one lobule, illustrated schematically by the cross section in Fig. 1, feeds into several dozen sinusoids that merge as they feed into the lobule's central vein (CV). Known hepatic features that are not needed to account for outflow profiles of the targeted data sets do not have corresponding components within the LOBULES. Examples include a separate hepatic arterial blood supply, the biliary system, and drug transport systems (into cells and into bile).

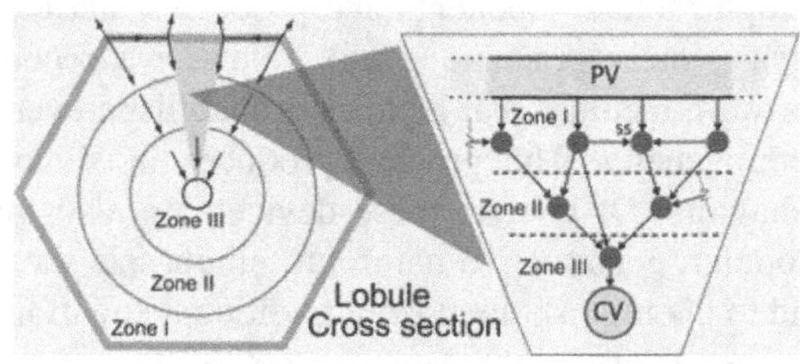

Fig. 1. A schematic of an idealized cross-section of a hepatic lobule showing half an acinus and the direction of flow between the terminal portal vein tract (PV), and the central hepatic vein (CV). SS: Sinusoidal Segment

2.4 Designing the in Silico Components

Computational Framework. Our devices are constructed within the Swarm framework (www.swarm.org). The methods do not require any particular formalism. But, the experimental framework is always formulated using Partially Ordered Sets; they are a generic way to specify concurrent processes with as few strictures as possible [9].

---

[1] When referring to the in silico counterpart of a biological component or process, such as "lobule," "endothelial cell," or "partition," we use SMALL CAPS.

Directed Graphs. A trace of flow paths within one lobule sketches a network that we represent by an interconnected, directed graph. Literature data [8] are used to constrain the accessible graph structures used. We only consider the subset of graphs that has more nodes connected to the portal vein tract (PV) (source) and fewer nodes connected to the CV (exit). Teutsch et al. [8] subdivide the lobule interior into concentric zones. For now, we impose a three-zone structure and require that each zone contain at least one node and that a shortest path from PV to CV will pass through at least one node and no more than one node per zone. The insert in Fig. 1 illustrates a portion of a graph that connects in silico PV outlets to the CV. Graph structure is specified by the number of nodes in each zone and the number of edges connecting those nodes. Edges are further identified as forming either inter-zone or intra-zone connections. The CV receives solute from the last node in each shortest path between PV and CV. Edges specify "flow paths" having zero length and containing no objects. A solute object exiting a parent node is randomly assigned to one of the available outgoing graph edges and appears immediately as input for the downstream child node. Randomly assigned intra-zone connections are allowed but are confined to Zones I and II. We randomly assign nodes to each of the zones so that the number of nodes in each zone is approximately proportional to the fraction of the total lobule volume found in that zone.

Sinusoidal Segments and the Fate of Solutes. Agents called sinusoidal segments (SSs) (Fig. 2) are placed at each graph node. There is one PV entrance (effectively covering the exterior of the LOBULE) and one CV exit for each LOBULE. A solute object is a passive representation of a chemical as it moves through the in silico environment. The PV creates solute objects, as dictated by the experimental dosage function, and distributes them to the SSs in Zone I. A solute object moving through the LOBULE represents molecules moving through the sinusoids of a lobule, and their behavior is dictated by rules specifying the relationships between solute location, proximity to other objects and agents, and the solute's physicochemical properties. Each solute has dose parameters and a scale parameter (molecules per solute object). The relative tendency of a solute object to move forward within a SS determines the effective flow pressure and this is governed by a parameter called *Turbo*. If there is no flow pressure (*Turbo* = 0), then solute movement is specified by a simple random walk. Increasing *Turbo* biases the random walk in the direction of the CV.

We have studied the behaviors of several sinusoidal segment designs and describe here the extensible design currently in use. Simpler designs generate behaviors that fail to meet our Similarity Measure criterion. Viewed from the center of perfusate flow out in Fig. 2, a SS is modeled as a tube with a rim surrounded by other layers. The tube and rim represent the sinusoidal space and its immediate borders. The tube

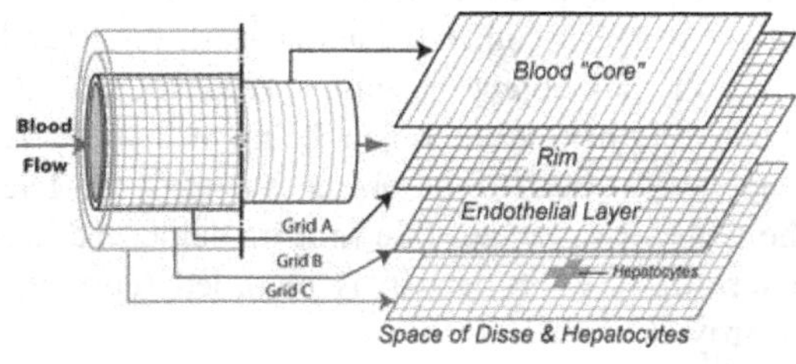

Fig. 2. Schematic of a sinusoidal segment (SS). Three types of SS are discussed in the text. One SS is placed at each node of the directed graph within each LOBULE

contains a fine-grained abstract Core space that represents blood flow. Grid A is the Rim. Grid B is wrapped around Grid A and represents the endothelial layer. Another fine-grained space (Grid C) is wrapped around Grid B to collectively represent the Space of Disse, hepatocytes, and bile canaliculi[2]. If needed, hepatocytes and connected features such as bile canaliculi can be moved to a fourth grid wrapped around Grid C. The properties of locations within each grid can be homogeneous or heterogeneous depending on the specific requirements and the experimental data. Objects can be assigned to one or more grid points. For example, a subset of Grid B points can represent one or more Kupffer cells. Objects that move from a location on a particular grid are subject to one or more lists of rules that are called into play at the next step. Within Grid B a parameter controls the size and prevalence of FENESTRATIONS[1] and currently 10% of Grid B in each SS is randomly assigned to FENESTRAE; the remaining 90% represents cells. Similarly, within the grid where some locations map to hepatocytes, there is a parameter that controls their relative density.

Classes of Sinusoidal Segments and Dynamics Within. To further enable accounting for sinusoidal heterogeneity, we defined two classes of SSs, S_A and S_B. Additional classes can be specified and used when needed. Relative to S_B the S_A have a shorter path length and a smaller surface-to-volume ratio, whereas the S_B have a longer path length and a larger surface-to-volume ratio. The circumference of each SS is specified by a random draw from a bounded uniform distribution. To reflect the observed relative range of real sinusoid path lengths, SS length is given by a random draw from a gamma distribution having a mean and variance specified by the three gamma function parameters, α, β, and γ.

Solute objects can enter a SS at either the Core or the Rim. At each step thereafter until it is METABOLIZED or collected it has several stochastic options, the aggregate properties of which are arrived at through Monte Carlo simulation. In the Rim or Core it can move within that space, jump from one space to the other, or exit the SS. From a Rim location it can also jump to Grid B or back to the Core. Within Grid B it can move within the space, jump back to Grid A or to Grid C. When it encounters an ENDOTHELIAL CELL within Grid B it may (depending on its properties) PARTITION into it. Once inside, it can move about, exit, bind or not. Within Grid C it can move within the space or jump back to Grid B. When a HEPATOCYTE is encountered the SOLUTE can (depending on its properties) PARTITION into it or move on. Once inside a HEPATOCYTE it can move about, exit, bind (and possibly get METABOLIZED) or not. Currently all objects within a HEPATOCYTE that bind can also METABOLIZE. The probability of a solute object being METABOLIZED depends on the object's properties. Once METABOLIZED the object is destroyed. The only other way to exit a SS is from the Core, Rim or into bile (not implemented here). When the SOLUTE exits a SS and enters the CV, its arrival is recorded (corresponding to being collected), and it is destroyed.

---

[2] Because we are building a normalized model there is no direct coupling between grid points within the fine-grained space and real measures such as hepatocyte volume or its dimensions in microns.

2.5 Similarity Measure

The *in situ* liver perfusion protocol is detailed in [4]. Briefly, a compound of interest is injected into the entering perfusate of the isolated liver. The entire outflow is collected at intervals and the fraction of the dose within is determined. The ^{14}C-sucrose outflow profile contains information only about features of the extracellular environments whereas the data for a drug contains information on those features as well as on intracellular environments in that same liver. When the results of in silico and *in situ* experiments are similar, an expert can inspect the two data sets and offer an opinion on the degree of their similarity. However, automated model generation and refinement requires having one or more Similarity Measure (SM) to substitute for the expert's judgment. A SM is a function which takes two sets of experimental data and returns a number as a measure of their similarity [7]. Classical regression approaches do not apply because we are comparing the outputs of two or more experiments.

There are two main contributors to intraindividual variability: methodological and biological. For replicate experiments in the same liver the coefficient of variation for fraction of dose within specific outflow collection intervals typically ranges between 10 and 40%. A coefficient of variation can define a continuous interval bounding the experimental data. Any new set of results that falls within those bounds and has essentially the same shape is defined as being experimentally indistinguishable. The same should hold even if the data comes from an in silico experiment, and that provides the basis for selecting and evaluating SMs.

The objective of the SM is to help select among device designs, not simply to specify a device and select among variations on that device. Hence, the successful SM must target the various features of the outflow profile that correlate with the generative structures and building blocks inside the device. However, for simplicity in these early studies we have assumed that the coefficients of variation of repeat observations within different regions of the curve are the same. In that way we can use a simple interval SM, which is what we do. A set of *in situ* outflow profiles, T, is used as training data. From this data, we calculate a distance, D, from a reference that will be the basis for a match. We then take two outflow profiles and pick one to be the reference profile, P^r. For each observation in P^r, create a lower, P^l, and an upper, P^u, bound by multiplying that observation by $(1 - D)$ and $(1 + D)$, respectively. The two curves P^l and P^u are the lower and upper bounds of a band around P^r. The two outflow profiles are deemed similar if the second profile, P, stays within the band. The distance D used for sucrose is one standard deviation of the array of relative differences between each repeat observation and the mean observations at that time. To calculate D for the training set T we choose experimental data on different subjects that were part of the same protocol [4, 5].

3 In Silico Experimental Results

To begin development of a new device we select an outflow profile and begin the process of finding the simplest design, given restrictions: one is that it be comprised of the minimum components needed to generate an acceptably similar profile. An initial unrefined parameterization is chosen based on available information. Components are added according to that initial parameterization. If the behavior of the resulting device is not satisfactory, any given piece of the device may be surplused and replaced, modified, or reparameterized with minimal impact on the other components within the device. This process continues until the device provides reasonable coverage of the targeted solution space. Once an acceptable parameterization is found, the parameter space is searched further [10] for additional solution sets. Bounds for the parameter space can be specified to indicate solution set regions for which the device validates (i.e., acceptable SM measures are obtained). Subsequently, for a second data set, a repeat of the first experiment or a data set for a second drug, we make only minimal adjustments and additions to the structure of the first device so that the resulting new device generates acceptable outflow profiles for both data sets.

The typical *in situ* outflow profile is an account of on the order of 10^{15} drug molecules percolating through several thousand lobules. The typical in silico dose for *one* run with one LOBULE is on the order of 5,000 drug objects, where each drug object can represent a number (• 1) of drug molecules. Thus, a resulting single outflow profile will be very noisy and will be inadequate to represent the referent *in situ* profile. Another independent run with that same device, parameter settings, and dose will produce a similar but uniquely different outflow profile. Changing the random number generator seed alters the specifics for all stochastic parameters (e.g., placement of SSs on the digraph), thus providing a unique, individual version of the LOBULE, analogous to the unique differences between two lobules in the same liver. A full in silico experiment is one that produces an outflow profile that is sufficiently smooth to use the SM, and typically combines the results from 20 or more independent runs using the same LOBULE.

One can identify several similar parameterizations that will yield outflow profiles that are experimentally indistinguishable from each other. That is because there is a region of device-structure space (model space) that will yield acceptable behaviors relative to a specific data set. Such a region can be viewed as a metaphor for the fact that all lobules are similar but not identical. We currently do not attempt to fully map acceptable regions of either model or parameter space. Our goal is simply to locate a region in each that meets our objectives.

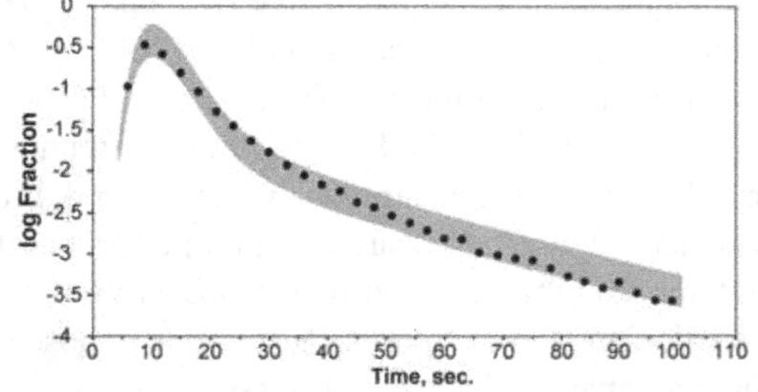

Fig. 3. An outflow profile for a device parameterized to match a sucrose outflow profile. Nodes per Zone: 55, 24 and 3 for Zones I, II and III, respectively; total edges: 60; intra-zone connections: Zone I = 10, Zone II = 8, Zone III = 0; inter-zone connections: I → II = 14, I → III = 4, II → III = 14; SSs: 50% S_A and S_B; Number of runs = 100

Figure 3 shows results from one parameterization of a LOBULE against a perfused liver sucrose outflow profile. The shaded region is a band enclosing the mean fraction of dose collected for each collection interval. The width of the band is ± 1 std about the mean. The filled circles are results obtained using the specified device parameter vector.

4 Higher and Lower Levels of Resolution

Different regions of a normal liver are indistinguishable. That is generally not the case for a diseased liver, where some lobules can be damaged or otherwise changed. So, to understand and account for such differences we need methods to shift levels of resolution without loss of information. The following summarizes how we are enabling resolution changes. To represent a whole diseased liver with heterogeneous properties we can first connect in parallel four to five different sized lobes, where each is a directed graph having multiple parallel, single node paths connecting portal and hepatic veins nodes (Fig. 4), and agents representing secondary units are placed at those nodes. A lobe is comprised of a large number of these units [8]. Each secondary unit can be similarly represented by a directed subgraph with LOBULES placed at each of its nodes (insert, Fig. 4). When subcellular networks within CELLS located in Grids B and C are needed, they may also be treated as directed graphs with nodes representing factors and with edges representing interactions and influences [11]. Having a mechanism for realizing networks allows us to replace sub-networks (at any location) with rules-based software modules.

5 Conclusion

We have tested and affirmed the hypothesis that perfused liver outflow data obtained following bolus administration of sucrose can, in conjunction with other data, be used to specify and parameterize a physiologically recognizable hepato-mimetic device that can generate outflow profiles that are experimentally indistinguishable from the original *in situ* data. Each device is *constructed* from software components that exhibit several essential properties including being designed to map logically to hepatic components at multiple levels of resolution, from subcellular to whole organ. This new technology is intended to provide powerful tools for optimizing the designs of real experiments. It will also help us challenge our understanding of how mammalian systems function in normal and diseased states, and when stressed or confronted with interventions.

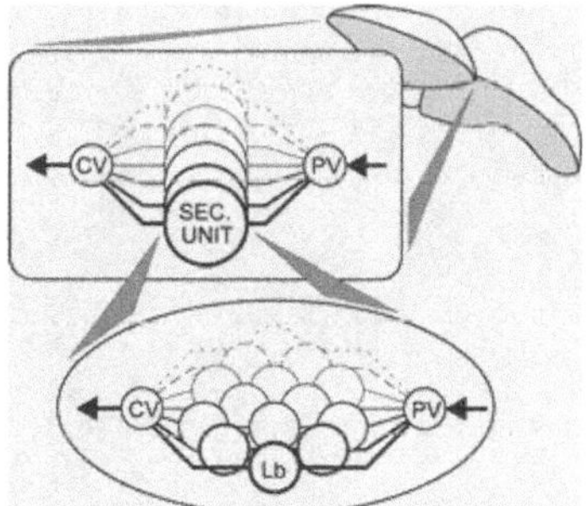

Fig. 4. An illustration of the hierarchical structure of an in silico liver. A lobe is comprised of a network of secondary units (SEC.UNIT) [8]; they, in turn are comprised of a network of lobules (Lb) as pictured in Fig. 2. PV: Portal vein. CV: Central hepatic vein

References

1. Roberts, M., Magnusson, B., Burczynski, F., Weiss, M.: Enterohepatic Circulation: Physiological, Pharmacokinetic and Clinical Implications. Clin. Pharmacokinet. 41(2002), 751-90

2. Noble D.: The Rise of Computational Biology. Nat. Rev. Mol. Cell. Bio. 3 (2002) 460-63
3. Noble, D.: The Future: Putting Humpty-Dumpty Together Again. Biochem. Soc. Transact. 31 (2003) 156-8
4. Roberts, M.S., Anissimov, Y.G.: Modeling of Hepatic Elimination and Organ Distribution Kinetics with the Extended Convection-Dispersion Model. J. Pharmacokin. Biopharm. 27 (1999) 343-382
5. Hung, D.Y., Chang, P., Weiss, M., Roberts, M.S.: Structure-Hepatic Disposition Relationships for Cationic Drugs in Isolated Perfused Rat Livers: Transmembrane Exchange and Cytoplasmic Binding Process. J. Pharmacol. Exper. Therap. 297 (2001) 780–89
6. Falkenhainer, B., Forbus, K.D.: Compositional Modeling: Finding the Right Model for the Job. Art. Intel. 51 (1991), 95-143
7. Santini, S., Jain, R.: Similarity Measures, IEEE Tran. Pattern Analysis and Machine Intelligence 21 (1999) 871-83
8. Teutsch, H.F., Schuerfeld, D., Groezinger, E.: Three-Dimensional Reconstruction of Parenchymal Units in the Liver of the Rat. Hepatology 29 (1999) 494-505
9. Burns, A., Davies, G.: Concurrent Programming. Addison-Wesley, Reading, MA (1993) 1-2
10. Sanchez, S.M., Lucas, T.W.: Exploring the World of Agent-Based Simulations: Simple Models, Complex Analyses. In: Yücesan, E., Chen, C.-H., Snowdon, J.L., Charnes, C.M., (eds.): Proceedings of the 2002 Winter Simulation Conference (2002) 116-26
11. Gumucio, J.J., Miller, D.L.: Zonal Hepatic Function: Solute-Hepatocyte Interactions Within the Liver Acinus. Prog. Liver Diseases 7 (1982) 17-30
12. Tyson, J.J., Chen, K., Novak, B.: Network Dynamics and Cell Physiology. Nat. Revs. Mol. Cell Biol. 2 (2001) 908-17

Building and Analysing an Integrative Model of HIV-1 RNA Alternative Splicing

A. Bockmayr[1], A. Courtois[1], D. Eveillard[1,2], and M. Vezain[1]

[1] LORIA, UMR 7503 (CNRS, INRIA, Universités de Nancy), BP 239
54506 Vandœuvre-lès-Nancy, France
[2] Laboratoire de Maturation des ARN et Enzymologie Moléculaire,
UMR 7567 CNRS-UHP, BP 239, 54506 Vandoeuvre-lès-Nancy, France
{bockmayr, acourtoi, eveillar, vezain}@loria.fr

Abstract. We present a multi-site model describing the alternative use of the RNA splicing sites A3, A4, A5 and A7 in the human immunodeficiency virus HIV-1. Our goal is to integrate experimental data obtained on individual splicing sites into a global model of HIV-1 RNA alternative splicing. We give a qualitative validation of our model, and analyse the possible impact of variations of regulatory protein concentrations on virus multiplication.

1 Introduction

The life cycle of the human immunodeficiency virus HIV-1 involves the production of different kinds of proteins. Among them the Tat and Nef proteins, which are lethal to the cell by inducing the apoptotic process. Their regulation is based on alternative splicing. Today, the control of the virus life cycle is only partially known. In order to improve existing knowledge on this finely regulated process, additional molecular biological experiments are needed. At present, the alternative splicing regulation is understood through experiments focusing on one particular splicing site. The global splicing behaviour is not studied experimentally, despite its importance for the HIV-1 life cycle. To overcome this difficulty, we propose an integrative modeling approach of alternative splicing regulation. Using an hybrid automaton with default reasoning, we construct a model that integrates the regulation at individual splicing sites into a multi-site model. Based on this modeling approach, we can study the impact of the regulation at one activator site, characterised by single-site experiments, on the production of the lethal HIV-1 proteins. In order to validate our approach, the model that covers different biological scales is analysed in a qualitative way.

The organisation of the paper is as follows. We start in Sect. 2 with a biological description of alternative splicing regulation. Based on a number of biological hypotheses in Sect. 3, we introduce in Sect. 4 an hybrid automaton model to describe the behaviour of the splicing process in a multi-site context. In Sect. 5, we analyse this model in a qualitative way, studying the impact of increasing the concentration of hnRNP A1 proteins. Based on biological queries and a model checking approach, the model is validated in Sect. 6 in a qualitative way. Sect. 7 contains some conclusions and perspectives for future work.

V. Danos and V. Schachter (Eds.): CMSB 2004, LNBI 3082, pp. 43–57, 2005.

Compared to our earlier work, the main contributions of this paper are as follows. While in [1, 2] we introduced our general modeling methodology, based on hybrid concurrent constraint programming, we develop here for the first time a realistic model of alternative splicing regulation in HIV-1, which integrates the sites A3, A4, A5, and A7. Furthermore, we give a qualitative analysis of this model to verify and discover different biological properties. Although further refinements are possible and desirable, this model can already be used to study *in silico* interesting biological questions, in particular the impact of variations of SR and hnRNP A1 regulatory protein concentrations on the virus life cycle.

2 Alternative Splicing: Regulation in Various Contexts

For eukaryotes and retroviruses such as HIV-1, the splicing reaction is a maturation process of the premessenger RNA (pre-mRNA), by elimination of non-coding sequences, introns, and junction of coding sequences, exons. The mature RNA (mRNA) can then be translated into a protein. The splicing reaction takes place in a large ribonucleoproteic complex called the spliceosome, which recognises signals on the pre-mRNA. Among them are the donor site (SD) and the acceptor site (SA), which border the non-coding intron [3]. Some pre-mRNAs are alternatively spliced, using competing splicing sites, which allows for the production of several different mRNAs from the same pre-mRNA. The selection of an alternative splicing site generally depends on a regulation ensured by the binding of an activator and/or an inhibitory protein. Members of the Serine-Arginine rich (SR) protein family (such as ASF/SF2 and SC35) are activator proteins, which help to recognise the regulated splicing site by the spliceosome. In contrast, the hnRNP A1 protein is well-known to inhibit splicing sites; in some cases, it acts by masking the site to the spliceosome [4].

For HIV-1 RNA splicing, the situation is more complex: the unique viral pre-mRNA contains 4 donor sites (D1 to D4) and 8 acceptor sites (A1 to A7), see Fig. 1. They can be used in different combinations to produce about 40 mRNAs, during 2 steps of the viral infection [5]. This controls the HIV-1 protein expression necessary to the production of a new virion. In the early phase, the viral pre-mRNA undergoes multiple splicing; the mature mRNAs produced have in common a splicing event between sites D4 and A7. Among them, *tat* mRNAs are also spliced at site A3 and contain two exons, coding an activator of the viral transcription, Tat. Other early mRNAs are the *nef* mRNAs spliced at site A5, and the *rev* mRNAs, spliced at any of the sites A4c, A4a, or A4b [5].

When enough Rev protein is synthesised, a shift from the early to the late phase of the infection occurs [6]. Indeed, the protein binds to the Rev Responsive Element (RRE) localised in the D4-A7 intron before this intron is spliced, and only incompletely spliced mRNAs characteristic of the late phase are produced. They all contain the D4-A7 intron and some of them are spliced at site A3, giving a Tat protein encoded by only one exon. Other sites (A4c, A4a, A4b and A5) are used to produce the *env* mRNAs [5].

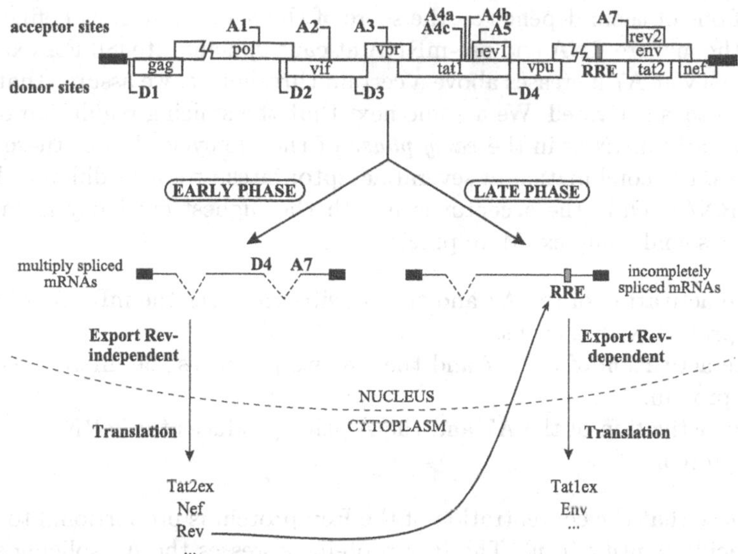

Fig. 1. Alternative splicing of HIV-1 RNA. Boxes delimitate the sequence of encoded proteins (ORF). In the mRNAs, spliced introns are designated by dotted lines, and exons by thick lines

The competing sites A3 to A5 are used to produce major viral proteins. Experimental studies show that these sites are regulated. Site A3 can be inhibited by the binding of the protein hnRNP A1 to the ESS2 (Exon Splicing Silencer) element, or activated by fixation of SR proteins to the ESEt (Exon Splicing Enhancer) element. SC35 proteins can also bind to the ESS2 element. They are expected to activate splicing at the site A3 by competing with the hnRNP A1 proteins binding to the ESS2 element.

3 Modeling Multi-site Alternative Splicing Regulation

We model the regulation by SR and hnRNP A1 proteins in the multi-site context (A3, A4, A5, A7) under several hypotheses. We consider the biological process of alternative splicing in an *in vitro* context, which corresponds to the experimental knowledge. Our model describes the dynamics of a closed system of nucleic acid concentrations. The state variables are the pre-mRNAs (*rna*) which become mRNAs (*tat1, tat2, rev, env, nef*) of different proteins. The splicing process depends on the SC35, ASF/SF2 and hnRNP A1 regulatory proteins. They activate four splicing acceptors sites A3, A4, A5, and A7. We neglect the regulation at the donor sites because these seem to be less regulated than the acceptor sites [7, 5]. The four acceptor sites are in competition for the regulatory protein allocation. We consider this regulation level only on a single-site scale. Thus we assume the local behaviour as a continuous competition between four acceptor sites. The choice of one of them depends on the score of the *splice efficiency* defined as the ratio of the mature RNA and pre-mRNA at each splicing site [8]. For example, if the efficiency at A7 increases above a certain threshold τ, we assume that the A7 acceptor site is activated. We assume next that the splicing regulation of HIV-1 produces early mRNAs in the *early phase* of the life cycle. Under these cellular conditions, the combination of several acceptor sites produces different kinds of mature RNAs. Only the acceptor site with the highest efficiency is chosen by the spliceosomal complex. More precisely,

- splice activation of the A7 and the A3 site produces the mRNA *tat2* for the Tat protein with 2 exons.
- splice activation of the A7 and the A4 site produces the mRNA *rev* for the Rev protein.
- splice activation of the A7 and the A5 site produces the mRNA *nef* for the Nef protein.

We suppose that the concentration of the Rev protein is proportional to the Rev nucleic acid quantity (*rev*). The Rev protein represses the A7 splicing site. We assume in our local model that the quantity of the Rev nucleic acid reduces the efficiency of the A7 splicing site. This hypothesis corresponds to a time-scale abstraction of the translation to proteins, in order to focus on the dynamics of the splicing regulation.

If the A7 efficiency decreases below a certain threshold, the A7 splicing site is deactivated. Now we consider the biological system to be in the *late phase*. Without splicing activation at the A7 site:

- splice activation at the A3 site produces the mRNA *tat1* for the Tat protein with 1 exon.
- splice activation at the A4 or A5 site produces the mRNA *env* for the Env protein.

The choice between different acceptor sites depends on the concentration of regulatory proteins that control the splicing site activity. Modeling the regulation of the A3 site [2] allowed us to extract the splice efficiency eff_3, defined as the ratio of the mature RNA and pre-mRNA at equilibrium, as an observer variable of the local behaviour at the acceptor site A3. Depending on the concentration of hnRNP A1, eff_3 is given by the formula

$$eff_3(hnRNPA1) = \frac{\eta \cdot P_{ESEt} \cdot (ASF+SC35) \cdot (k_{ESS2a} \cdot (k_{ESS2b}+SC35)+hnRNPA1 \cdot k_{ESS2b})}{\eta' \cdot (k_{ESE}+ASF+SC35) \cdot P_{ESS2} \cdot hnRNPA1 \cdot k_{ESS2b}} \quad (1)$$

A similar approach for the acceptor site A7 yields an efficiency function eff_7, integrating the effects of SR and Rev proteins:

$$eff_7(hnRNPA1, rev) = \frac{\frac{P_{ESE3} \cdot ASF}{k_{ESE3b} \cdot (1+\frac{hnRNPA1}{k_{ESE3}})+ASF} + \frac{P_{ESS3ab} \cdot SC35}{k_{ESS3ab}+SC35}}{\frac{P_{ISS} \cdot hnRNPA1}{k_{ISS}+hnRNPA1} + \frac{P_{ESS3b} \cdot hnRNPA1}{k_{ESS3b}+hnRNPA1} + \frac{P_{RRE} \cdot rev}{k_{RRE}+rev}} \quad (2)$$

In both formulas, $P_x > 0$ (resp. $k_x > 0$) are Michaelis-Menten constants that correspond to the maximal affinity (resp. half-saturation coefficient) at the binding site x. The concentrations of the regulatory proteins ASF/SF2, SC35, and hnRNP A1 are denoted by ASF, $SC35$, $hnRNPA1$, respectively. Finally, η, η' are kinetic constants.

The A4 and A5 acceptor sites are not well-studied yet, due to the complexity of the corresponding nucleic acid domain [9]. Their local behaviour is captured by the splice efficiencies eff_4 and eff_5, which are defined as interval constants. To model the relative interaction between those sites, we use default reasoning. Following [10], we describe A5 as a preferential activated site in the presence of regulatory proteins, and A4 as the default splicing site.

4 Integrative Model

To model the global alternative splicing regulation according to the hypotheses introduced in Sect. 3, we use an hybrid automaton with default reasoning, called $\mathcal{A}$, which is given in Fig. 2. There are five states corresponding to the production (after splicing) of the mRNAs *rev, env, nef, tat1, tat2* for the proteins Rev, Env, Nef, Tat with 1 exon, and Tat with 2 exons, respectively. In each state, the dynamics is described by a set of differential equations, and a law for matter conservation.

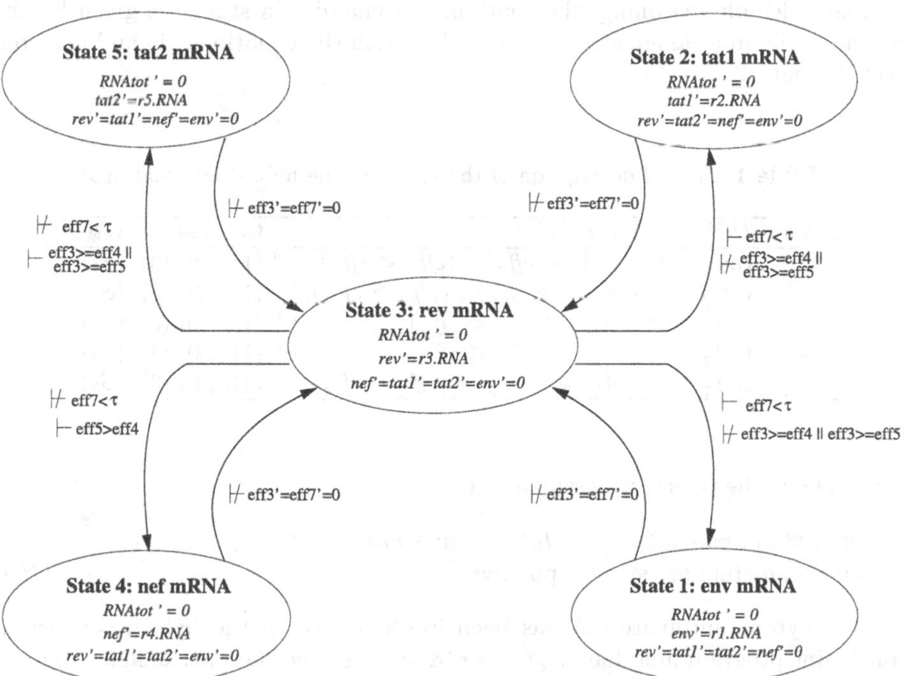

Fig. 2. Hybrid automaton $\mathcal{A}$ for the alternative splicing regulation in a multi-site context

The transitions from one state to another depend on constraints. We will consider two types of transitions from a state i to a state j:

- a transition $i \xrightarrow{\vdash c} j$ occurs if in state i the condition c holds.
- a transition $i \xrightarrow{\nvdash c} j$ occurs if in state i the condition c does not hold. This may have two reasons: either $\neg c$ holds or c is unknown (default reasoning).

  Equivalently, we will say that $\neg c$ holds *by default*, and then write $i \xrightarrow{\vdash^d \neg c} j$

For reasons of readability, we give in Fig. 2 only the default transitions. A complete description of the automaton can be found in Tab. 1.

To be in state $i \in \{1, \ldots, 5\}$, the constraint $\mathcal{D}_i(hnRNPA1, rev)$ from Tab. 1 has to be satisfied. Here, $eff_3(hnRNPA1)$ (resp. $eff_7(hnRNPA1, rev)$) are abbreviated by eff_3 (resp. eff_7). The vector $\boldsymbol{u} = (env, tat1, rev, nef, tat2)^T \in \mathbb{R}^5_{\geq 0}$ evolves in state i according to the system of differential equations

$$du_i/dt = r_i \cdot rna, \quad du_j/dt = 0, \text{ for } j \neq i \qquad (\mathcal{S}_i)$$

with rate constants $r_1, \ldots, r_5 > 0$. We also assume that the total quantity of RNA, $rna_{tot} = rna + env + tat1 + tat2 + nef + rev$ is invariant, i.e.

$$drna_{tot}/dt = 0 \qquad (3)$$

Using default reasoning, the continuous dynamics in state i is given by the system of differential equations $\mathcal{S}_i$, together with the equations (1) and (2) from Sect. 3, and (3).

Table 1. Formal description of the states of the hybrid automaton $\mathcal{A}$

State i	$\mathcal{D}_i(hnRNPA1, rev)$		Continuous system
1	$(eff_7 < \tau) \wedge (eff_3 < eff_5) \wedge (eff_3 < eff_4)$	$\vdash^d$	(1), (2), (3), $(\mathcal{S}_1)$
2	$(eff_7 < \tau) \wedge [(eff_3 \geq eff_5) \vee (eff_3 \geq eff_4)]$	$\vdash^d$	(1), (2), (3), $(\mathcal{S}_2)$
3	$(eff_7 \geq \tau) \wedge (eff_3 < eff_5 < eff_4)$	$\vdash^d$	(1), (2), (3), $(\mathcal{S}_3)$
4	$(eff_7 \geq \tau) \wedge (eff_3 < eff_4 \leq eff_5)$	$\vdash^d$	(1), (2), (3), $(\mathcal{S}_4)$
5	$(eff_7 \geq \tau) \wedge [(eff_3 \geq eff_5) \vee (eff_3 \geq eff_4)]$	$\vdash$	(1), (2), (3), $(\mathcal{S}_5)$

Throughout the paper, we assume that

- $\forall t \in \mathbb{R}_{\geq 0} : rna_{tot} > tat_1 + tat_2 + rev + env + nef$
- all parameters are strictly positive $\qquad (\mathcal{H}_{\mathcal{P}})$

The hybrid automaton $\mathcal{A}$ has been implemented in the hybrid concurrent constraint programming language `Hybrid cc` [11], see the part `GENERIC MODEL` in the program given in Fig. 3. In `Hybrid cc`, the pair of constraints $c \vdash^d A, \neg c \vdash B$ is noted as `unless` $\neg c$ A `else` B. Furthermore, f' denotes the time derivative

```
hnRNPA1=...; tat2=...; tat1=...; rev=...; env=...; nef=...;
      /*** VARIABLE INITIALISATION at t=0 with interval constants ***/

always{                                              /*** GENERIC MODEL ***/
  unless (eff7 >= tau) {
    tat2'=0;  rev'=0;  nef'=0;
    unless (eff3 >= eff5 || eff3 >= eff4) {                    //State 1//
      tat1'=0;  env'=r1*(ARNtot-tat1-tat2-rev-env-nef);
    }
    else {                                                     //State 2//
      tat1'=r2*(ARNtot-tat1-tat2-rev-env-nef);
      env'=0;
    };
  }
  else {
    env'=0;  tat1'=0;
    unless(eff3 >= eff4 || eff3 >= eff5) {
      tat2'=0;
      unless (eff5 >= eff4) {                                  //State 3//
        rev'=r3*(ARNtot-tat1-tat2-rev-env-nef);  nef'=0;
      }
      else {                                                   //State 4//
        nef'=r4*(ARNtot-tat1-tat2-rev-env-nef); rev'=0;
      };
    }
    else {                                                     //State 5//
      tat2'=r5*(ARNtot-tat1-tat2-rev-env-nef);  rev'=0;  nef'=0;
    };
  };
  eff3=(eta*PESEt*(ASF+SC35)*...;
  eff4=[v1,v3];                                         // 0 <= v1 <= v3
  eff5=[w1,w3];                                         // 0 <= w1 <= w3
  eff7=(PESE3*ASF/(KESE3b*...;
}

always {                                         /*** SPECIALISED MODEL ***/
  eff4=[v2,v2];  eff5=[w2,w2];             //v2 in [v1,v3], w2 in [w1,w3]
  hnRNPA1'=kh*hnRNPA1;
}
```

Fig. 3. Hybrid cc program implementing the hybrid automaton

df/dt of f. As argued in [1], constraint programming with *default logic* is well-suited for modeling biological systems because it allows us to handle partial or incomplete information. Each constraint gives one piece of information on the system that is studied. Using logical combinators, the hybrid automaton uses this information for the transition from one state to the other.

5 Model Analysis

Integrated modeling of local and global behaviours allows us to study alternative splicing regulation for different parameter variations. We may introduce in the hybrid automaton variations of hnRNP A1, ASF/SF2 or SC35 proteins based on the corresponding production rates. In order to validate our integrative model, we apply an infinite-time analysis to the hybrid automaton. This kind of analysis aims at answering biological queries about the system, without knowing a priori about numerical parameterisation and initial values. For the purpose of this paper, the generic model in Sect. 4 has been specialised by introducing a function *hnRNPA1*(t) describing the variation of hnRNP A1:

$$d\,hnRNPA1(t)/dt = k_h \cdot hnRNPA1(t), \quad \text{with} \quad k_h > 0 \tag{4}$$

Other definitions of *hnRNPA1*(t) would be possible. For our analysis, we only need that *hnRNPA1*(t) is strictly growing, and that *hnRNPA1*$(0) > 0$. Indeed, the sign of the partial derivative $\partial eff_3/\partial hnRNPA1$ depends only on the sign of the derivative $hnRNPA1'$. So $hnRNPA1' > 0$ implies $\partial eff_3/\partial hnRNPA1 < 0$. The sign of $\partial^2 eff_7/\partial hnRNPA1\, \partial rev$ depends on the signs of $hnRNPA1'$ and rev'. So $hnRNPA1' > 0$ and $rev' \geq 0$ implies $\partial^2 eff_7/\partial hnRNPA1\, \partial rev < 0$. As eff_3 and eff_7 are monotonic, Tab. 2 below will not change, and the same analysis can be performed.

Note that since *hnRNPA1* is now totally defined on $\mathbb{R}_{\geq 0}$, the default operator $\vdash^d$ is not needed anymore in the specialised program, i.e. we could replace the statement `unless` $\neg c\, A$ `else` B, with `if` c `then` A `else` B.

Our analysis of the hybrid automaton proceeds in several steps. To determine the possible behaviour in each individual state, we first analyse the signs of the (partial) derivatives and the limits of eff_3 and eff_7. From that we construct a transition table, which is then used to obtain transition graphs. Finally, we eliminate indeterminism by adding constraints that have been inferred during the analysis. Applications to biological queries about the system will be given in Sect. 6.

Variations of eff_3 and eff_7. Let $x \nearrow y$ (resp. $\searrow, \rightarrow$) denote that a given function is strictly increasing (resp. decreasing, constant), with greatest lower bound x and least upper bound y. Let h_0 be the concentration of *hnRNPA1* at time $t = 0$, and r_0 (resp. r_{t_k}) the concentration of rev at time $t = 0$ (resp. $t = t_k$). Based on an analysis of the signs of the (partial) derivatives of eff_3 and eff_7 and using the lower and upper bounds

$$\begin{aligned}
\alpha &= eff_3(h_0) \\
\beta &= lim_{hnRNPA1 \to +\infty} eff_3 = \frac{(ASF+SC35)\cdot P_{ESEt}\cdot \eta}{\eta'\cdot(k_{ESE}+ASF+SC35)\cdot P_{ESS2}} \\
\gamma &= eff_7(h_0, r_0)
\end{aligned}$$

$$\begin{aligned}\delta_{t_k} &= lim_{hnRNPA1 \to +\infty} eff_7(r_{t_k}) \\ &= \frac{P_{ESS3ab} \cdot SC35 \cdot (k_{RRE} + r_{t_k})}{(k_{ESS3ab} + SC35)(P_{ISS} \cdot k_{RRE} + P_{ISS} \cdot r_{t_k} + P_{ESS3b} \cdot k_{RRE} + P_{ESS3b} \cdot r_{t_k} + P_{RRE} \cdot r_{t_k})}\end{aligned}$$

$$\varepsilon \quad = lim_{(hnRNPA1, rev) \to (+\infty, +\infty)} eff_7 = \frac{P_{ESS3ab} \cdot SC35}{(P_{RRE} + P_{ISS} + P_{ESS3b}) \cdot (k_{ESS3ab} + SC35)}$$

the qualitative behaviour in the different states is given in Tab. 2.

Table 2. Qualitative behaviour in the different states

	t	*hnRNPA1*	*rev*	eff_3	eff_7
States1, 2, 4, 5	$0 \nearrow +\infty$	$h_0 \nearrow +\infty$	$r_{t_k} \to r_{t_k}$	$\alpha \searrow \beta$	$\gamma \searrow \delta_{t_k}$
State3			$r_0 \nearrow +\infty$		$\gamma \searrow \varepsilon$

For all $(hnRNPA1, rev) \in \mathbb{R}_{>0} \times \mathbb{R}_{\geq 0}$, we have $\varepsilon < eff_7(hnRNPA1, rev) < \gamma$ and $\varepsilon < \delta_{t_k} < \gamma$.

Using the table above, we can determine under which conditions the automaton will stay definitely in a given state i, denoted $i \circlearrowleft_\infty$:

State 1: Since eff_7 and eff_3 decrease, the system will stay in state 1 once it has reached it.
State 2: The system will stay in state 2, if $\beta \geq \max\{eff_4, eff_5\}$.
State 3: The system will stay in state 3, if $\varepsilon \geq \tau$.
State 4: The system will stay in state 4, if $\delta_{t_{k'}} \geq \tau$, for some $t_{k'} > 0$.
State 5: The system will stay in state 5, if $\delta_{t_{k''}} \geq \tau$, for some $t_{k''} > 0$, and $\beta \geq \max\{eff_4, eff_5\}$.

Transitions from state i to state j. Next we analyse which transitions are possible in the hybrid automaton. Tab. 3 summarises these results. Col. 1 indicates the transition $i \to j$ in question. Col. 2 gives necessary conditions for this transition, for a given time t. Cols. 3 to 5 recall the behaviour of rev, eff_3, eff_7 in state i. As a conclusion, Col. 6 indicates whether or not transition $i \to j$ is possible. Failing conditions in Col. 2 are underlined.

The expression τ^- means that the limit τ is reached from below. eff_3 and eff_7 are abbreviations for $eff_3(hnRNPA1)$ and $eff_7(hnRNPA1, rev)$.

Finally we observe that for $i \in \{1, .., 4\}$, transitions $i \to i+1$ are not allowed. So $\delta_{t_{k''}}$ and $\delta_{t_{k'}}$ are unique if they exist, and $\delta_{t_{k''}} \geq \delta_{t_{k'}}$.

Transition graphs. To classify the qualitative behaviour of the hybrid automaton, we introduce the following five constraints, which have been obtained during the analysis given before:

Table 3. Table of transitions

$\mathrm{T}_{i\,\rightarrow j}$	Conditions on eff_3 and eff_7 for $i \rightarrow j$ transition	rev^i	eff_3^i	eff_7^i	possible
1 $\circlearrowleft_\infty$	$(\mathit{eff}_7 < \tau) \wedge (\mathit{eff}_3 < \mathit{eff}_5) \wedge (\mathit{eff}_3 < \mathit{eff}_4)$	$\rightarrow$	$\searrow$	$\searrow$	yes
$1 \rightarrow 2$	$\underline{\mathit{eff}_3 \nearrow} \wedge (\mathit{eff}_3 = \mathit{eff}_5^- \vee \mathit{eff}_3 = \mathit{eff}_4^-)$	$\rightarrow$	$\searrow$	$\searrow$	no
$1 \rightarrow 3$	$\underline{\mathit{eff}_7 \nearrow} \wedge (\mathit{eff}_7 = \tau^-)$	$\rightarrow$	$\searrow$	$\searrow$	no
$1 \rightarrow 4$	$\underline{\mathit{eff}_7 \nearrow} \wedge (\mathit{eff}_7 = \tau^-)$	$\rightarrow$	$\searrow$	$\searrow$	no
$1 \rightarrow 5$	$\underline{\mathit{eff}_7 \nearrow} \wedge \underline{\mathit{eff}_3 \nearrow} \wedge (\mathit{eff}_7 = \tau^-)$ $\wedge (\mathit{eff}_3 = \mathit{eff}_5^- \vee \mathit{eff}_3 = \mathit{eff}_4^-)$	$\rightarrow$	$\searrow$	$\searrow$	no
$2 \rightarrow 1$	$\mathit{eff}_3 \searrow \wedge (\mathit{eff}_3 = \mathit{eff}_5 \vee \mathit{eff}_3 = \mathit{eff}_4)$	$\rightarrow$	$\searrow$	$\searrow$	yes
2 $\circlearrowleft_\infty$	$(\mathit{eff}_7 < \tau) \wedge (\mathit{eff}_3 \geq \mathit{eff}_5 \vee \mathit{eff}_3 \geq \mathit{eff}_4)$ $\wedge (\beta \geq \max\{\mathit{eff}_4, \mathit{eff}_5\})$	$\rightarrow$	$\searrow$	$\searrow$	yes
$2 \rightarrow 3$	$\underline{\mathit{eff}_7 \nearrow} \wedge \mathit{eff}_3 \searrow \wedge (\mathit{eff}_7 = \tau^-)$ $\wedge (\mathit{eff}_3 = \mathit{eff}_5 \vee \mathit{eff}_3 = \mathit{eff}_4)$	$\rightarrow$	$\searrow$	$\searrow$	no
$2 \rightarrow 4$	$\underline{\mathit{eff}_7 \nearrow} \wedge \underline{\mathit{eff}_3 \nearrow} \wedge (\mathit{eff}_7 = \tau^-)$ $\wedge (\mathit{eff}_3 = \mathit{eff}_5 \vee \mathit{eff}_3 = \mathit{eff}_4)$	$\rightarrow$	$\searrow$	$\searrow$	no
$2 \rightarrow 5$	$\underline{\mathit{eff}_7 \nearrow} \wedge (\mathit{eff}_7 = \tau^-) \wedge (\mathit{eff}_3 \geq \mathit{eff}_5 \vee \mathit{eff}_3 \geq \mathit{eff}_4)$	$\rightarrow$	$\searrow$	$\searrow$	no
$3 \rightarrow 1$	$\mathit{eff}_7 \searrow \wedge (\mathit{eff}_7 = \tau) \wedge (\mathit{eff}_3 < \mathit{eff}_4)$	$\nearrow$	$\searrow$	$\searrow$	yes
$3 \rightarrow 2$	$\mathit{eff}_7 \searrow \wedge \underline{\mathit{eff}_3 \nearrow} \wedge (\mathit{eff}_7 = \tau) \wedge (\mathit{eff}_3 = \mathit{eff}_4^-)$	$\nearrow$	$\searrow$	$\searrow$	no
3 $\circlearrowleft_\infty$	$(\mathit{eff}_7 \geq \tau) \wedge (\mathit{eff}_3 < \mathit{eff}_4 < \mathit{eff}_5) \wedge (\varepsilon \geq \tau)$	$\nearrow$	$\searrow$	$\searrow$	yes
$3 \rightarrow 4$	$\underline{\mathit{eff}_4 \nearrow} \vee \underline{\mathit{eff}_5 \searrow}$	$\nearrow$	$\searrow$	$\searrow$	no
$3 \rightarrow 5$	$\underline{\mathit{eff}_3 \nearrow} \wedge (\mathit{eff}_7 \geq \tau) \wedge (\mathit{eff}_3 = \mathit{eff}_4^-)$	$\nearrow$	$\searrow$	$\searrow$	no
$4 \rightarrow 1$	$\mathit{eff}_7 \searrow \wedge (\mathit{eff}_7 = \tau) \wedge (\mathit{eff}_3 < \mathit{eff}_5)$	$\rightarrow$	$\searrow$	$\searrow$	yes
$4 \rightarrow 2$	$\mathit{eff}_7 \searrow \wedge \underline{\mathit{eff}_3 \nearrow} \wedge (\mathit{eff}_7 = \tau) \wedge (\mathit{eff}_3 = \mathit{eff}_5^-)$	$\rightarrow$	$\searrow$	$\searrow$	no
$4 \rightarrow 3$	$(\mathit{eff}_7 \geq \tau) \wedge (\underline{\mathit{eff}_4 \searrow} \vee \underline{\mathit{eff}_5 \nearrow})$	$\rightarrow$	$\searrow$	$\searrow$	no
4 $\circlearrowleft_\infty$	$(\mathit{eff}_7 \geq \tau) \wedge (\mathit{eff}_3 < \mathit{eff}_5 \leq \mathit{eff}_4) \wedge (\delta_{k'} \geq \tau)$	$\rightarrow$	$\searrow$	$\searrow$	yes
$4 \rightarrow 5$	$\underline{\mathit{eff}_3 \nearrow} \wedge (\mathit{eff}_7 \geq \tau) \wedge (\mathit{eff}_3 = \mathit{eff}_5^-)$	$\rightarrow$	$\searrow$	$\searrow$	no
$5 \rightarrow 1$	$\mathit{eff}_7 \searrow \wedge \mathit{eff}_3 \searrow \wedge (\mathit{eff}_7 = \tau)$ $\wedge (\mathit{eff}_3 = \mathit{eff}_5 \vee \mathit{eff}_3 = \mathit{eff}_4)$	$\rightarrow$	$\searrow$	$\searrow$	yes
$5 \rightarrow 2$	$\mathit{eff}_7 \searrow \wedge (\mathit{eff}_7 = \tau) \wedge (\mathit{eff}_3 \geq \mathit{eff}_5 \vee \mathit{eff}_3 \geq \mathit{eff}_4)$	$\rightarrow$	$\searrow$	$\searrow$	yes
$5 \rightarrow 3$	$\mathit{eff}_3 \searrow \wedge (\mathit{eff}_7 \geq \tau) \wedge (\mathit{eff}_3 = \mathit{eff}_4) \wedge (\mathit{eff}_4 < \mathit{eff}_5)$	$\rightarrow$	$\searrow$	$\searrow$	yes
$5 \rightarrow 4$	$\mathit{eff}_3 \searrow \wedge (\mathit{eff}_7 \geq \tau) \wedge (\mathit{eff}_3 = \mathit{eff}_5) \wedge (\mathit{eff}_5 \leq \mathit{eff}_4)$	$\rightarrow$	$\searrow$	$\searrow$	yes
5 $\circlearrowleft_\infty$	$(\mathit{eff}_7 \geq \tau) \wedge (\mathit{eff}_3 \geq \mathit{eff}_5 \vee \mathit{eff}_3 \geq \mathit{eff}_4)$ $\wedge (\delta_{t_{k''}} \geq \tau) \wedge (\beta \geq \max\{\mathit{eff}_4, \mathit{eff}_5\})$	$\rightarrow$	$\searrow$	$\searrow$	yes

$$c_1 \equiv (\mathit{eff}_4 < \mathit{eff}_5), \qquad c_2 \equiv (\beta \geq \max\{\mathit{eff}_4, \mathit{eff}_5\}), \qquad c_3 \equiv (\varepsilon \geq \tau),$$
$$c_4 \equiv (\delta_{t_{k'}} \geq \tau), \qquad c_5 \equiv (\delta_{t_{k''}} \geq \tau)$$

Each of these constraints may be satisfied or not, depending on the initial values and the parameterisation. This yields $2^5 = 32$ theoretically possible behaviours. Since $\delta_{t_{k''}} \geq \delta_{t_{k'}}$, we have $c_4 \Rightarrow c_5$. Furthermore, if c_1 holds, then c_4 is not relevant. Therefore 20 different behaviours remain, see Tab. 4. Here, the relation symbols $<, \geq$ indicate whether a constraint or its negation holds. An irrelevant value is marked by "–". After simplification, we obtain 11 possible transition graphs, which are given in Fig. 4.

Table 4. The 20 behaviours depending on constraints. *Def states* means that the system will stay in the corresponding state, once it has reached it

c_1	c_2	c_3	c_4	c_5	Def states	c_1	c_2	c_3	c_4	c_5	Def states	c_1	c_2	c_3	c_4	c_5	Def states
$<$	$<$	$\geq$	$-$	$\geq$	1, 3	$<$	$<$	$\geq$	$-$	$<$	1, 3	$<$	$<$	$<$	$-$	$\geq$	1
$<$	$<$	$<$	$-$	$<$	1	$<$	$\geq$	$\geq$	$-$	$\geq$	1, 2, 3, 5	$<$	$\geq$	$\geq$	$-$	$<$	1, 2, 3
$<$	$\geq$	$<$	$-$	$\geq$	1, 2, 5	$<$	$\geq$	$<$	$-$	$<$	1, 2	$\geq$	$<$	$\geq$	$\geq$	$\geq$	1, 4
$\geq$	$<$	$\geq$	$<$	$\geq$	1	$\geq$	$<$	$\geq$	$<$	$<$	1	$\geq$	$<$	$<$	$\geq$	$\geq$	1, 4
$\geq$	$<$	$<$	$<$	$\geq$	1	$\geq$	$<$	$<$	$<$	$<$	1	$\geq$	$\geq$	$\geq$	$\geq$	$\geq$	1, 2, 4, 5
$\geq$	$\geq$	$\geq$	$<$	$\geq$	1, 2, 5	$\geq$	$\geq$	$\geq$	$<$	$<$	1, 2	$\geq$	$\geq$	$<$	$\geq$	$\geq$	1, 2, 4, 5
$\geq$	$\geq$	$<$	$<$	$\geq$	1, 2, 5	$\geq$	$\geq$	$<$	$<$	$<$	1, 2						

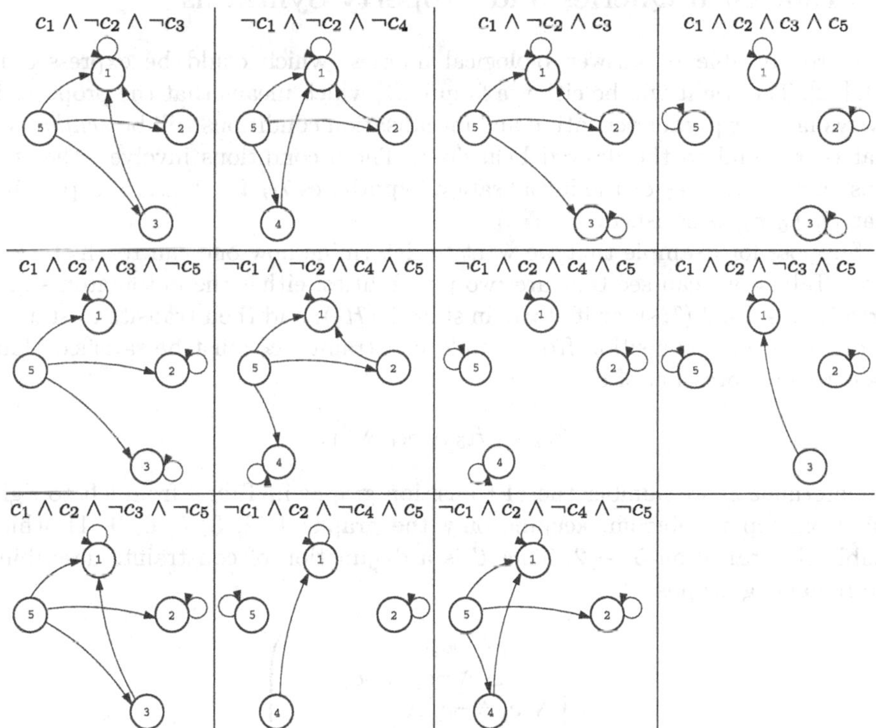

Fig. 4. Transition graphs

Adding new constraints. Note that 7 graphs are indeterministic. If the automaton is initialised in state 5, the successor state could be 1, 2, 3 or 4. In order to simplify the queries in Sect. 6, we eliminate this indeterminism by introducing additional constraints. Define
$t_m = \min\{t \in \mathbb{R}_{\geq 0} \mid \mathit{eff}_3(\mathit{hnRNPA1}(t)) = \mathit{eff}_4$ or $\mathit{eff}_3(\mathit{hnRNPA1}(t)) = \mathit{eff}_5\}$, if the minimum exists, and $t_m = -1$, otherwise. Given the constraints

$$c_6 \equiv (t_m \geq 0 \wedge \mathit{eff}_7(\mathit{hnRNPA1}(t_m), r_0) \geq \tau) \text{ and}$$
$$c_7 \equiv (t_m \geq 0 \wedge \mathit{eff}_7(\mathit{hnRNPA1}(t_m), r_0) = \tau),$$

$$\text{the successor state of 5 is } \begin{cases} 1, \text{ if } c_7 \\ 2, \text{ if } \neg c_6 \\ 3 \text{ or } 4, \text{ if } c_6 \wedge \neg c_7 \end{cases}$$

Together with $c_1, \ldots, c_5$, the constraints c_6, c_7 allow us to define $25 = 3 \times 7 + 4$ deterministic transition graphs, which cover the possible behaviours in the deterministic case.

6 Biological Queries and Property Synthesis

Now we are able to answer biological queries, which could be expressed in CTL [12]. The result will be either a failure $\varnothing$, which means that the property is always false, or parameterisation and initialisation conditions will be *synthesised* that correspond to the desired behaviour. These conditions involve either the constraints $c_1, \ldots, c_7$ or the initialisation hypotheses $\mathcal{H}_i, i = 1, \ldots, 5$, expressing that $\mathcal{D}_i(h_0, r_0)$ is satisfied.

Suppose for example that we want to determine how one can reach state 2. From Tab. 4, we can see that are two possibilities: either the automaton starts already in state 2 ($\mathcal{H}_2$), or it starts in state 5 ($\mathcal{H}_5$), and then transits to state 2. In order to get a transition from 5 to 2, constraint $\neg c_6$ must be satisfied. This leads to an expression of the form

$$\mathcal{H}_2 \vee [\, \mathcal{H}_5 \wedge \neg c_6 \wedge \mathcal{C}] \,.$$

To determine $\mathcal{C}$, we number the 11 transition graphs in Tab. 4 from left to right and from top to bottom, keeping only the graphs 1, 2, 3, 5, 6, 9, 11 which enable the transition $5 \rightarrow 2$. Thus $\mathcal{C}$ is a disjunction of constraints describing the remaining graphs :

$$\mathcal{C} = \begin{pmatrix} c_1 \wedge \neg c_2 \wedge \neg c_3 \\ \vee \neg c_1 \wedge \neg c_2 \wedge \neg c_4 \\ \vee\, c_1 \wedge \neg c_2 \wedge c_3 \\ \vee\, c_1 \wedge c_2 \wedge c_3 \wedge \neg c_5 \\ \vee \neg c_1 \wedge \neg c_2 \wedge c_4 \wedge c_5 \\ \vee\, c_1 \wedge c_2 \wedge \neg c_3 \wedge \neg c_5 \\ \vee \neg c_1 \wedge c_2 \wedge \neg c_4 \wedge \neg c_5 \end{pmatrix}$$

After simplification of $\mathcal{C}$, we obtain

$$\mathcal{H}_2 \vee \left[\mathcal{H}_5 \wedge \neg c_6 \wedge \begin{pmatrix} c_1 \wedge \neg c_5 \\ \vee \neg c_2 \wedge c_5 \\ \vee \neg c_4 \wedge \neg c_5 \end{pmatrix} \right]$$

Further examples of biological queries and answers are given in Tab. 5.

Table 5. Synthesising conditions to satisfy the goals

Query/Goal	Conditions
1) Reach state 3	$[\mathcal{H}_5 \wedge c_1 \wedge (\neg c_2 \vee \neg c_5) \wedge c_6 \wedge \neg c_7] \vee \mathcal{H}_3$
1') Reach and exit from state 3	$\neg c_3 \wedge [(\mathcal{H}_5 \wedge c_1 \wedge (\neg c_2 \vee \neg c_5) \wedge c_6 \wedge \neg c_7) \vee \mathcal{H}_3]$
2) Transit from {1,2} to {3,4,5}	$\varnothing$
3) Reach state 2	$\mathcal{H}_2 \vee [\mathcal{H}_5 \wedge \neg c_6 \wedge (c_1 \wedge \neg c_5 \vee \neg c_2 \wedge c_5 \vee \neg c_4 \wedge \neg c_5)]$

Query 1 yields the well-known qualitative biological conditions to be in the state which produces the *rev* mRNA [13]. In order to be validated, our model should give a positive answer to this query and produce some sufficient conditions.

Query 2 corresponds to a question where we expect a negative answer. Similarly to Query 1, the result is a priori well-known. It concerns the crucial switch between the early and late phase in the virus life cycle. In a validated model, the virus cannot switch back from the late to the early phase. Thus the negative answer agrees with biological knowledge.

Correct answers to such queries give a partial validation of the integrative model. Furthermore, the hybrid automaton allows us to synthesise sufficient conditions for a given property.

Since our model is in accordance with existing biological knowledge, we may also ask a different type of query, where we do not know the answer yet. Query 3 can be seen as a biological exploration based on a formal model. The answer shows that, given the production of hnRNP A1 proteins, state 2 can be reached only from state 5. Thus, *tat1* mRNA production is only possible after *tat2* mRNA production. Our model suggests that by increasing the concentration of the repressor proteins hnRNP A1, the system may switch from the early to the late phase. This observation corresponds to a new biological hypothesis, which should be verified experimentally.

7 Conclusions and Perspectives

Our approach combines single-site and multi-site modeling approaches of the alternative splicing regulation. The integration is achieved by an hybrid automaton with default reasoning, in accordance with available biological knowledge. The splice efficiency is used as a time-scale abstraction of the local behaviour at one site inside a more global multi-site model. For the experimental biologist, this integrative model serves as a computational tool to study a fine-grained biological process on different scales. In particular, one can analyze the effect of the local regulation at one site on the global regulation involving different sites. On the one hand, the above queries may validate the biological hypotheses underlying the model. On the other hand, they may suggest future experiments by focusing on one particular process of interest, which remains difficult for classical approaches.

Our integrative model is partially validated by biological queries on its qualitative behaviour. An hybrid automaton appears to be an efficient model to represent the switching conditions between the early and the late phase, despite the lack of numerical values.

The present qualitative analysis concerns only one instantiation of the generic hybrid automaton, which consists in increasing the concentration of hnRNP A1 proteins. Thus our model is validated only in this situation. In order to generalise our approach, we could study the effect of other regulatory proteins in order to extract novel biological hypotheses, which may initiate new experimental work.

In a virological study of the HIV-1 life cycle, Hammond [14] describes the variation of the proteins translated from mRNAs. His study does not take into account the effect of alternative splicing. In future work, we plan to introduce the alternative splicing regulation in Hammond's model of the virus life cycle. The variation of the viral proteins can be represented by additional constraints in the hybrid automaton. Using this approach, our goal is to develop a more dynamic model of alternative splicing regulation, and to quantify the effects of this complex biological process in the HIV-1 life cycle.

References

1. Bockmayr, A., Courtois, A.: Using hybrid concurrent constraint programming to model dynamic biological systems. In: 18th International Conference on Logic Programming, ICLP'02, Copenhagen. Springer LNCS 2401 (2002) 85–99
2. Eveillard, D., Ropers, D., Jong, H.d., Branlant, C., Bockmayr, A.: Multiscale modeling of alternative splicing regulation. In: Computational Methods in Systems Biology, CMSB'03, Rovereto, Italy. Springer LNCS 2602 (2003) 75–87. Long version to appear in *Theoretical Computer Science*.
3. Moore, M., Query, C., Sharp, P.: Splicing of precursors to messenger RNAs by the spliceosome. In: The RNA World. Cold Spring Harbor Laboratory Press (1993)
4. Smith, C.W., Valcarcel, J.: Alternative pre-mRNA splicing: the logic of combinatorial control. Trends in Biochemical Sciences **25** (2000) 381–388
5. Purcell, D., Martin, M.: Alternative splicing of human immunodeficiency virus type 1 mRNA modulates viral protein expression, replication and infectivity. J. Virol. **67** (1993) 6365–78
6. Hope, T.: The ins and outs of HIV rev. Arch. Biochem. Biophys. **365** (1999) 186–91
7. O'Reilly, M., McNally, M., Beemon, K.: Two strong 5' splice sites and competing, suboptimal 3' splice sites involved in alternative splicing of human immunodeficiency virus type 1 RNA. Virology **213** (1995) 373–85
8. Si, Z., Amendt, B.A., Stoltzfus, C.M.: Splicing efficiency of human immunodeficiency virus type 1 tat RNA is determined by both a suboptimal 3' splice site and a 10 nucleotide exon splicing silencer element located within tat exon 2. Nucleic Acids Res **25** (1997) 861–7
9. Swanson, A.K., Stoltzfus, C.M.: Overlapping *cis* sites used for splicing of HIV-1 env/nef and rev mRNAs. J. Biol. Chem. **273** (1998) 34551–7
10. Schaal, H., Freund, M., Kammler, S., Asang, C., Caputi, M.: A bidirectional SR protein-dependent exonic splicing enhancer regulates *rev*, *env*, *vpu* and *nef* gene expression. In: Eukaryotic mRNA Processing. (2003)

11. Gupta, V., Jagadeesan, R., Saraswat, V.: Computing with continuous change. Science of Computer Programming **30** (1998) 3–49
12. Chabrier, N., Fages, F.: Symbolic model checking of biochemical networks. In: Computational Methods in Systems Biology, CMSB'03, Rovereto, Italy. Springer LNCS 2602 (2003) 149–162
13. Pongoski, J., Asai, K., Cochrane, A.: Positive and negative modulation of human immunodeficiency virus type 1 Rev function by *cis* and *trans* regulators of viral RNA splicing. J. Virol. **76** (2002) 5108–20
14. Hammond, B.J.: Quantitative study of the control of HIV-1 gene expression. J. Theor. Biol. **163** (1993) 199-221

Graph-Based Modeling of Biological Regulatory Networks: Introduction of Singular States

Adrien Richard, Jean-Paul Comet, and Gilles Bernot

La.M.I UMR 8042, CNRS & Université d'Évry
Boulevard François Mitterrand, 91025 Évry cedex-France
{arichard, comet, bernot}@lami.univ-evry.fr

Abstract. In the field of biological regulation, models extracted from experimental works are usually complex networks comprising intertwined feedback circuits. R. Thomas and coworkers introduced a qualitative description of the dynamics of such regulatory networks, called the generalized logical analysis, and used the concept of circuit-characteristic states to identify all steady states and functional circuits. These characteristic states play an essential role on the dynamics of the system, but they are not represented in the state graph. In this paper we present an extension of this formalism in which all singular states including characteristic ones are represented. Consequently, the state graph contains all steady states. Model checking is then able to verify temporal properties concerning singular states. Finally, we prove that this new modeling is coherent with R. Thomas' modeling since all paths of R. Thomas' dynamics are represented in the new state graph, which in addition shows the influence of singular states on the dynamics.

1 Introduction

Biological regulatory systems are often complex networks comprising several intertwined feedback circuits. The behavior of such systems is extremely anti-intuitive and cannot be solved without adequate formalization. They can be accurately described by non-linear ordinary differential equations [1, 2, 3, 4] which, however, cannot be solved analytically and use kinetic parameters which are most often unknown. The generalized logical analysis developed by R. Thomas and coworkers [5, 6, 7, 8] to describe biological regulatory networks extracts the essential qualitative features of the dynamics of such systems by logical parameters [9, 5] which can take a finite number of values. But some states, the *singular states*, are not explicitly represented in the state graph obtained with this formalism whereas they can be steady. Even if the steady singular states can be detected with the concept of circuit-characteristic states [10, 11], it is not possible to use model checking for verifying temporal properties concerning singular states. This paper provides, in section 2, our extension of R. Thomas' modeling. This new formalism considers the singular states and consequently represents all the steady states of a classic continuous description of regulatory networks. This

V. Danos and V. Schachter (Eds.): CMSB 2004, LNBI 3082, pp. 58–72, 2005.

continuous description is also the ground of the work of R. Thomas. Section 3 presents this description and shows, by the introduction of a discretisation map, why our qualitative modeling extracts similarly its essential qualitative features. Then we study in section 4 how the introduction of singular states gives a new light on the properties of characteristic states of feedback circuits. Finally conclusions and perspectives are presented.

2 Qualitative Dynamics of Regulatory Networks

In our qualitative approach, the entities of a biological regulatory network, often macromolecules or genes, have discrete expression levels defined as *qualitative values.*

Definition 1 (Qualitative Values). *A* qualitative value, *denoted by* $|a,b|$, *is a couple of integers* ($|a,b| \in \mathbb{N}^2$) *where* $a \leq b$. *The relations* $=,<,>,\subseteq$ *are defined for two qualitative values* $|a,b|$ *and* $|c,d|$ *by:*

- $|a,b| = |c,d|$ *if* $a = c$ *and* $b = d$.
- $|a,b| < |c,d|$ *if* ($b < c$) *or* ($b = c$ *and* ($a < b$ *or* $c < d$))
- $|a,b| > |c,d|$ *if* $|c,d| < |a,b|$
- $|a,b| \subseteq |c,d|$ *if* ($|a,b| = |c,d|$) *or* ($a = b$ *and* $c < a$ *and* $b < d$) *or* ($a < b$ *and* $c \leq a$ *and* $b \leq d$).

Intuitively, if $a < b$ then $|a,b|$ is said *singular* and represents the open interval $]a,b[$. Otherwise, if $a = b$ then $|a,b|$ is said *regular* and represents the closed interval $[a,b]$ which only contains the integer a. Then two qualitative values are comparable if the corresponding intervals are not overlapping and the relation $\subseteq$ is simply the inclusion relation between these intervals. To shorten the notation of the qualitative values we denote by $|a|$ the regular qualitative value $|a,a|$.

Interactions between biological entities are classically represented by directed graphs, where vertices abstracts biological entities and edges their interactions. In the sequel we denote by $\#S$ the cardinal of a set S and by $G^-(v)$ (resp. $G^+(v)$) the set of predecessors (resp. successors) of a vertex v in a graph G.

Definition 2 (*QRN*). *A* qualitative regulatory network *(QRN for short) is a labelled directed graph* $N = (V,E)$ *where:*

- *each vertex* $v \in V$, *called* variable, *represents a biological entity. The set* Q_v *of all* possible qualitative expression levels *of* v *is defined as* $Q_v = \{|0|, |0,1|, |1|, ..., |q-1|, |q-1,q|, |q|, ..., |\#N^+(v)|\}$.
- *each edge* $u \to v \in E$, *called* interaction, *is labelled by a couple* (α_{uv}, q_{uv}) *where* α_{uv} *is the* sign *of the interaction* ($\alpha_{uv} = +$ *(resp.* $\alpha_{uv} = -$) *if* $u \to v$ *is an activation (resp. inhibition)) and where* q_{uv} *is an integer in* $\{1, 2, ..., \#N^+(u)\}$ *such that* $q_{uv} \neq q_{uw}$ *for all* $w \in N^+(u)$ *distinct from* v. *The* threshold t_{uv} *of the interaction is defined as* $t_{uv} = |q_{uv} - 1, q_{uv}|$.

At a given time, the data made of the expression level of each variable is called the state of the network.

Definition 3 (States of a *QRN*). *Let* $N = (V, E)$ *be a* QRN. *A* state x *of* N *is a vector* $\mathrm{x} = (\mathrm{x}_v)_{v \in V}$ *such that* $\mathrm{x}_v \in Q_v$ *for all* $v \in V$. *A state is said* singular *if one of its component is singular and* regular *otherwise.*

As a majority of biological interactions behave in a cooperative way and have a sigmoid nature, they are in a *QRN* labelled by thresholds and they model switch-like reactions: at a given state x, an interaction $u \to v$ is said effective when $\mathrm{x}_u > t_{uv}$, not effective when $\mathrm{x}_u < t_{uv}$ and uncertain when $\mathrm{x}_u = t_{uv}$. Thus $\mathrm{x}_u = |q|$ means that u is an effective regulator for q of its successors and $\mathrm{x}_u = |q, q+1|$ means that u is an effective regulator for the same q successors and it is an uncertain regulator for the successor v such that $t_{uv} = \mathrm{x}_u$ (v exists inevitably).

Definition 4 (Resources). *Let* $N = (V, E)$ *be a* QRN, v *be a variable of* N *and* x *be a state of* N. *The sets of* regular resources $R_v(\mathrm{x})$ *and* singular resources $S_v(\mathrm{x})$ *of* v *at the state* x *are given by:*

- $R_v(\mathrm{x}) = \{u \in N^-(v) \mid (\mathrm{x}_u > t_{uv} \text{ and } \alpha_{uv} = +) \text{ or } (\mathrm{x}_u < t_{uv} \text{ and } \alpha_{uv} = -)\}$
- $S_v(\mathrm{x}) = \{u \in N^-(v) \mid \mathrm{x}_u = t_{uv}\}$

A regular resource of v is a variable which acts positively on v, that is to say an effective activator or a non effective inhibitor of v. A singular resource is just an uncertain regulator.

Definition 5 (Qualitative Model). *A* qualitative model M *of a* QRN $N = (V, E)$ *is a couple* $M = (N, K)$ *where* $K = \{K_{v,\omega} \mid v \in V \text{ and } \omega \subseteq N^-(v)\}$ *is a set of integers, called* qualitative parameters, *such that :*

- *if* $\omega = \emptyset$ *then* $K_{v,\omega} = 0$ *and* $K_{v,\omega} \in \{0, 1, ..., \#N^+(v)\}$ *otherwise.*
- *if* $\omega \subseteq \omega'$ *then* $K_{v,\omega} \leq K_{v,\omega'}$.

At a given state x, the expression level of a variable v evolves toward a qualitative value according to its regular and singular resources. This qualitative value, called attractor and noted $A_v(\mathrm{x})$, is defined with two parameters indexed by the regular and singular resources of v:

Definition 6 (Attractors). *Let* $M = (N, K)$ *be a qualitative model and* x *be a state of* N. *The* attractor $A_v(\mathrm{x})$ *of* $v \in V$ *at the state* x *is :*

$$A_v(\mathrm{x}) = \left| K_{v,R_v(\mathrm{x})}, K_{v,R_v(\mathrm{x}) \cup S_v(\mathrm{x})} \right|$$

At a given state x, if v does not have singular resources ($S_v(\mathrm{x}) = \emptyset$) then it evolves toward the qualitative value $|K_{v,R_v(\mathrm{x})}|$. Otherwise ($S_v(\mathrm{x}) \neq \emptyset$) v has singular resources and evolves toward an expression level greater than $|K_{v,R_v(\mathrm{x})}|$, i.e. the case where all the singular resources are not regarded as regular resources, and less than $|K_{v,R_v(\mathrm{x}) \cup S_v(\mathrm{x})}|$, i.e. the case where all the singular resources are regarded as regular resources: $|K_{v,R_v(\mathrm{x})}| < |K_{v,R_v(\mathrm{x})}, K_{v,R_v(\mathrm{x}) \cup S_v(\mathrm{x})}| < |K_{v,R_v(\mathrm{x}) \cup S_v(\mathrm{x})}|$ if $|K_{v,R_v(\mathrm{x})}| < |K_{v,R_v(\mathrm{x}) \cup S_v(\mathrm{x})}|$. Naturally, if $\mathrm{x}_v < A_v(\mathrm{x})$ then v tends to increase, if $\mathrm{x}_v > A_v(\mathrm{x})$ then v tends to decrease, and otherwise

($\mathrm{x}_v \subseteq A_v(\mathrm{x})$) v is steady. A steady state is thus a state where all the variables are steady, that is, a state x such that for all variable $v \in V$:

$$|K_{v,R_v(\mathrm{x})}| = \mathrm{x}_v = |K_{v,R_v(\mathrm{x}) \cup S_v(\mathrm{x})}| \quad \text{or} \quad |K_{v,R_v(\mathrm{x})}| < \mathrm{x}_v < |K_{v,R_v(\mathrm{x}) \cup S_v(\mathrm{x})}| \quad (1)$$

In section 4, we show how these static constraints (in which the dynamics of the system does not matter) can be used for the detection of homeostasis and multistationnarity. It can be proved that these constraints are equivalent to those given by E. H. Snoussi and R. Thomas in [10].

To sum up, we deduce from a model the tendencies of variables at each state, which is sufficient to define its dynamics with the following state graph.

Definition 7 (State Graph). *The* state graph *of a qualitative model $M = (N, K)$, is a directed graph where the set of vertices is the set of states of N, and where $\mathrm{x} \to \mathrm{y}$ is an edge, called* transition, *if there is a variable v verifying :*

$$\textit{for all } u \neq v,\ \mathrm{y}_u = \mathrm{x}_u \quad \textit{and} \quad \begin{cases} \mathrm{y}_v = \Delta^+(\mathrm{x}_v) \textit{ if } \mathrm{x}_v < A_v(\mathrm{x}) \\ \mathrm{y}_v = \Delta^-(\mathrm{x}_v) \textit{ if } \mathrm{x}_v > A_v(\mathrm{x}) \end{cases}$$

with Δ^+ and Δ^- the evolution operators *defined by:*

$$\Delta^+(\alpha) = \begin{cases} |q, q+1| \textit{ if } \alpha = |q| \\ |q| \textit{ if } \alpha = |q-1, q| \end{cases} \quad \textit{and} \quad \Delta^-(\alpha) = \begin{cases} |q-1, q| \textit{ if } \alpha = |q| \\ |q| \textit{ if } \alpha = |q, q+1| \end{cases}$$

In this definition two variables cannot evolve simultaneously towards their respective attractors, the state graph is thus asynchronous and can be deduced from a synchronous one as in [6]. Indeed, when several variables tend to evolve at a given state, additional information (time delays associated to each transition [8]) is needed to select which one first changes. As this information is most often unknown, all possible transitions are considered. Thus the system is non deterministic and can translate the stochastic character of biological interactions. Consequently, a state for which n variables tend to evolve has n successors. In particular, if $n = 0$ then the state is steady and does not have any successor.

A qualitative model in R. Thomas' approach can be defined in the same way but the state graph deduced from it just gives transitions between regular states. Indeed, in R.Thomas' state graph of a model $M = (N, K)$ the vertices are all the regular states of N and $x \to y$ is a transition if there is a variable v verifying:

$$\text{for all } u \neq v,\ \mathrm{y}_u = \mathrm{x}_u \quad \text{and} \quad \begin{cases} \mathrm{y}_v = |q+1| \text{ if } \mathrm{x}_v < A_v(\mathrm{x}) \\ \mathrm{y}_v = |q-1| \text{ if } \mathrm{x}_v > A_v(\mathrm{x}) \end{cases} \quad \text{with } |q| = \mathrm{x}_v$$

As we can see in figure 1, for a given model, R. Thomas' state graph is present in our state graph: our state graph can be viewed as a refinement of R. Thomas' one formally expressed in theorem 3.

We now illustrate our formalism with the *QRN* $N = (V, E)$ whose representation is given in figure 1. It represents a small genetic network controlling the mucus production of *Pseudomonas aeruginosa* [12, 13, 14]. We have $V = \{u, v\}$ and

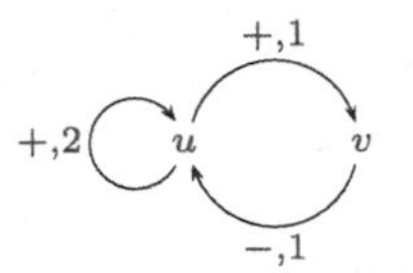

states x		symbolic attractors		attractors		tendencies	
x_u	x_v	$A_u(x)$	$A_v(x)$	$A_u(x)$	$A_v(x)$	u	v
$\|0\|$	$\|0\|$	$\|K_{u,\{v\}}\|$	$\|K_{v,\emptyset}\|$	$\|2\|$	$\|0\|$	$\nearrow$	$\rightsquigarrow$
$\|0\|$	$\|0,1\|$	$\|K_{u,\emptyset}, K_{u,\{v\}}\|$	$\|K_{v,\emptyset}\|$	$\|0,2\|$	$\|0\|$	$\nearrow$	$\searrow$
$\|0\|$	$\|1\|$	$\|K_{u,\emptyset}\|$	$\|K_{v,\emptyset}\|$	$\|0\|$	$\|0\|$	$\rightsquigarrow$	$\searrow$
$\|0,1\|$	$\|0\|$	$\|K_{u,\{v\}}\|$	$\|K_{v,\emptyset}, K_{v,\{u\}}\|$	$\|2\|$	$\|0,1\|$	$\nearrow$	$\nearrow$
$\|0,1\|$	$\|0,1\|$	$\|K_{u,\emptyset}, K_{u,\{v\}}\|$	$\|K_{v,\emptyset}, K_{v,\{u\}}\|$	$\|0,2\|$	$\|0,1\|$	$\rightsquigarrow$	$\rightsquigarrow$
$\|0,1\|$	$\|1\|$	$\|K_{u,\emptyset}\|$	$\|K_{v,\emptyset}, K_{v,\{u\}}\|$	$\|0\|$	$\|0,1\|$	$\searrow$	$\searrow$
$\|1\|$	$\|0\|$	$\|K_{u,\{v\}}\|$	$\|K_{v,\{u\}}\|$	$\|2\|$	$\|1\|$	$\nearrow$	$\nearrow$
$\|1\|$	$\|0,1\|$	$\|K_{u,\emptyset}, K_{u,\{v\}}\|$	$\|K_{v,\{u\}}\|$	$\|0,2\|$	$\|1\|$	$\rightsquigarrow$	$\nearrow$
$\|1\|$	$\|1\|$	$\|K_{u,\emptyset}\|$	$\|K_{v,\{u\}}\|$	$\|0\|$	$\|1\|$	$\searrow$	$\rightsquigarrow$
$\|1,2\|$	$\|0\|$	$\|K_{u,\{v\}}, K_{u,\{u,v\}}\|$	$\|K_{v,\{u\}}\|$	$\|2\|$	$\|1\|$	$\nearrow$	$\nearrow$
$\|1,2\|$	$\|0,1\|$	$\|K_{u,\emptyset}, K_{u,\{u,v\}}\|$	$\|K_{v,\{u\}}\|$	$\|0,2\|$	$\|1\|$	$\rightsquigarrow$	$\nearrow$
$\|1,2\|$	$\|1\|$	$\|K_{u,\emptyset}, K_{u,\{u\}}\|$	$\|K_{v,\{u\}}\|$	$\|0,2\|$	$\|1\|$	$\rightsquigarrow$	$\rightsquigarrow$
$\|2\|$	$\|0\|$	$\|K_{u,\{u,v\}}\|$	$\|K_{v,\{u\}}\|$	$\|2\|$	$\|1\|$	$\rightsquigarrow$	$\nearrow$
$\|2\|$	$\|0,1\|$	$\|K_{u,\{u\}}, K_{u,\{u,v\}}\|$	$\|K_{v,\{u\}}\|$	$\|2\|$	$\|1\|$	$\rightsquigarrow$	$\nearrow$
$\|2\|$	$\|1\|$	$\|K_{u,\{u\}}\|$	$\|K_{v,\{u\}}\|$	$\|2\|$	$\|1\|$	$\rightsquigarrow$	$\rightsquigarrow$

QRN

+,1
+,2 u v
−,1

R. Thomas' state graph

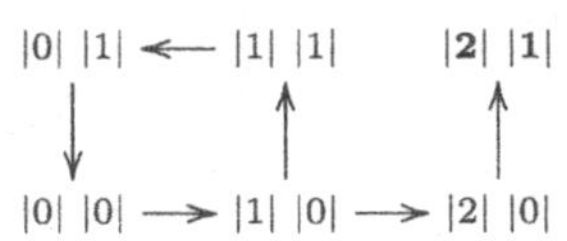

state graph

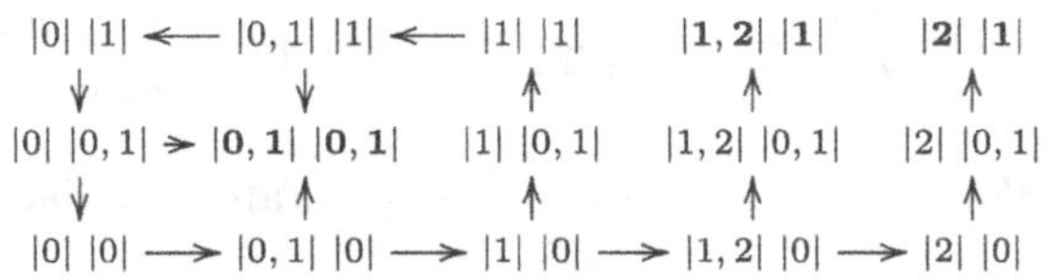

Fig. 1. The table gives, all the states of the QRN, the attractors of variables at each state, the values of the attractors deduced from the qualitative model where $K_{u,\emptyset} = 0$, $K_{u,\{v\}} = 2$, $K_{u,\{u\}} = 2$, $K_{u,\{u,v\}} = 2$, $K_{v,\emptyset} = 0$ and $K_{v,\{u\}} = 1$, and finally the corresponding tendencies ($\nearrow$ if $x_u < A_u(x)$, $\searrow$ if $x_u > A_u(x)$ and $\rightsquigarrow$ if $x_u \subseteq A_u(x)$). These tendencies allow us to construct our state graph and R. Thomas' one

$E = \{u \to u, u \to v, v \to u\}$. Variable u activates v and itself ($\alpha_{uu} = \alpha_{uv} = +$) when its expression level respectively reaches the thresholds $t_{uv} = |0,1|$ and $t_{uu} = |1,2|$. In return, variable v inhibits u ($\alpha_{vu} = -$) when its expression level reaches the threshold $t_{vu} = |0,1|$. Consequently, the possible expression levels of u are $Q_u = \{|0|, t_{uv}, |1|, t_{uu}, |2|\}$ and those of v are $Q_v = \{|0|, t_{vu}, |1|\}$. Thus 15 states are associated to the network, 6 are regular and 9 are singular (see the table of figure 1). The qualitative parameters corresponding to N are $K_{u,\emptyset}, K_{u,\{u\}}, K_{u,\{v\}}, K_{u,\{u,v\}} \in \{0,1,2\}$ and $K_{v,\emptyset}, K_{v,\{u\}} \in \{0,1\}$. The attractors expressed with the qualitative parameters are given in the table of figure 1. For a given model, the values of attractors allows us to deduce the tendencies of each variable at each state and to build, for both formalisms, the corresponding state graphs. For example, at the state $([1|, |0|)$ both u and v have an expression level less than their attractors (respectively equal to $|2|$ and $|1|$) and thus both variables tend to increase. Consequently $([1|, |0|)$ has two successors:

- in our state graph we have $(|1|, |0|) \to (|1,2|, |0|)$ for the increase of u and $(|1|, |0|) \to (|1|, |0,1|)$ for the increase of v.
- in R. Thomas' state graph we have $(|1|, |0|) \to (|2|, |0|)$ for the increase of u and $(|1|, |0|) \to (|1|, |1|)$ for the increase of v.

One can notice that our state graph contains two more steady states than R. Thomas' one (they are both singular states).

3 Discretization Map

R. Thomas' approach has been built as a discretization of the continuous approach presented in this section. Our formalism can also be viewed as a discretization of this description, which gives the dynamics of the regulatory networks defined as follow.

Definition 8 (*RN*)**.** *A* regulatory network *(RN for short) is a labelled directed graph* $\mathcal{N} = (V, E)$ *where :*

- *each vertex* v *of* V*, called* variable, *represents a biological entity,*
- *each edge* $u \to v$ *of* E*, called* interaction, *is labelled by a couple* $(\alpha_{uv}, \theta_{uv})$ *where* α_{uv} *is the* sign *of the interaction and where* $\theta_{uv} \in \mathbb{R}_+$ *is its* threshold.

To each variable v of $\mathcal{N} = (V, E)$ is associated a continuous variable $x_v \in \mathbb{R}_+$ which represents its expression level. At a given time, each variable x_v has a unique expression level and the vector $x = (x_v)_{v \in V}$ defines the state of the *RN*. The continuous dynamics of $\mathcal{N}$ can be given by the following system of piecewise-linear differential equations [9]:

$$\frac{\mathrm{d}x_v}{\mathrm{d}t} = \mathcal{S}_v(x) - \lambda_v x_v \quad \forall v \in V \quad \text{with} \quad \mathcal{S}_v(x) = \sum_{u \in \mathcal{N}^-(v)} \mathcal{I}^{\alpha_{uv}}(x_u, \theta_{uv}) \qquad (2)$$

where $\lambda_v > 0$ is the degradation coefficient of v, $\mathcal{S}_v(x)$ is its synthesis rate and $\mathcal{I}^{\alpha_{uv}}$ is a step function (figure 2) describing the effect of u on the synthesis rate of v:

$$\mathcal{I}^+(x_u, \theta_{uv}) = \begin{cases} 0 & \text{if } x_u < \theta_{uv} \\ k_{uv} & \text{if } x_u > \theta_{uv} \end{cases} \qquad \mathcal{I}^-(x_u, \theta_{uv}) = \begin{cases} k_{uv} & \text{if } x_u < \theta_{uv} \\ 0 & \text{if } x_u > \theta_{uv} \end{cases}$$

With such a definition, $\mathcal{I}^{\alpha_{uv}}$ is undefined for $x_u = \theta_{uv}$. A state in which there is at least one variable on a threshold is thus called a *singular* state. To define the system (2) for the singular states E. H. Snoussi and R. Thomas proposed in [10] to represent the *uncertain* influence of u on v when $x_u = \theta_{uv}$ by an open interval: $\mathcal{I}^{\alpha_{uv}}(\theta_{uv}, \theta_{uv}) =]0, k_{uv}[$. This interval represents the set of possible effects of u on v strictly included between the case where u acts on v ($x_u > \theta_{uv}$) and the case where it does not ($x_u < \theta_{uv}$). Then the system has to be seen as a system of differential inclusions [15]:

$$\frac{\mathrm{d}x_v}{\mathrm{d}t} \in \mathcal{S}_v(x) - \lambda_v x_v \quad \forall v \in V \quad \text{with} \quad \mathcal{S}_v(x) = \sum_{u \in \mathcal{N}^-(v)} \mathcal{I}^{\alpha_{uv}}(x_u, \theta_{uv}) \qquad (3)$$

Definition 9 (Model). *A* model $\mathcal{M}$ *is a tuple* $\mathcal{M} = (\mathcal{N}, k, \lambda)$ *where* $\mathcal{N} = (V, E)$ *is a RN,* $k = \{k_{uv}\}_{u \to v \in E}$ *is the set of parameters associated to each interaction, and* $\lambda = \{\lambda_v\}_{v \in V}$ *is the set of degradation rates associated to each variable.*

Definition 10 (Discretization Map). *Let* $\mathcal{N}$ *be a RN and* u *a variable. The* discretization map $d_u : 2^{\mathbb{R}_+} \setminus \emptyset \to \mathbb{N}^2$ *is defined for all non empty open intervals* $I =]\alpha, \beta[$ *and for all singletons* $I = [\alpha, \beta]$ *with* $\alpha = \beta$ *by :*

$$d_u(I) = \Big| \ \#\{\theta \in \Theta_u \mid \theta < \alpha\} \ , \ \#\{\theta \in \Theta_u \mid \theta \leq \beta\} \ \Big|$$

where $\Theta_u = \{\theta_{uv} \mid v \in \mathcal{N}^+(u)\}$ *is the* set of out-thresholds *of* u.

Let us highlight some properties of d_u which are useful in the sequel. The order relations $<$, $>$ and $\subseteq$ on non empty open intervals and singletons are defined similarly to the corresponding order relations on qualitative values. Thus, d_u is an increasing function: if $A < B$ then $d_u(A) \leq d_u(B)$. Then if $d_u([\alpha, \alpha]) = |a|$ and $d_u([\beta, \beta]) = |b|$ with $\alpha \leq \beta$ we have $d_u(]\alpha, \beta[) = |a, b|$. Finally $d_u(A) \subseteq d_u(B)$ iff $A \subseteq B$, in particular $d_u([\alpha, \alpha]) \subseteq d_u(B)$ iff $\alpha \in B$. In the remainder, for a singleton $[\alpha, \alpha]$, $d_u(\alpha)$ denotes $d_u([\alpha, \alpha])$ by abuse of notation.

Definition 11 (Qualitative Form). *The* qualitative form *of a RN* $\mathcal{N} = (V, E)$ *is the QRN* $N = (V, E)$ *such that each interaction* $u \to v$ *has the sign of the corresponding interaction in* $\mathcal{N}$ *and is such that* $t_{uv} = d_u(\theta_{uv})$.

According to the previous definition, a *RN* has a qualitative form iff it has no variable which acts on two successors with the same threshold, but it is a marginal case since thresholds are real values a priori different.

Let $\mathcal{N}$ be a *RN* and N its qualitative form. The discretization $d_u(x_u)$ of a continuous expression level x_u is a qualitative expression level of u in N:

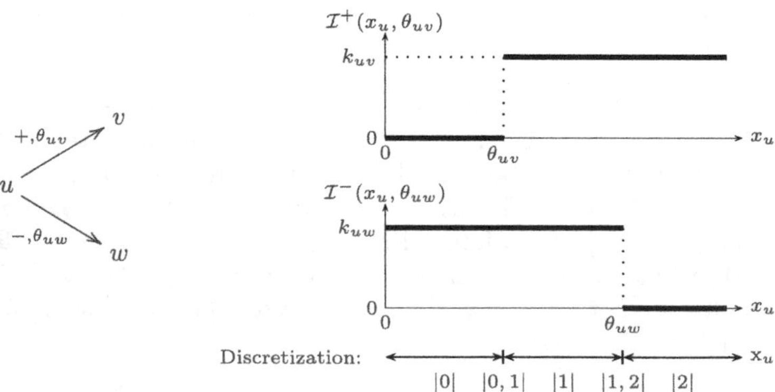

Fig. 2. Step functions associated to the interactions $u \to v$ and $u \to w$ with $\theta_{uv} < \theta_{uw}$ and discretization of the expression levels of u

$d_u(x_u) \in Q_u$ (figure 2). Thus each continuous state of $\mathcal{N}$ corresponds to one qualitative state of N but a qualitative state of N can correspond to an infinity of continuous states of $\mathcal{N}$. To link the states of N with those of $\mathcal{N}$ we define $\mathcal{D}_v : Q_v \to 2^{\mathbb{R}_+}$ and $\mathcal{D} : \prod_{v \in V} Q_v \to \prod_{v \in V} 2^{\mathbb{R}_+}$ by:

$$\mathcal{D}_v(\mathrm{x}_v) = \{x_v \in \mathbb{R}_+ \mid d_v(x_v) = \mathrm{x}_v\} \qquad \text{and} \qquad \mathcal{D}(\mathrm{x}) = (\mathcal{D}_v(\mathrm{x}_v))_{v \in V}$$

$\mathcal{D}_v(\mathrm{x}_v)$ and $\mathcal{D}(\mathrm{x})$ are respectively called the *domains* of x_v and x. Let $\mathcal{M} = (\mathcal{N}, k, \lambda)$ be a model. The differential equation system (3) has one analytic solution on each domain $\mathcal{D}(\mathrm{x})$ where x is regular. For the initial state of the system $x^0 \in \mathcal{D}(\mathrm{x})$, the solution is:

$$x_v(t) = \mathcal{A}_v(x^0) - (\mathcal{A}_v(x^0) - x_v^0)e^{-\lambda_v t} \quad \forall v \in V \quad \text{with} \quad \mathcal{A}_v(x) = \frac{\mathcal{S}_v(x)}{\lambda_v}$$

Thus all continuous states of the domain $\mathcal{D}(\mathrm{x})$ tend to the same constant state $\mathcal{A}(x^0) = (\mathcal{A}_v(x^0))_{v \in V}$ called the *attractor* of the domain $\mathcal{D}(\mathrm{x})$. If $\mathcal{A}(x^0) \in \mathcal{D}(\mathrm{x})$, all states of $\mathcal{D}(\mathrm{x})$ will never go out of the domain $\mathcal{D}(\mathrm{x})$ and they will reach (in $+\infty$) the continuous steady state $\mathcal{A}(x^0)$. Otherwise, if $\mathcal{A}(x^0) \notin \mathcal{D}(\mathrm{x})$, then a state x of $\mathcal{D}(\mathrm{x})$ will evolve towards $\mathcal{A}_v(x^0)$ until it goes out of the domain $\mathcal{D}(\mathrm{x})$. Outside the domain, the solution of the system is not the same and the attractor is modified. In such a case the state $\mathcal{A}_v(x^0)$ can never be reached. To sum up, at the state x if $x_v < \mathcal{A}_v(x)$ (resp. $x_v > \mathcal{A}_v(x)$) then v tends to increase (resp. decrease) and if $x_v = \mathcal{A}_v(x)$ then v is steady.

If $x \in \mathcal{D}(\mathrm{x})$ with x a singular state then x is also singular and there is at least one variable u such that $x_u = \theta_{uv}$ with $v \in \mathcal{N}^+(u)$. Thus $\mathcal{A}_v(x)$ is an open interval and the tendencies of v are defined in the same way except that v is considered steady if $x_v \in \mathcal{A}_v(x)$.

Definition 12 (Qualitative Form of a Model). *The* qualitative form of a model $\mathcal{M} = (\mathcal{N}, k, \lambda)$ *is the qualitative model* $M = (N, K)$ *such that* N *is the qualitative form of* $\mathcal{N}$ *and such that for all parameters* $K_{v,\omega}$ *of* K *we have* $|K_{v,\omega}| = d_v(\sum_{u \in \omega} \frac{k_{uv}}{\lambda_v})$.

Notice that $\mathcal{M}$ has a qualitative form iff $(\sum_{u \in \omega} \frac{k_{uv}}{\lambda_v}) \notin \Theta_v$ for all v and ω which is a reasonable hypothesis. By setting down $|K_{v,\omega}| = d_v(\sum_{u \in \omega} \frac{k_{uv}}{\lambda_v})$ we respect the constraints given in the definition of qualitative parameters. Indeed, if $\omega = \emptyset$ then $K_{v,\omega} = 0$ and $d_v(\sum_{u \in \omega} \frac{k_{uv}}{\lambda_v}) = d_v(0) = |0|$. Otherwise, $K_{v,\omega} \in \{0, 1, ..., \#N^+(v)\}$ and since $(\sum_{u \in \omega} \frac{k_{uv}}{\lambda_v}) \notin \Theta_v$ we have $d_v(\sum_{u \in \omega} \frac{k_{uv}}{\lambda_v}) \in \{|0|, |1|, ..., |\#N^+(v)|\}$. Then, if $\omega \subseteq \omega'$ we have $\sum_{u \in \omega} \frac{k_{uv}}{\lambda_v} \leq \sum_{u \in \omega'} \frac{k_{uv}}{\lambda_v}$ and since d_v is an increasing map we have $d_v(\sum_{u \in \omega} \frac{k_{uv}}{\lambda_v}) \leq d_v(\sum_{u \in \omega'} \frac{k_{uv}}{\lambda_v})$, that is $K_{v,\omega} \leq K_{v,\omega'}$.

Theorem 1. *Let* $\mathcal{M}$ *be a model and* M *its qualitative form. For all variable* u*, for all qualitative state* x *of* M *and for all continuous state* $x \in \mathcal{D}(\mathrm{x})$ *of* $\mathcal{M}$*, we have* $A_u(\mathrm{x}) = d_u(\mathcal{A}_u(x))$.

Proof. According to the definition of d_u, we have $\mathrm{x}_u > t_{uv}$ iff $x_u > \theta_{uv}$, $\mathrm{x}_u < t_{uv}$ iff $x_u < \theta_{uv}$ and $\mathrm{x}_u = t_{uv}$ iff $x_u = \theta_{uv}$. Thus $u \in R_v(\mathrm{x})$ iff $\mathcal{I}^{\alpha_{uv}}(x_u, \theta_{uv}) = k_{uv}$ and $u \in S_v(\mathrm{x})$ iff $\mathcal{I}^{\alpha_{uv}}(x_u, \theta_{uv}) =]0, k_{uv}[$. Thus, if $S_v(\mathrm{x}) = \emptyset$ we have $A_v(\mathrm{x}) = |K_{v,R_v(\mathrm{x})}|$ and $\mathcal{A}_v(x) = \sum_{u \in R_v(\mathrm{x})} \frac{k_{uv}}{\lambda_v}$. Since $\sum_{u \in R_v(\mathrm{x})} \frac{k_{uv}}{\lambda_v} \notin \Theta_v$ we have $A_u(\mathrm{x}) = d_u(\mathcal{A}_u(x))$. If $S_v(\mathrm{x}) \neq \emptyset$ we have $A_v(\mathrm{x}) = |K_{v,R_v(\mathrm{x})}, K_{v,R_v(\mathrm{x}) \cup S_v(\mathrm{x})}|$ and $\mathcal{A}_v(x) = \sum_{u \in R_v(\mathrm{x})} \frac{k_{uv}}{\lambda_v} + \sum_{u \in S_v(\mathrm{x})} \frac{]0, k_{uv}[}{\lambda_v}$. Thus $\mathcal{A}_v(x)$ is equal to the open interval $]\alpha, \beta[=]\sum_{u \in R_v(\mathrm{x})} \frac{k_{uv}}{\lambda_v}, \sum_{u \in R_v(\mathrm{x}) \cup S_v(\mathrm{x})} \frac{k_{uv}}{\lambda_v}[$. Since α and β are not in Θ_v we have $A_u(\mathrm{x}) = d_u(\mathcal{A}_u(x))$.

Consequently we have, for all $x \in \mathcal{D}(\mathrm{x})$ and for each variable v, $\mathrm{x}_v < A_v(\mathrm{x})$ iff $x_v < \mathcal{A}_v(x)$ and $\mathrm{x}_v > A_v(\mathrm{x})$ iff $x_v > \mathcal{A}_v(x)$. Thus if v is not steady, it tends to evolve in the same way at the state x and at the state $x \in \mathcal{D}(\mathrm{x})$. Moreover, $\mathrm{x}_v \subseteq A_v(\mathrm{x})$ iff there exists $x \in \mathcal{D}(\mathrm{x})$ such that $x_v = \mathcal{A}_v(x)$ or $x_v \in \mathcal{A}_v(x)$. So if v is steady at the state x then v can be steady in $\mathcal{D}(\mathrm{x})$. Thus, a qualitative state x is steady iff there is a continuous steady state in $\mathcal{D}(\mathrm{x})$. We can sketch these properties by saying that the dynamics of M extracts the essential qualitative features of the dynamics of $\mathcal{M}$, and in particular both kinds of dynamics have the same number of steady states.

In practice, the values of the kinetic parameters k and λ are most often unknown. If we want to carry out a continuous modeling of a biological system, an infinity of models has to be considered corresponding to all the possible values of real parameters. The qualitative parameters K which define the dynamics of a *QRN* (definition 5) are also most often unknown but they can take a finite number of qualitative values. Then we can use the following fruitful exhaustive strategy to model a system: *to generate all the models with the aim to select those which give a dynamics coherent with the experimental knowledge of the system.* We have developed a software, called `SMBioNet` [16], which automatically carries out this generation and selection of models using three approaches : feedback circuit functionality, temporal logic and model checking. It has been used successfully to model the mucoidy and the cytotoxicity of *Pseudomonas aeruginosa* [13, 14].

4 Circuit Characteristic States

The most important generalized logical analysis concepts are certainly those of positive and negative circuits, which respectively generate multistationarity and cycle in the state graph, when the corresponding circuit is functional [17, 18, 19, 20, 21]. These concepts are especially important when modeling biological systems where differentiation and homeostasis need to be represented [13, 14]. A circuit is said positif (resp. negatif) when it contains an even (resp. odd) number of inhibitions. It is said functionnal when there is a steady circuit characteristic state (which is singular by definition) associated to it [10]. Consequently, the steady singular states play an essential role on the dynamics of a system. A characteristic state of a circuit is a singular state in which the set of uncertain interactions is equal to the set of edges of the circuit. This notion of characteristic state can be extended to the union of disjoint circuits.

Definition 13 (Characteristic State). *Let N be a QRN and C be a circuit of N. A state* x *is a* characteristic state *of C if $C = \bigcup_{v \in V}\{u \to v \mid u \in S_v(\mathrm{x})\}$.*

Some examples of circuits with their characteristic states are given in Figure 3. E. H. Snoussi and R. Thomas proved for their formalism that a singular state can

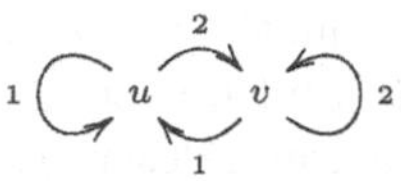

$C_1 = \{u \to u\}$		$C_2 = \{v \to v\}$		$C_3 = \{u \to v, v \to u\}$		$C_1 \cup C_2$	
x_u	x_v	x_u	x_v	x_u	x_v	x_u	x_v
$\lvert 0,1\rvert$	$\lvert 0\rvert$	$\lvert 0\rvert$	$\lvert 1,2\rvert$	$\lvert 1,2\rvert$	$\lvert 0,1\rvert$	$\lvert 0,1\rvert$	$\lvert 1,2\rvert$
$\lvert 0,1\rvert$	$\lvert 1\rvert$	$\lvert 1\rvert$	$\lvert 1,2\rvert$				
$\lvert 0,1\rvert$	$\lvert 2\rvert$	$\lvert 2\rvert$	$\lvert 1,2\rvert$				

Fig. 3. Characteristic states of all the circuits and unions of disjoint circuits in a *QRN* (C_1 and C_2 have three characteristic states and the others only one). The interactions of are not labelled by any sign since they do not play a role in the definition of characteristic states

be steady only if it characterizes a feedback circuit. This property is preserved in our qualitative modeling.

Theorem 2. *Among singular states, only characteristic ones can be steady.*

Proof. Let $N = (V, E)$ be a *QRN* and let x a non characteristic singular state. Then there is an edge $v \to w$ such that v is a singular resource of w and such that all resources of v are regular. Then x_v is a singular value and the attractor $A_v(\mathrm{x})$ is a regular qualitative value: $A_v(\mathrm{x}) = |K_{v,R_v(\mathrm{x})}|$ since $S_v(\mathrm{x}) = \emptyset$. A singular qualitative value cannot be contained in a regular one, thus $\mathrm{x}_v \not\subseteq A_v(\mathrm{x})$ and x cannot be steady.

We now compare the dynamics of models for which some circuits are functional in both modelings with the aim to hightlight how the presence of singular states makes more explicit the functionality. Let us start with the example of figure 4. In both formalisms, homeostasis induced by the stationarity of the negative circuit characteristic state is represented. But the steady characteristic state towards which tends the softened oscillation, representing the homeostasis, in the continuous description is represented in our state graph while the homeostasis is reflected as an infinite oscillation in R. Thomas' one. Thus, our state graph extracts more precisely the qualitative features of the continuous formalism. Notice that the paths of R. Thomas' state graph do not correspond to paths between regular states in our state graph. The presence a of steady characteristic state of a negative loop is the only case where R. Thomas' state graph is not "included" in our state graph (see theorem 3 for formal explanation). In a general way, the softening generated by the functionality of negative circuits is not represented in R. Thomas' modeling. That can lead to a confusion about the interpretation

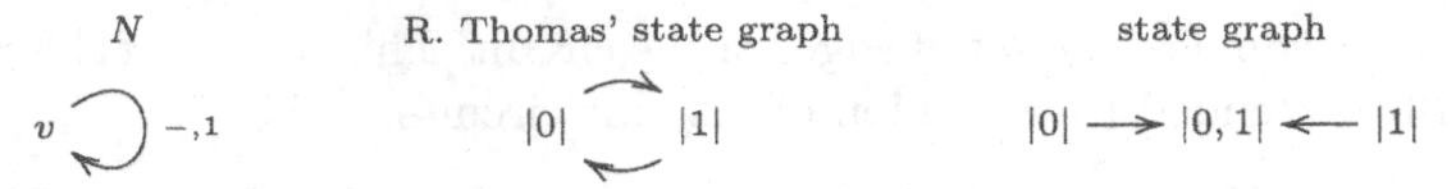

Fig. 4. State graphs, in the two approaches, deduced from the model $M = (N, \{K_{v,\emptyset} = 0, K_{v,\{v\}} = 1\})$ which makes functional the negative circuit N

of the circuit functionality. Let us consider the *QRN* of figure 1 containing a negative and a positive circuit. The model presented in the same figure makes functional both circuits, there is two steady singular states (equation 1) which are characteristic of both circuits, and thus multistationarity and homeostasis are predicted. For the homeostasis let us remark that our state graph describes a dynamics in which infinite and softened oscillations are possible. For the multistationnarity, in R. Thomas' state graph there is only one steady state and from each state it is possible to reach it. The state graph does not really illustrate the multistationarity. In our state graph, the presence of all the steady states (two singular steady states and the previous regular one) makes more explicit the multistationarity. In both state graphs, the paths between regular states are coherent. Indeed, each transition $\mathrm{x} \to \mathrm{z}$ of R. Thomas' state graph corresponds, in our state graph, to a path $\mathrm{x} \to \mathrm{y} \to \mathrm{z}$ where y is the singular state adjacent to the regular states x and z. Note that, according to the following theorem, for all models deduced from this network, R. Thomas' state graph is "included" in our one since the network does not contain a negative loop.

Theorem 3. *Let* $\mathrm{x} \to \mathrm{z}$ *be a transition of R. Thomas' state graph deduced from a qualitative model, and let* v *be the only variable which evolves during the transition* $\mathrm{x} \to \mathrm{z}$ $(\mathrm{x}_v \neq \mathrm{z}_v)$*. Let* y *be the singular state adjacent to* x *and* z *defined by :* $\mathrm{y}_u = \mathrm{x}_u = \mathrm{z}_u$ *for all* $u \neq v$ *and, setting down* $\mathrm{x}_v = |q|$, $\mathrm{y}_v = |q, q+1|$ *if* $\mathrm{z}_v = |q+1|$ *and* $\mathrm{y}_v = |q-1, q|$ *if* $\mathrm{z}_v = |q-1|$*. Then our state graph contains the path* $\mathrm{x} \to \mathrm{y} \to \mathrm{z}$ *if* $\mathrm{y}_v \not\subseteq A_v(\mathrm{y})$ $(\mathrm{y}_v \subseteq A_v(\mathrm{y})$ *imposes that* y *is a characteristic state of the negative circuit* $v \to v)$*.*

Proof. As x is a regular state, we have $A_v(\mathrm{x}) = |K_{v,R_v(\mathrm{x})}|$. Let us suppose that $\mathrm{x}_v = |q| < |K_{u,R_u(\mathrm{x})}|$. We have $\mathrm{y}_v = |q, q+1|$ and so $\mathrm{x} \to \mathrm{y}$ is a transition of our state graph. Moreover, $\mathrm{z}_v = |q+1|$ since $\mathrm{x} \to \mathrm{z}$ is a transition of R. Thomas' state graph.

As y_v is the only component of y which is singular, y is not a characteristic state if v does not regulate itself or if it regulates itself with a threshold not equal to y_v. In this case, $S_v(\mathrm{y}) = \emptyset$ and $A_v(\mathrm{y}) = |K_{v,R_v(\mathrm{y})}| = |K_{v,R_v(\mathrm{x})}|$. Thus $\mathrm{y} \to \mathrm{z}$ is a transition of our state graph since $|q| < |K_{u,R_u(\mathrm{x})}|$ implies that $|q, q+1| < |K_{u,R_u(\mathrm{x})}|$.

If y is a characteristic state (v regulates itself and $t_{vv} = \mathrm{y}_v$) then $S_v(\mathrm{y}) = \{v\}$. Thus, if $\alpha_{vv} = +$ then $A_v(\mathrm{y}) = |K_{v,R_v(\mathrm{y})}, K_{v,R_v(\mathrm{y})\cup\{v\}}| = |K_{v,R_v(\mathrm{x})}, K_{v,R_v(\mathrm{x})\cup\{v\}}|$. So $\mathrm{y}_v = |q, q+1| < |K_{v,R_v(\mathrm{x})}, K_{v,R_v(\mathrm{x})\cup\{v\}}|$ and $\mathrm{y} \to \mathrm{z}$ is a transition of our state graph (if $\alpha_{vv} = +$ then v cannot be steady at the state y). Otherwise, if $\alpha_{vv} = -$ then $A_v(\mathrm{y}) = |K_{v,R_v(\mathrm{y})}, K_{v,R_v(\mathrm{y})\cup\{v\}}| = |K_{v,R_v(\mathrm{x})\setminus\{v\}}, K_{v,R_v(\mathrm{x})}|$. So, if

v is not steady at the state y we have $\mathrm{y}_v = |q, q+1| \not\subseteq |K_{v,R_v(\mathrm{x})\setminus\{v\}}, K_{v,R_v(\mathrm{x})}|$ which implies that $q < K_{v,R_v(\mathrm{x})\setminus\{v\}}$ equivalent to $q+1 \leq K_{v,R_v(\mathrm{x})\setminus\{v\}}$. Thus $|q, q+1| < |K_{v,R_v(\mathrm{x})\setminus\{v\}}, K_{v,R_v(\mathrm{x})}|$ and $\mathrm{y} \rightarrow \mathrm{z}$ is a transition of our state graph.

In the other case, if $\mathrm{x}_v = |q| > |K_{u,R_u(\mathrm{x})}|$, the proof is similar.

5 Conclusions and Perspectives

In this paper we present a new qualitative modeling based on the generalized logical analysis of R. Thomas which allows us to represent the singular states in the dynamics. Both modeling are built as a dicretization of a piecewise-linear differential equations system but our modeling, taking into account the singular states, permits us to represent all the steady states of the continuous dynamics. In spite of an exponential increase in the number of states, there is not an increase in the number of models associated to a network. Moreover, the state graph reflects the softening of the negative functional circuits and it is a refinement of the dynamics of R. Thomas.

The representation of all steady states is essential to confront with precision the models to biological knowledge. The concepts of circuit functionality allow us to select models which present homeostasis and/or multistationarity with only static constraints, that is inequality constraints for the steadiness of singular states.

To still go further such static conditions must be reinforced by properties on the dynamics. To achieve the selection of the acceptable models (with temporal properties coherent with all available biological knowledge) we will take advantage of the corpus of formal methods. We have already implemented a user-friendly software, `SMBioNet` [13, 14, 16] (Selection of Models of Biological Networks), which allows one to select models of given regulatory networks according to their temporal properties. The software takes as input a QRN (with a graphical interface) and some temporal properties expressed as CTL formulae and a set of functional circuits. It generates all models associated to the network which makes functional the specified circuits and gives as output those whose corresponding R. Thomas' state graph satisfy the specified temporal properties (using the NuSMV model checker [22]). Then, the selected models can be used to make and formally test hypotheses or to run simulations. The input of `SMBioNet` not consists to a complex file with several reaction-rules or parameters assignment as in several other tools using a qualitative approach to model biological regulatory networks [23, 24, 25, 26, 27]. Indeed, R. Thomas' formalism catches the qualitative structure of a system in a simple graphical object (a QRN) easily extractable from present biological data. Nevertheless, it is difficult to represent with this approach a physical change of state of biological entities after an interaction or the formation/breakage of complexes. However, the effectiveness of the generalized logical analysis has been demonstrated in the study of a number of genetic regulatory systems [28, 29, 5, 30, 31, 32, 13, 14].

Naturally, a short term perspective is to introduce in this software our new formalism. The generation of the models making some circuits functional will remains the same but the increase in the number of states will makes more difficult to check a CTL formula. However, this formula will be able to express temporal properties concerning regular and singular states. To struggle against this increase of states, we can already propose to automatically remove from the state graph some singular states, for example those whose the set of successors is reduced to a single regular state (the states $(|1|, |0,1|)$, $(|1,2|, |0,1|)$ and $(|2|, |0,1|)$ can be removed from the state graph of figure 1).

More generally the formal methods can be applied in the field of biological regulatory networks in order to explicit some behaviors or to take into account not yet modelled biological knowledge. Let us mention for example that the introduction of transitions in the regulatory graph could help to specify how the different regulators cooperate for inducing or repressing their common target [33]. One can also separate inhibitors from regulators [34] to increase the readability of the approach, or take into account time delays [8] between the beginning of the activation order and the synthesis of the product and conversely for the turn-off delays. These constitute ongoing or future works of our genopole and G^3 research groups.

Acknowledgement

We thank genopole-research in Evry for constant supports. In particular we gratefully acknowledge the members of the genopole working group "*observability*" and G^3 for stimulating interactions.

References

1. Sveiczer, A., Csikasz-Nagy, A., Gyorffy, B., Tyson, J., Novak, B.: Modeling the fission yeast cell cycle: quantized cycle times in wee1- cdc25delta mutant cells. Proc. Natl. Acad. Sci. U S A. **97** (2000) 7865–70
2. Chen, K., Csikasz-Nagy, A., Gyorffy, B., Val, J., Novak, B., Tyson, J.: Kinetic analysis of a molecular model of the budding yeast cell cycle. Mol. Biol. Cell. **11** (2000) 369–91.
3. Tyson, J., Novak, B.: Regulation of the eukaryotic cell cycle: molecular antagonism, hysteresis, and irreversible transitions. J. Theor. Biol. **210** (2001) 249–63
4. Hasty, J., McMillen, D., Collins, J.: Engineered gene circuits. Nature **420** (2002) 224–30
5. Thomas, R., d'Ari, R.: Biological Feedback. CRC Press (1990)
6. Thomas, R.: Regulatory networks seen as asynchronous automata : A logical description. J. Theor. Biol. **153** (1991) 1–23
7. Thomas, R., Kaufman, M.: Multistationarity, the basis of cell differentiation and memory. I. structural conditions of multistationarity and other nontrivial behavior. Chaos **11** (2001) 170–179
8. Thomas, R., Kaufman, M.: Multistationarity, the basis of cell differentiation and memory. II. logical analysis of regulatory networks in terms of feedback circuits. Chaos **11** (2001) 180–195

9. Snoussi, E.: Qualitative dynamics of a piecewise-linear differential equations : a discrete mapping approach. Dynamics and stability of Systems **4** (1989) 189–207
10. Snoussi, E., Thomas, R.: Logical identification of all steady states : the concept of feedback loop caracteristic states. Bull. Math. Biol. **55** (1993) 973–991
11. Thomas, R., Thieffry, D., Kaufman, M.: Dynamical behaviour of biological regulatory networks - I. biological role of feedback loops an practical use of the concept of the loop-characteristic state. Bull. Math. Biol. **57** (1995) 247–76
12. Pérès, S., Comet, J.P.: Contribution of computation tree logic to biological regulatory networks : example from pseudomonas aeruginosa. In Priami, C., ed.: Proceedings of the 1st Intern. Workshop CMSB'2003. LNCS 2602, Springer-Verlag (2003) 47–56
13. Bernot, G., Comet, J.P., Richard, A., Guespin, J.: A Fruitful Application of Formal Methods to Biological Regulatory Networks, extending Thomas' asynchronous logical approach with temporal logic. J. Theor. Biol., in press (2004)
14. Guespin, J., Bernot, G., Comet, J.P., Mriau, A., Richard, A., , Hulen, C., Polack, B.: Epigenesis and dynamic similarity in two regulatory networks in pseudomonas aeruginosa. Acta Biotheoretica, in press (2004)
15. Filippov, A.: Differential Equations with Discontinuous Righthand Sides. Kluwer Academic Publishers (1988)
16. Richard, A., Comet, J.P., Bernot, G.: SMBioNet : Selection of Models of Biological Networks. (http://smbionet.lami.univ-evry.fr)
17. Thomas, R.: The role of feedback circuits: positive feedback circuits are a necessary condition for positive real eigenvalues of the Jacobian matrix. Ber. Bunsenges. Phys. Chem. **98** (1994) 1148–1151
18. Plahte, E., Mestl, T., Omholt, S.W.: Feedback loop, stability and multistationarity in dynamical systems. J. Biol. Syst. **3** (1995) 569–577
19. Demongeot, J.: Multistationarity and cell differentiation. J. Biol. Syst **6** (1998) 11–15
20. Snoussi, E.: Necessary conditions for multistationarity and stable periodicity. J. Biol. Syst. **6** (1998) 3–9
21. Soulé, C.: Graphic requirements for multistationarity. ComPlexUs **1** (2003)
22. Cimatti, A., Clarke, E.M., Giunchiglia, E., Giunchiglia, F., Pistore, M., Roveri, M., Sebastiani, R., Tacchella, A.: Nusmv 2: An opensource tool for symbolic model checking. In: Proceeding of International Conference on Computer-Aided Verification (CAV 2002), Copenhagen, Denmark (2002)
23. Heidtke, K., Schulze-Kremer, S.: Design and implementation of a qualitative simulation model of λ phage infection. Bioinformatics **14** (1998) 81–91
24. Heidtke, K., Schulze-Kremer, S.: BioSim : A New Qualitative Simulation Environment for Molecular Biology. In: Proceedings of the 6th International Conference in Intelligent Systems for Molecular Biology, AAAI Press (1998) 85–94
25. De Jong, H., Geiselmann, J., Hernandez, C., Page, M.: Genetic network analyzer: qualitative simulation of genetic regulatory networks. Bioinformatics **19** (2003) 336–44.
26. Chaouiya, C., Remy, E., Mossé, B., Thieffry, D.: Qualitative analysis of regulatory graphs: a computation tools based an a discrete formal framwork. Lecture Notes on Control and Information Sciences (2003) Accepted.
27. Chabrier-Rivier, N., Chiaverini, M., Danos, V., Fages, F., Schächter, V.: Modeling and querying biochemical networks. Theoretical Computer Science (2004) To appear.
28. Thieffry, D., Thomas, R.: Dynamical behaviour of biological regulatory networks - ii. immunity control in bacteriophage lambda. Bull. Math. Biol. **57** (1995) 277–97

29. Thomas, R., Gathoye, A., Lambert, L.: A complex control circuit. regulation of immunity in temperate bacteriophages. Eur. J. Biochem. **71** (1976) 211–27
30. Sánchez, L., van Helden, J., Thieffry, D.: Establishement of the dorso-ventral pattern during embryonic development of drosophila melanogaster: a logical analysis. J. Theor. Biol. **189** (1997) 377–89.
31. Sánchez, L., Thieffry, D.: A logical analysis of the drosophila gap-gene system. J. Theor. Biol. **211** (2001) 115–41
32. Mendoza, L., Thieffry, D., Alvarez-Buylla, E.: Genetic control of flower morphogenesis in arabidopsis thaliana: a logical analysis. Bioinformatics. **15** (1999) 593–606.
33. Bassano, V., Bernot, G.: Marked regulatory graphs: A formal framework to simulate biological regulatory networks with simple automata. In: Proc. of the 14th IEEE International Workshop on Rapid System Prototyping, RSP 2003, San Diego, California, USA (2003)
34. Bernot, G., Cassez, F., Comet, J.P., Delaplace, F., Müller, C., Roux, O., Roux, O.: Semantics of Biological Regulatory Networks. In: BioConcur, Workshop on Concurrent Models in Molecular Biology. ENTCS series (2003)

IMGT-Choreography: Processing of Complex Immunogenetics Knowledge

Denys Chaume[1], Véronique Giudicelli[1], Kora Combres[1], Chantal Ginestoux[1], and Marie-Paule Lefranc[1,2]

[1] Laboratoire d'ImmunoGénétique Moléculaire LIGM, Université Montpellier II, UPR CNRS 1142, Institut de Génétique Humaine IGH, 141 rue de la Cardonille 34396 Montpellier, Cedex 5, France, Phone: +33 4 99 61 99 65, Fax: +33 4 99 61 99 01
[2] Institut Universitaire de France, 103 Boulevard Saint-Michel, 75005 Paris, France
{Denys.Chaume, Kora.Combres}@igh.cnrs.fr
{giudi, chantal, lefranc}@ligm.igh.cnrs.fr
http://imgt.cines.fr

Abstract. IMGT, the international ImMunoGeneTics information system® (http://imgt.cines.fr) is a high quality integrated knowledge resource specialized in immunoglobulins (IG), T cell receptors (TR), major histocompatibility complex (MHC) and related proteins of the immune system (RPI) of human and other vertebrates. IMGT provides a common access to standardized data from genome, proteome, genetics and tridimensional structures. The accuracy and the consistency of IMGT data are based on IMGT-ONTOLOGY, a semantic specification of terms used in immunogenetics and immunoinformatics. IMGT-ONTOLOGY is formalized using XML schemas (IMGT-ML) for interoperability with other information systems. We are developing Web services to automatically query IMGT databases and tools. This is the first step towards IMGT-Choreography which will trigger and coordinate dynamic interactions between IMGT Web services, to process complex significant biological and clinical requests. IMGT-Choreography will further increase the IMGT leadership in immunogenetics medical research (repertoire analysis in autoimmune diseases, AIDS, leukemias,...), biotechnology related to antibody engineering and therapeutical approaches.

1 Introduction

Genome and proteome analysis interpretation represents the current great challenge, as a huge quantity of data is produced by many scientific fields, including fundamental, clinical, veterinary and pharmaceutical research. In particular, the number of sequences and related data published in the immunogenetics fields is growing exponentially. Indeed, the molecular synthesis and genetics of the immunoglobulin (IG) and T cell receptor (TR) chains is particularly complex and unique as it includes biological mechanisms such as DNA molecular rearrangements in multiple loci (three for IG and four for TR in humans) located on different chromosomes (four in humans), nucleotide deletions and insertions at the

V. Danos and V. Schachter (Eds.): CMSB 2004, LNBI 3082, pp. 73 – 84, 2005.

rearrangement junctions (or N-diversity), and somatic hypermutations in the IG loci [for review 1,2]. The number of potential protein forms of IG and TR is almost unlimited.

IMGT, the international ImMunoGeneTics information system® (http://imgt.cines.fr) [3], created in 1989, by the Laboratoire d'ImmunoGénétique Moléculaire (LIGM) (Université Montpellier II and CNRS) at Montpellier, France, is a high quality integrated knowledge resource specialized in IG, TR, major histocompatibility complex (MHC) and related proteins of the immune systems (RPI) of human and other vertebrates [4-11]. IMGT consists of four sequence databases, one genome database and one tridimensional (3D) structure database, interactive tools for sequence, genome and 3D structure analysis, Web resources ("IMGT Marie-Paule page") comprising 8,000 HTML pages of synthesis (IMGT Repertoire), knowledge (IMGT Scientific chart, IMGT Education, IMGT Index) and external links (IMGT Bloc-notes and IMGT other accesses). IMGT is the international reference in immunogenetics and immunoinformatics [12].

The accuracy and the consistency of the IMGT data, as well as the coherence between the different IMGT components (databases, tools and Web resources) are based on IMGT-ONTOLOGY, a semantic specification of terms used in immunogenetics and immunoinformatics [13]. IMGT-ONTOLOGY is formalized using XML schemas (IMGT-ML) [14,15]. In order to allow any IMGT component to be automatically queried, we are developing Web services, the corresponding input and output being formalized according to IMGT-ML. It is the first step towards the implementation of IMGT-Choreography. IMGT-Choreography corresponds to the connection of treatments performed by the IMGT components. In particular, it will trigger and coordinate dynamic interactions between IMGT Web services to process complex significant biological and clinical requests, according to either genomic, genetic or structural approaches.

2 IMGT-ONTOLOGY Concepts

IMGT has developed a formal specification of the terms to be used in the domain of immunogenetics and immunoinformatics to ensure accuracy, consistency and coherence in IMGT. This has been the basis of IMGT-ONTOLOGY [13], the first ontology in the domain, which allows the management of the immunogenetics knowledge for human and other vertebrate species. IMGT-ONTOLOGY comprises five main concepts: IDENTIFICATION, CLASSIFICATION, DESCRIPTION, NUMEROTATION and OBTENTION [13]. Standardized keywords, standardized sequence annotation, standardized IG and TR gene nomenclature, the IMGT unique numbering, and standardized origin/methodology were defined, respectively, based on these five main concepts.

The IDENTIFICATION concept: standardized keywords. IMGT standardized keywords have been assigned to all entries of IMGT/LIGM-DB, the first and largest IMGT sequence database containing more than 82,000 IG and TR nucleotide sequences in April 2004 [16]. They include (i) general keywords: essential for sequence identification, they are described in an exhaustive and non redundant list, for the assignment of species, molecule type, receptor type, chain type, gene type, structure, functionality and specificity, and (ii) specific keywords: they are more specifically associated with particularities of the sequences (orphon, transgene, etc.) or with diseases (leukemia, lymphoma, myeloma, etc.) [17].

The DESCRIPTION concept: standardized labels and annotations. Two hundred fifteen feature labels are necessary in IMGT/LIGM-DB to describe all structural and functional subregions that compose the IG and TR sequences [17], whereas only seven of them are available in EMBL, GenBank or DDBJ [18-20]. Annotation of sequences with these labels constitutes the main part of the expertise and provides a detailed and accurate description. Prototypes represent the organizational relationship between labels and give information on their order and their expected length (in number of nucleotides) [17, 21]. Following a similar approach, one hundred seventy two additional labels were more recently defined for the IG, TR, MHC and RPI amino acid sequences and domain structures in the IMGT protein database, IMGT/PROTEIN-DB, and in the IMGT structure database, IMGT/3D structure-DB [22].

The CLASSIFICATION concept: standardized IG and TR gene nomenclature. The CLASSIFICATION concept has been used to set up a unique nomenclature of human IG and TR genes, which was approved by the Human Genome Organization (HUGO) Nomenclature Committee (HGNC) in 1999 [23] and has become the community standard. The complete list of the human IG and TR gene names [1, 2] has been entered by the IMGT Nomenclature Committee in the IMGT genome database IMGT/GENE-DB [24], Genome DataBase GDB (Canada), LocusLink at NCBI (USA), and GeneCards. The complete list of the mouse IG and TR gene names was sent by IMGT, in July 2002, to the Mouse Genome Informatics MGI Mouse Genome Database MGD, LocusLink at NCBI, and HGNC.

The NUMEROTATION concept: the IMGT unique numbering. A uniform numbering system for IG and TR sequences of all species has been established to facilitate sequence comparison and cross-referencing between experiments from different laboratories whatever the antigen receptor (IG or TR), the chain type, the domain (variable V or constant C), or the species [25-27]. This numbering results from the analysis of more than 5000 IG and TR variable region sequences of vertebrate species from fish to human. It takes into account and combines the definition of the framework (FR) and complementarity determining regions (CDR) [28], structural data from X-ray diffraction studies [29] and the characterization of the hypervariable loops [30].

The OBTENTION concept: standardized origin/methodology. The OBTENTION concept is a set of standardized terms that precise the origins of the sequence (the 'origin' concept) and the conditions in which the sequences have been obtained (the 'methodology' concept). The 'origin' concept comprises the subsets of 'cell, tissue or organ', 'auto-immune diseases', 'clonal expansion diseases' (such as leukemia, lymphoma, myeloma), whereas the 'methodology' concept comprises the subsets related to the 'hybridoma', to the experimental conditions (sequences amplified by 'PCR'), to the obtention from 'libraries' (genomic, cDNA, combinatorial, etc.) or from 'transgenic' organisms (animal, plant). At this stage of development, the exhaustive definition of the concepts of obtention and of their instances is still in progress.

3 IMGT Scientific Chart, IMGT-ML and Web Services

IMGT-ONTOLOGY concepts are formalized, for the biologists and IMGT users, in the IMGT Scientific chart and, for the computing scientists, in IMGT-ML.

3.1 IMGT Scientific Chart

IMGT Scientific chart is a section of the IMGT Web resources which provide the controlled vocabulary and the annotation rules defined by IMGT-ONTOLOGY for the identification, the description, the classification and the numbering of the IG, TR, MHC and RPI data of human and other vertebrates. These HTML pages are devoted to biologists, IMGT users and IMGT annotators. All IMGT data are expertly annotated according to the IMGT Scientific chart.

3.2 IMGT-ML

IMGT-ML represents the formal specification of IMGT-ONTOLOGY [13]. IMGT-ONTOLOGY main concepts have been formalized in XML schemas [14, 15]. By making data portable, XML is useful both internally for the integration of data and externally for sharing data with other information systems. Because of this data integration ability, XML has become the underpinning for Web-related computing. IMGT-ML defines XML schemas to encode data with XML tags which respect the IMGT-ONTOLOGY concepts (XML schemas are available from IMGT index>IMGT-ML and at http://oriel.igh.cnrs.fr/IMGT-ML.xhtml). IMGT-ML schemas are used for distributive data using the Web services (http://www.w3.org/2002/ws/) technology.

3.3 IMGT Web Services

IMGT Web services are developed using the Apache Tomcat server (http://jakarta.apache.org/tomcat) and are implemented with Axis (for Java programs) (http://ws.apache.org/) or with SOAP:lite (for Perl programs). IMGT Web service data exchanges are based on valid IMGT-ML streams. IMGT Web services can easily be interconnected since IMGT-ML is the unique language used for both Web services input (query) and output (response).

The IMGT/LIGM-DB Web service is the first Web service currently developed and implemented with Axis. It includes the "queryKnowledge" and "querySeqData" services. The queryKnowledge service provides the lists of instances for the IMGT-ONTOLOGY concepts, for example the list of chain types, functionalities, specificities defined in the IDENTIFICATION concept, the lists of groups and subgroups defined in the CLASSIFICATION concept, or the list of labels defined in the DESCRIPTION concept. The querySeqData service allows the retrieval of any sequence related data, identified, classified, described according to the IMGT concepts, such as the nucleotide sequence, the description labels, the literature references, the metadata, etc. The querySeqData input has the form of an incomplete IMGT-ML data entry. The given values are used as criteria to query the database. The result is then a list of data entries, in IMGT-ML format, sharing these given values.

4 IMGT-Choreography Modelling

IMGT-Choreography, which will combine and join several database queries and analysis tools, includes three main types of approaches: genomic, genetic and structural. A schematic representation of IMGT-Choreography with the three approaches is shown in Figure 1, with examples of the IMGT databases and tools.

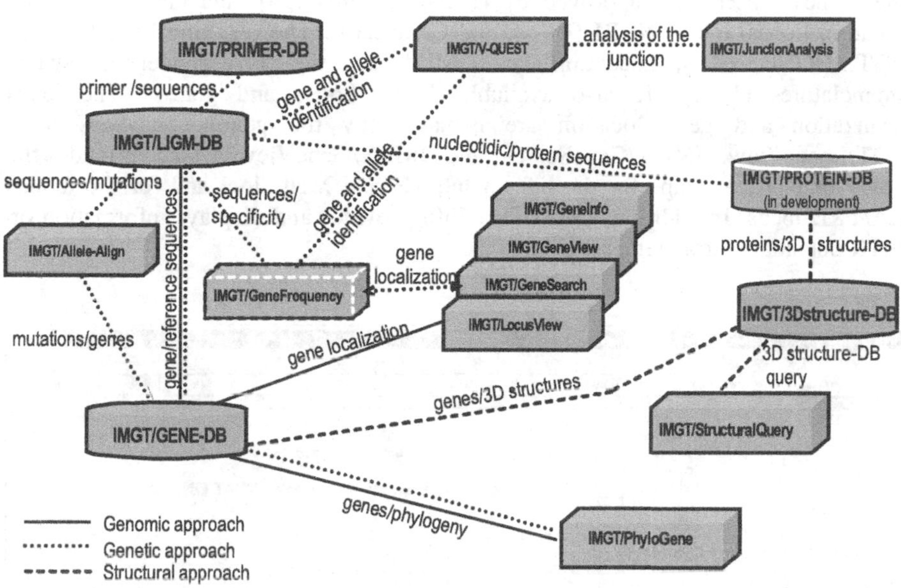

Fig. 1. IMGT-Choreography. Examples of interactions between IMGT components (databases and tools) in the 3 main approaches: genomic, genetic and structural. The IMGT/MHC sequence databases hosted at EBI [4] and the IMGT Repertoire Web resources (8000 HTML pages) [10] are not shown

Potential interactions between IMGT components are technically almost infinite. A rigorous analysis of the biologist requests and of the clinician needs are required to keep only significant approaches. The detailed interactions between IMGT components are currently being be carefully modelled in UML [31] collaboration diagrams, in order to identify the next Web services which need to be developed.

4.1 The Genomic Approach

The genomic approach is mainly orientated towards the study of the genes within their loci and on the chromosomes. The corresponding Web resources are compiled in IMGT Repertoire. The "Locus and genes" section includes chromosomal localizations, locus representations, locus description, gene tables, potential germline repertoires and the complete lists of human and mouse IG and TR genes. All the human and mouse IG and TR genes are managed in IMGT/GENE-DB, which is the comprehensive IMGT genome database [24]. Queries can be performed according to

IG and TR gene classification criteria and IMGT reference sequences have been defined for each allele of each gene based on one or, whenever possible, several of the following criteria: germline sequence, first sequence published, longest sequence, mapped sequence [21]. In April 2004, IMGT/GENE-DB contained 1,375 genes and 2,201 alleles (673 IG and TR genes and 1,024 alleles from *Homo sapiens*, and 702 IG and TR genes and 1,177 alleles from *Mus musculus, Mus cookii, Mus pahari, Mus spretus, Mus saxicola, Mus minutoides*). As mentioned earlier, all the human IMGT gene names [1, 2] were approved by HGNC in 1999 [25], and entered in GDB (Canada), LocusLink at NCBI (USA), and GeneCards. The resulting links between IMGT, HGNC, GDB, LocusLink and OMIM, and the correspondence between nomenclatures [1, 2] are also available from "Locus and genes". The locus organization and gene location are managed by the genome analysis tools IMGT/LocusView, IMGT/GeneSearch and IMGT/GeneView which provide the display of physical maps for the human IG, TR and MHC loci and for the mouse TRA/TRD locus. In addition, IMGT/GeneInfo provides and displays information on the TR potential rearrangements [32].

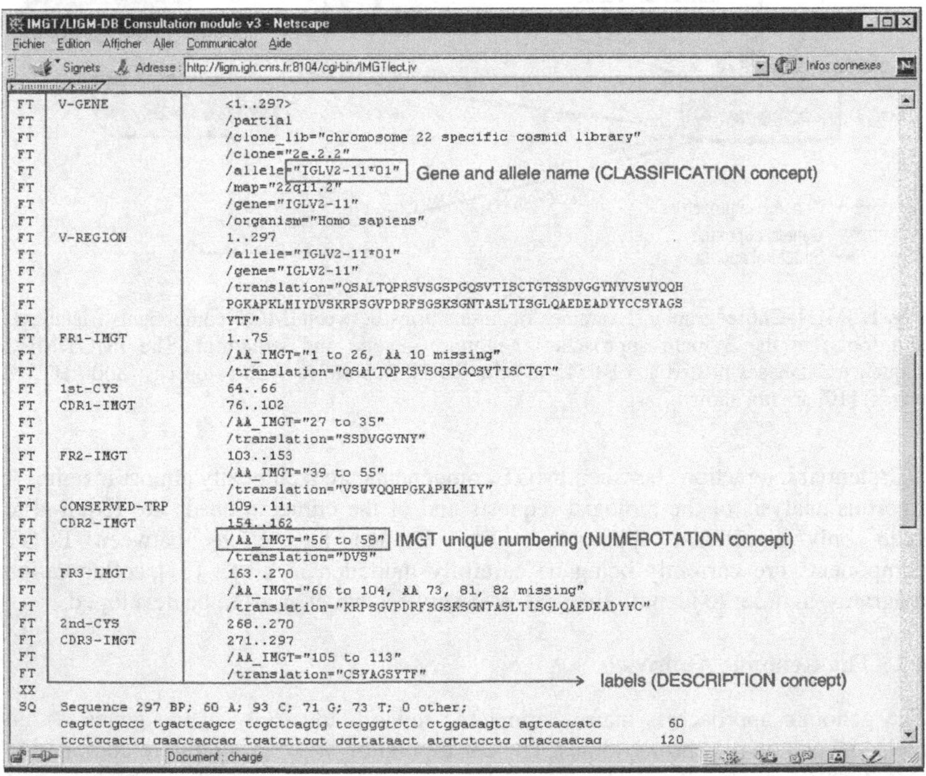

Fig. 2. An example of fully annotated sequence by IMGT/Automat [33]: gene and allele names, labels, and IMGT unique numbering are according to the CLASSIFICATION, DESCRIPTION, and NUMEROTATION concepts of IMGT-ONTOLOGY, respectively [13]

IMGT/GENE-DB interacts dynamically with IMGT/LIGM-DB to download and display gene-related sequence data. IMGT/LIGM-DB is the comprehensive IMGT database of IG and TR nucleotide sequences from human and other vertebrate species, with translation for fully annotated sequences, created in 1989 by LIGM, Montpellier, France, on the Web since July 1995 [17]. The unique source of data for IMGT/LIGM-DB is EMBL [18] which shares data with the other two generalist databases GenBank and DDBJ [19,20]. Based on expert analysis, specific detailed annotations are added to IMGT flat files in a second step [16]. An internal tool, IMGT/Automat, has been developed to automatically perform the annotation of the rearranged cDNA sequences in IMGT/LIGM-DB [33] (Figure 2). The Web interface allows searches according to immunogenetic specific criteria and is easy to use without any knowledge in a computing language. IMGT/LIGM-DB gene and allele name assignment and sequence annotation are performed according to the IMGT Scientific chart rules, based on the CLASSIFICATION, DESCRIPTION and NUMEROTATION concepts of IMGT-ONTOLOGY [13] (Figure 2).

4.2 The Genetic Approach

The genetic approach refers to the study of the genes in relation with their polymorphisms, their evolution and their specificity. The genetic approach heavily relies on the IMGT unique numbering [27]: in this numbering, conserved amino acids from frameworks always have the same number whatever the IG or TR variable sequence, and whatever the species they come from (as examples: Cysteine 23 (in FR1-IMGT), Tryptophan 41 (in FR2-IMGT), Leucine 89 and Cysteine 104 (in FR3-IMGT). Based on the IMGT unique numbering, standardized 2D graphical representations, designated as IMGT Colliers de Perles [21], are available in IMGT Repertoire. The IMGT unique numbering represents a big step forward in the analysis of the IG and TR variable region (V-REGION) sequences of all vertebrate species [34]. Moreover, it gives insight into the structural configuration of the variable domain [27, 35]. Note that the IMGT unique numbering opens interesting views on the evolution of the proteins belonging to the "immunoglobulin superfamily" (IgSF) [36]. It has been applied with success to all the sequences of domains belonging to the IgSF V-set, designated as V-LIKE-DOMAINs in IMGT (human CD4 Xenopus CTXg1, etc.) and in invertebrates (drosophila amalgam, etc.) [25-27]. This standardized approach has been applied to the constant domain (C-DOMAIN) of the IG and TR, and to the C-LIKE-DOMAINs of proteins other than IG and TR. An IMGT unique numbering has also been implemented for the groove domain (G-DOMAIN) [37] of the MHC class I and II chains, and for the G-LIKE-DOMAINs of proteins other than MHC.

The study of the gene polymorphisms is performed using IMGT/Allele-Align, the analysis of the gene evolution using IMGT/Phylogene [38], and the characterization of the gene specificity using IMGT/V-QUEST [9, 39], IMGT/JunctionAnalysis [40], and IMGT/GeneFrequency. Figure 1 shows the current interactions of the IMGT genetic analysis tools with the IMGT/LIGM-DB sequence database and with the IMGT/GENE-DB genome database. IMGT/Allele-Align allows the comparison of two alleles highlighting the nucleotide and amino acid differences. IMGT/Phylogene is an easy to use tool for phylogenetic analysis of variable region (V-REGION) and constant domain (C-DOMAIN) sequences. This tool is particularly useful in developmental and comparative immunology. The users can analyse their own sequences by comparison with the IMGT standardized reference sequences for human and mouse IG and TR [38].

IMGT/V-QUEST (V-QUEry and STandardization) is an integrated software for IG and TR [9, 39]. This tool, easy to use, analyses an input IG or TR germline or rearranged variable nucleotide sequence. IMGT/V-QUEST results comprise the identification of the V, D and J genes and alleles and the nucleotide alignments by comparison with sequences from the IMGT reference directory, the delimitations of the FR-IMGT and CDR-IMGT based on the IMGT unique numbering, the protein translation of the input sequence, the identification of the JUNCTION, and the two-dimensional Collier de Perles representation of the V-REGION. IMGT/JunctionAnalysis [40] is a tool, complementary to IMGT/V-QUEST, which provides a thorough analysis of the V-J and V-D-J junction of IG and TR rearranged genes. IMGT/JunctionAnalysis identifies the D-GENEs and alleles involved in the IGH, TRB and TRD V-D-J rearrangements by comparison with the IMGT reference directory, and delimits precisely the P, N and D regions (IMGT/JunctionAnalysis output results). Several hundreds of junction sequences can be analysed simultaneously.

IMGT/GeneFrequency is a new IMGT interactive tool which dynamically computes and represents histograms for the gene expression in response to an antigen selected by the user, by querying the IMGT/LIGM-DB sequences. IMGT/GeneFrequency combines IMGT Web services in order to answer complex requests on the relationships between specificity, gene expression and gene location in the locus. It will be available from the IMGT Home page at http://imgt.cines.fr within the year 2004. In a next step, dynamic interactions will be developed with IMGT/GENE-DB and the genome analysis tools, and with the IMGT/PROTEIN-DB and IMGT/PRIMER-DB [41] sequence databases.

4.3 The Structural Approach

The structural approach focuses on the 3D structures of the IG, TR, MHC and RPI, in IMGT/3Dstructure-DB [22], and on their interactions with the antigens or ligands, using the IMGT/StructuralQuery tool [22]. IMGT/3Dstructure-DB is the IMGT 3D structure database for IG, TR, MHC and RPI of human and other vertebrate species, created by LIGM, on the Web since November 2001 [22]. IMGT/3Dstructure-DB comprises IG, TR, MHC and RPI with known 3D structures. Coordinate files are extracted from the Protein Data Bank (PDB) [42], and IMGT annotations are added according to the IMGT Scientific chart rules, based on the IMGT-ONTOLOGY concepts [13]. An IMGT/3Dstructure-DB card provides IMGT gene and allele identification (based on the CLASSIFICATION concept), domain delimitations (based on the DESCRIPTION concept), amino acid positions according to the IMGT unique numbering [27] (based on the NUMEROTATION concept). IMGT/StructuralQuery allows to retrieve the IMGT/3Dstructure-DB entries, based on specific structural characteristics: phi and psi angles, accessible surface area (ASA), amino acid type, distance in angstrom between amino acids, CDR-IMGT lengths [22]. IMGT/StructuralQuery is currently available for the V-DOMAINs.

5 Conclusion

IMGT databases and tools are extensively queried and used by scientists from academic and industrial laboratories working in fundamental research, in medical research (repertoire in normal and pathological situations: autoimmune diseases, infectious diseases, AIDS, leukemias, lymphomas, myelomas), in veterinary research, in genome diversity and genome evolution studies of the adaptive immune responses, in biotechnology related to antibody engineering (single chain Fragment variable (scFv), phage displays, combinatorial libraries), in diagnostics and in therapeutical approaches (clonalities, grafts, immunotherapy, detection and follow up of residual diseases). By its high quality and its data distribution based on IMGT-ONTOLOGY, IMGT has an important role to play in the development of immunogenetics Web services. The design of IMGT-Choreography and the creation of dynamic interactions between the IMGT databases and tools, using the Web services and IMGT-ML, represent novel and major developments of IMGT, the international reference in immunogenetics and immunoinformatics.

Acknowledgments

We thank S. Jeanjean for the implementation of IMGT/GeneFrequency, M. Lemmatize, M.-C. Beckers, G. Folch and D. Valette from EUROGENTEC S.A., Belgium, for their scientific contribution to IMGT/PRIMER-DB. We are deeply grateful to the IMGT team for its expertise and constant motivation and specially to our curators for their hard work and enthusiasm. IMGT is funded by the European Union's 5th PCRDT programme (QLG2-2000-01287), the Centre National de la Recherche Scientifique (CNRS), the Ministère de l'Education Nationale and the Ministère de la Recherche. The ORIEL project is funded by the European Union IST programme (ST-2001-32688). Subventions have been received from Association pour la Recherche sur le Cancer (ARC) and the Région Languedoc-Roussillon.

References

1. Lefranc, M.-P., Lefranc, G.: The Immunoglobulin FactsBook. Academic Press, London, UK, 458 pages (2001)
2. Lefranc, M.-P., Lefranc, G.: The T cell receptor FactsBook. Academic Press, London, UK, 398 pages (2001)
3. Lefranc, M.-P.: IMGT, the international ImMunoGeneTics database®. Nucleic Acids Res. 31 (2003) 307-310
4. Ruiz, M., Giudicelli, V., Ginestoux, C., Stoehr, P., Robinson, J., Bodmer, J., Marsh, S.G., Bontrop, R, Lemaitre, M., Lefranc, G., Chaume, D., Lefranc, M.-P.: IMGT, the international ImmunoGeneTics database. Nucleic Acids Res. 28 (2000) 219-221
5. Lefranc, M.-P.: IMGT, the international ImMunoGeneTics database. Nucleic Acids Res. 29 (2001) 207-209
6. Lefranc, M.-P.: IMGT, the international ImMunoGeneTics database: a high-quality information system for comparative immunogenetics and immunology. Dev. Comp. Immunol. 26 (2002) 697-705

7. Lefranc, M.-P.: IMGT® databases, web resources and tools for immunoglobulin and T cell receptor sequence analysis, http://imgt.cines.fr. Leukemia 17 (2003) 260-266
8. Lefranc, M.-P., Giudicelli, V., Ginestoux, C., Chaume, D.: IMGT, the international ImmunoGeneTics information system®, http://imgt.cines.fr: the reference in immunoinformatics. Proceeding of the Medical Informatics Europe 2003, MIE 2003, IOS Press, Amsterdam (2003) 74-79
9. Lefranc, M.-P.: IMGT, the international ImMunoGeneTics information system® (http://imgt.cines.fr). In: Antibody Engineering: Methods and Protocols. 2nd edition (ed. Lo B.K.C.) Methods in Molecular Biology. Humana Press, Totowa, NJ, USA, vol. 248, chap. 3 (2003) 27-49
10. Lefranc, M.-P., Giudicelli, V., Ginestoux, C., Bosc, N., Folch, G., Guiraudou, D., Jabado-Michaloud, J., Magris, S., Scaviner, D., Thouvenin, V., Combres, K., Girod, D., Jeanjean, S., Protat, C., Yousfi Monod, M., Duprat, E., Kaas, Q., Pommié, C., Chaume, D., Lefranc, G.: IMGT-ONTOLOGY for Immunogenetics and Immunoinformatics (http://imgt.cines.fr). In Silico Biology 4 (2003) 0004. http://www.bioinfo.de/isb/2003/04/2004/
11. Lefranc, M.-P.: IMGT-ONTOLOGY and IMGT databases, tools and web resources for immunogenetics and immunoinformatics. Molecul. Immunol. 40 (2004) 647-659
12. Warr, G.W., Clem, L.W., Söderhall, K., Editorial : The International ImMunoGeneTics database IMGT. Dev. Comp. Immunol. 27 (2003) 1-2
13. Giudicelli, V., Lefranc, M.-P.: Ontology for Immunogenetics: the IMGT-ONTOLOGY. Bioinformatics 12 (1999) 1047-1054
14. Chaume, D., Giudicelli, V., Lefranc, M.-P.: IMGT-ML a language for IMGT-ONTOLOGY and IMGT/LIGM-DB data. In: CORBA and XML: Towards a bioinformatics integrated network environment, Proceedings of NETTAB 2001, Network tools and applications in biology (2001) 71-75
15. Chaume, D., Giudicelli, V., Combres, K., Lefranc, M.-P. : IMGT-ONTOLOGY and IMGT-ML for Immunogenetics and immunoinformatics. European Congress in Computational Biology ECCB'2003, Sequence databases and Ontologies satellite event, Paris (2003)
16. Chaume, D., Giudicelli, V., Lefranc, M.-P.: IMGT/LIGM-DB. In : The Molecular Biology Database Collection: 2004 update (Galperin M.Y. ed.). Nucleic Acids Res. 32 (2004) http://www3.oup.co.uk/nar/database/summary/504
17. Giudicelli, V., Chaume, D., Bodmer, J., Müller, W., Busin, C., Marsh, S., Bontrop, R., Lemaitre, M., Malik, A., Lefranc, M.-P.: IMGT, the international ImMunoGeneTics database. Nucleic Acids Res. 25 (1997) 206-211
18. Stoesser, G., Baker, W., Van den Broek, A., Garcia-Pastor, M., Kanz, C., Kulikova, T., Leinonen, R., Lin, Q., Lombard, V., Lopez, R., Mancuso, R., Nardone, F., Stoehr, P., Tuli, M.A., Tzouvara, K., Vaughan, R.: The EMBL nucleotide sequence database: major new developments. Nucleic Acids Res. 31 (2003) 17-22
19. Benson, D., A., Karsch-Mizrachi, I., Lipman, D.J., Ostell, J., Wheeler, D.L.: GenBank. Nucleic Acids Res. 31 (2003) 23-27
20. Miyazaki, S., Sugawara, H., Gojobori, T., Tateno, Y.: DNA Data Bank of Japan (DDBJ) in XML. Nucleic Acids Res. 31 (2003) 13-16
21. Lefranc, M.-P., Giudicelli, V., Ginestoux, C., Bodmer, J., Müller, W., Bontrop, R., Lemaitre, M., Malik, A., Barbié, V., Chaume, D.: IMGT, the international ImMunoGeneTics database. Nucleic Acids Res. 27 (1999) 209-212

22. Kaas, Q., Ruiz, M., Lefranc, M.-P.: IMGT/3Dstructure-DB and IMGT/StructuralQuery, a database and a tool for immunoglobulin, T cell receptor and MHC structural data. Nucleic Acids Res. 32 (2004) D208-D210
23. Wain, H.M., Bruford, E.A., Lovering, R.C., Lush, M.J., Wright, M.W., Povey, S.: Guidelines for human gene nomenclature. Genomics 79 (2002) 464-470
24. Giudicelli, V., Lefranc, M.-P.: IMGT/GENE-DB. In : The Molecular biology Database Collection: 2004 update (Galperin M.Y. ed.). Nucleic Acids Res. 32 (2004) http://www3.oup.co.uk/nar/database/summary/503
25. Lefranc, M.-P.: Unique database numbering system for immunogenetic analysis. Immunol. Today 18 (1997) 509
26. Lefranc, M.-P.: The IMGT unique numbering for Immunoglobulins, T cell receptors and Ig-like domains. The Immunologist 7 (1999) 132-136
27. Lefranc, M.-P., Pommié, C., Ruiz, M., Giudicelli, V., Foulquier, E., Truong, L., Thouvenin-Contet, V., Lefranc, G.: IMGT unique numbering for immunoglobulin and T cell receptor variable domains and Ig superfamily V-like domains. Dev. Comp. Immunol. 27 (2003) 55-77
28. Kabat, E.A., Wu, T.T.: Identical V region amino acid sequences and segments of sequences in antibodies of different specificities. Relative contributions of VH and VL genes, minigenes, and complementarity-determining regions to binding of antibody-combining sites. J. Immunol. 147 (1991) 1709-1719
29. Satow, Y., Cohen, G.H., Padlan, E.A., Davies, D.R.: Phosphocholine binding immunoglobulin Fab McPC603. J. Mol. Biol. 190 (1986) 593-604
30. Chothia, C., Lesk, A. M.: Canonical structures for the hypervariable regions of immunoglobulins. J. Mol. Biol. 196 (1987) 901-917
31. Cranefield, S.: UML and the Semantic Web. Proceedings of the International Semantic Web Working Symposium (SWWS) (2001)
32. Baum, T.P., Pasqual, N., Thuderoz, F., Hierle, V., Chaume, D., Lefranc, M.-P., Jouvin-Marche, E., Marche, P.,N., Demongeot, J.: IMGT/GeneInfo: enhancing V(D)J recombination database accessibility. Nucleic Acids Res. (2004) D51-D54
33. Giudicelli, V., Protat, C., Lefranc, M.-P,: The IMGT strategy for the automatic annotation of IG and TR cDNA sequences: IMGT/Automat. Poster DKB_31. ECCB'2003, European Conference on Computational Biology (2003) http://www.inra.fr/Internet/Departements/bia/recherche/colloques/ECCB-2003/posters/pdf/Annot_Giudicelli_20030528_160703.pdf
34. Pommié, C., Sabatier, S., Lefranc, G., Lefranc, M.-P.: IMGT standardized criteria for statistical analysis of immunoglobulin V-REGION amino acid properties J. Mol. Recognit. 17 (2004) 17-32
35. Ruiz, M., Lefranc, M.-P.: IMGT gene identification and Colliers de Perles of human immunoglobulins with known 3D structures. Immunogenetics 53 (2002) 857-883
36. Williams, A.F., Barclay A.N.: The immunoglobulin family: domains for cell surface recognition. Annu. Rev. Immunol. 6 (1988) 381-405
37. Duprat, E., and Lefranc, M.-P.: IMGT standardization and analysis of V-LIKE-, C-LIKE- and G-LIKE-DOMAINs. Poster PS_32. ECCB'2003, European Conference on Computational Biology (2003) http://www.inra.fr/Internet/Departements/bia/recherche/colloques/ECCB-2003/posters/pdf/Molec_Duprat_20030604_163520.ps
38. Elemento, O., Lefranc, M.-P.: IMGT/PhyloGene: an on-line tool for comparative analysis of immunoglobulin and T cell receptor genes. Dev. Comp. Immunol. 27(2003)763-779
39. Giudicelli, V., Chaume, D., Lefranc, M.-P.: IMGT/V-QUEST, an integrated software program for immunoglobulin and T cell receptor V-J and V-D-J rearrangement analysis. Nucleic Acids Res. Web issue (2004) in press

40. Yousfi Monod, M., Giudicelli, V., Chaume, D., Lefranc, M.-P.: IMGT/JunctionAnalysis: the first tool for the analysis of the immunoglobulin and T cell receptor complex V-J and V-D-J JUNCTIONs. Bioinformatics (2004) in press
41. Folch G., Bertrand J., Lemaitre M. and Lefranc M.-P.: IMGT/PRIMER-DB. In : The Molecular Biology Database Collection: 2004 update (Galperin M.Y. ed.). Nucleic Acids Res. 32 (2004) http://www3.oup.co.uk/nar/database/summary/505
42. Berman, H.M., Westbrook, J., Feng, Z., Gilliland, G., Bhat, T.N., Weissig, H., Shindyalov, I.N., Bourne, P.E.: The Protein Data Bank. Nucleic Acids Res. 28 (2000) 235-242

Address for Correspondence

Professor Marie-Paule Lefranc
Laboratoire d'ImmunoGénétique Moléculaire LIGM, UPR CNRS 1142, Institut de Génétique Humaine IGH, 141 rue de la Cardonille 34396 Montpellier Cedex 5, France
Phone: +33 4 99 61 99 65, Fax: +33 4 99 61 99 01
e-mail: lefranc @ligm.igh.cnrs.fr.

Model Checking Biological Systems Described Using Ambient Calculus*

Radu Mardare, Corrado Priami, Paola Quaglia, and Oleksandr Vagin

Dipartimento di Informatica e Telecomunicazioni,
Università di Trento, Italy

Abstract. We propose a way of performing model checking analysis for biological systems. The technics were developed for a CTL* logic built upon Ambient Calculus.

We introduce *labeled syntax trees* for ambient processes and use them as possible worlds in a Kripke structure developed for a propositional branching temporal logic. The accessibility relation over labeled syntax trees is generated by the reduction over corresponding Ambient Calculus processes.

Providing the algorithms for calculating the accessibility relation between states, we open the perspective of using model checking algorithms developed for temporal logics in analyzing any phenomena described in Ambient Calculus.

1 Introduction

Ambient Calculus [13] is a useful tool to construct mathematical models for complex systems because of its facilities in expressing hierarchies of locations and their mobility. At the same time properties as "the protein has split", or "there is a path of computation where the complexAB precedes the proteinA" are not expressible inside the calculus. Only a logic built on top of it can handle such properties.

Modal logics and especially temporal logics have emerged in many domains as a good compromise between expressiveness and abstraction. Many of them support useful computational applications as model checking. For the particular case of temporal logics, these technics were developed up to the construction of some tools able to perform such analysis (see, e.g. SMV [4], NuSMV [2], HyTech [1], VIS [5]).

This paper presents a propositional branching temporal logic for Ambient Calculus that is the representative calculus for the paradigm of calculi expressing hierarchies of locations and their mobility. We believe that the same sort of logic can be constructed for other calculi in this paradigm like, e.g., BioAmbients Calculus [7], or Brane Calculi [9].

* Work partially supported by the FET project IST-2001-32072 DEGAS under the pro-active initiative on Global Computing.

V. Danos and V. Schachter (Eds.): CMSB 2004, LNBI 3082, pp. 85–103, 2005.

The main feature of our logic is that the final state of any computation can be reconstructed by just having information about the initial state and the history of the computation. The spatial structure of a state is fully described by a set of atomical propositions, while the possible states are described using, in addition, a temporal modality. In this respect our approach is different from those used in Ambient Logic [12, 11], or Spatial Logic [10], giving us the advantages of simplicity and expressivity that a CTL* logic have w.r.t. the cited modal logics.

The rest of the paper is organized as follows. We first present a couple of simple case studies coming from biology. They are used to comment on the advantages of applying temporal logics to the Ambient Calculus specification of phenomena related to life sciences. Section 3 introduces the theoretical underpinning of our logic: *labeled syntax trees.* In Section 4 we define a branching temporal logic for Ambient Calculus, and show how to run simple reachability properties on our case studies. The final section concludes the presentation with a description of an implementation of our logic. The platform actually consists in the development of a suitable interface to NuSMV [2].

2 Case Studies from Biology

The advantage of using a temporal logic is relevant in the representation of biological phenomena because it gives us the power to predict over the future. Consider the model of *the trimetric GTP binding proteins (G-proteins)* that plays an important role in the signal transduction pathway for numerous hormones and neurotransmitters [3, 6]. It consists of five processes: a regulatory molecule RM, a receptor R, and three domains that are bound together composing the protein α, β and γ. Data sent by RM to R determine a communication between the receptor R and the protein that causes the breakage of the boundary of α, β and γ. We ca express this in Ambient Calculus by the following specification:

$$RM \stackrel{def}{=} open\ n.RM,\ R \stackrel{def}{=} n[\langle GTP\rangle|R],$$

$$Protein \stackrel{def}{=} (GDP)(\alpha|\beta|\gamma), \text{ where } GDP \text{ is a name that appear in } \alpha \text{ only, bounded by the input prefix}$$

$$\begin{aligned} RM|R|Protein \equiv open\ n.RM \mid n[\langle GTP\rangle|R] \mid (GDP)(\alpha|\beta|\gamma) \rightarrow \\ RM \mid R \mid \langle GTP\rangle \mid (GDP)(\alpha|\beta|\gamma) \rightarrow \\ RM|R|(\alpha|\beta|\gamma)(GDP/GTP) \rightarrow \\ RM|R|(\alpha)(GDP/GTP)|\beta|\gamma \end{aligned}$$

where we denoted by $(\alpha)(GDP/GTP)$ the process obtained by substituting GDP with GTP inside α.

W.r.t the above example we are interested to express properties like, e.g., *for all possible future paths, sometime in the future, we will have the interaction that will generate the split of the protein.* One could also want to express that the protein will not be split before the interaction between R and RM will be performed (*a property will not be satisfied until an other one will be*). Both properties above are examples of 'temporal' properties which cannot be expressed using other modal logics.

Consider now the interaction between a Virus and a Macrophage. The Macrophage is recognizing the Virus by its characteristics. Once it is recognized, the Virus is moved by Macrophage inside itself were it is destroyed. We decided to describe the Macrophage as an ambient named n that contains a process *Digest* able to destroy the virus; the virus is an ambient k' that contains inside a process *Infect*. The Macrophage recognizes the virus by the name k' and by its structure (i.e. Macrophage knows the names k, k'' that define the structure of the virus). Using these information, Macrophage manages to put in parallel the processes *Infect* and *Digest* and in this way annihilates the action of the virus. We can describe this action in Ambient Calculus in a way similar with the description of the action of a firewall:

$$Macrophage \stackrel{def}{=} n[k[out\ n.in\ k'.in\ n.0] | open\ k'.open\ k''.Digest]$$

$$Virus \stackrel{def}{=} k'[open\ k.k''[Infect]]$$

$$Virus | Macrophage \equiv$$

$$k'[open\ k.k''[Infect]] | n[k[out\ n.in\ k'.in\ n.0] | open\ k'.open\ k''.Digest]$$

$$\rightarrow^* k'[open\ k.k''[Infect]] | k[in\ k'.in\ n.0] | n[open\ k'.open\ k''.Digest]$$

$$\rightarrow^* k'[open\ k.k''[Infect] | k[in\ n.0]] | n[open\ k'.open\ k''.Digest]$$

$$\rightarrow^* k'[k''[Infect] | in\ n.0] | n[open\ k'.open\ k''.Digest]$$

$$\rightarrow^* n[k'[k''[Infect]] | open\ k'.open\ k''.Digest]$$

$$\rightarrow^* n[k''[Infect] | open\ k''.Digest]$$

$$\rightarrow^* n[Infect | Digest]$$

For this situation, we can be interested if our system succeeds, in all possible time paths, to achieve the state where the processes *Infect* and *Digest* are in parallel inside the ambient n (that represent the Macrophage), such that the virus to be annihilated. If our system succeeds to do this, we can say that is an appropriate one, otherwise we have to reconsider our approach. Such properties, we will argue further, can be naturally expressed in a temporal logic.

Another reason for using temporal logics to model Ambient Calculus is the possibility of performing model checking for our calculus, by reusing some software already developed for these logics such as SMV [4], NuSMV [2], HyTech [1], VIS [5].

3 The Construction of the Labeled Syntax Trees

In order to define the temporal logic, we reorganize the spatio-temporal information contained by an ambient process. This will be done by defining a special labelling function for the syntax trees of Ambient Calculus.

A syntax tree $S = (S, \rightarrow_S)$ for a process is a graph with $S = \Pi \cup \Gamma \cup \Omega = (\Pi_P \cup \Pi_A) \cup \Gamma \cup \Omega$ where

Π is a set that contain all the unspecified process nodes (hereafter atomical processes[1] and collected in the subset Π_P) and the ambient nodes (collected in the subset Π_A);

Γ is the set of capability nodes (we include here the input nodes and the nodes of variables over capabilities as well); and

Ω is the set of syntactical operator nodes (this set contains the parallel operators $|$ and the prefix operators, $\bullet$). We identify the subset $\Omega' = \{\bullet_1 \in \Omega \mid \bullet_1 \rightarrow_S | \} \subseteq \Omega$ of the prefix nodes that are immediately followed in the syntax tree by the parallel operator because they play an important role in the spatial structure of the ambient process [2].

We consider also the possibility of having circular branches in our trees, when recursive definitions are involved. All the further discussion is including these cases as well.

The intuition behind the construction of a labeled syntax tree is to associate to each node of the syntax tree some labeles by two functions: *id* that gives to each node an identity, and *sp* that registers the spatial position of the node.

The identity function *id* associates a label (urelement or $\emptyset$):

1. to each unspecified process and to each ambient; this label will identify the node and will help us further to distinguish between processes that have the same name
2. to each capability, the identity of the process in front of which this capability is placed
3. $\emptyset$, to each syntactical node

The spatial function *sp* associates:

1. to each ambient the set of the identities of its children[3], while to unspecified processes associates the *id*-label.
2. to each capability, a natural number that counts the position of this capability in the chain of capabilities (if any) belonging to the same process
3. to each syntactical node the spatial function associates 0, except for the nodes in Ω' to which the function *sp* will associate the set of identities of the processes connected by the main parallel operator in the compound process that this point is prefixing. For example in the situation $c.(P|Q)$, $sp(\bullet) = \{id(P), id(Q)\}$.

---

[1] We use these to denote unspecified processes found inside an ambient process; this is a necessary requirement in developing model checking for Ambient Calculus because we have to recognize and distinguish, over time, unspecified processes inside the target process. For instance P is an unspecified process in $n[in\ m.P]$.

[2] These point operators are those that connect a capability with a process formed by a parallel composition of other processes bounded together by brackets, hereafter *complex processes*, as in $c.(P|Q)$

[3] We use the terms *parent* and *child* about processes, meaning the immediate parent and immediate child in Ambient Calculus processes.

We recall here some basic definitions of Set Theory and Graph Theory that are needed to formally define the functions *id* and *sp* above.

We choose to work inside Zermelo-Fraenkel system of Set Theory ZFC with the Foundation Axiom (FA), as being a fertile field that offers many tools for analyzing structures, as argued in [8]. This approach allows us to describe the spatial structure of ambient processes as equations in set theory, each such equation being then used as atomical proposition in our logic. In this way we will not use a modality in describing the hierarchy of locations, but only in describing the evolution of the hierarchy in time. Hereafter, we assume a class $\mho$ of urelements, set-theoretical entities which are not sets (they do not have elements) but can be elements of sets. The urelements together with the empty set $\emptyset$ will generate all the sets we will work with (sometimes sets of sets).

Definition 1. *A set a is* transitive *if all the elements of a set b, which is an element of a, also belong to a: $\forall b \in a$ if $c \in b$ then $c \in a$.*

The transitive closure *of a, denoted by $TC(a)$ is the smallest transitive set including a. The existence of $TC(a)$ could be justified as follows:*
$TC(a) = \cup\{a, \cup a, \cup\cup a, ...\}$

Definition 2. *The* support *of a set a, denoted by $supp(a)$ is $TC(a) \cap \mho$. The elements of $supp(a)$ are the urelements that are* somehow involved *in a.*

Definition 3. *If $a \subseteq \mho$ then $V(a) \stackrel{def}{=} \{b \mid b \text{ is a set and } supp(b) \subseteq a\}$. $V(a)$ is the class of all sets in which the only urelements that are somehow involved are the urelements of a.*

Definition 4. *Let $S_P = (S, \rightarrow_S)$ be the syntax tree associated with the ambient process P. We call* the structure graph *associated with P, the graph obtained by restricting the edge relation of the syntax tree to $\Pi \cup \Omega'$, i.e. the graph $T_P = (\Pi \cup \Omega', \rightarrow_T)$ defined by:*
for $n, m \in \Pi \cup \Omega'$ we have $n \rightarrow_T m$ iff $n \rightarrow_S^* m$ and $\nexists p \in \Pi \cup \Omega'$ such that $n \rightarrow_S^* p \rightarrow_S^* m$

Intuitively, the structure graph of a process is obtained by restricting the edge relation of its syntax tree to Π.

Definition 5. *A* decoration *of a graph $G = (G, \rightarrow_G)$ is an injective function $e : G \rightarrow V(\mho) \cup \mho$ such that for all $a \in G$ we have:*

- *if $\nexists b \in G$ such that $a \rightarrow_G b$ then $e(a) \in \mho$*
- *if $\exists b \in G$ such that $a \rightarrow_G b$ then $e(a) = \{e(b) \mid \text{ for all } b \text{ such that } a \rightarrow_G b\}$.*

We now introduce a set of auxiliary functions that are the building blocks for *id* and *sp* (for the application of these and the following definitions see the Appendix).

Definition 6. *Let the next functions be defined on the subsets of nodes of the syntax tree $(S, \rightarrow)$ as follows:*

- *Let $sp_\Pi : \Pi \cup \Omega' \rightarrow V(\mho) \cup \mho$ be a decoration of the structure graph associated with our syntax tree.*
- *Let $id_\Pi : \Pi \rightarrow \mho$ be an injective function such that $id_\Pi(P) = sp_\Pi(P)$ for all $P \in \Pi_P$. Consider $U_P \stackrel{def}{=} id_\Pi(\Pi_P) \subset \mho$, $U_A \stackrel{def}{=} id_\Pi(\Pi_A) \subset \mho$*
- *Let $sp_\Omega : \Omega \rightarrow \mho \cup V(\mho) \cup N$ defined by (N is the class of natural numbers)*

$$sp_\Omega(s) = \begin{cases} sp_\Pi(s) \text{ iff } s \in \Omega' \\ 0 \quad\quad\ \text{ iff } s \in \Omega \setminus \Omega' \end{cases}$$

Consider $O \stackrel{def}{=} sp_\Omega(\Omega') \subset V(\mho)$
- *Let $id_\Omega : \Omega \rightarrow V(\mho) \cup \mho$ defined by*

$$id_\Omega(s) = \emptyset$$

- *Let $sp_\Gamma : \Gamma \rightarrow N$ such that*

$$sp_\Gamma(c) = \begin{cases} 1 & \text{iff } | \rightarrow \bullet \rightarrow c \text{ or } n \rightarrow \bullet \rightarrow c \text{ with } n \in \Pi \\ k+1 & \text{iff } \bullet_1 \rightarrow \bullet_2 \rightarrow c \text{ and } \bullet_1 \rightarrow c' \in \Gamma \text{ with } sp_\Gamma(c') = k \end{cases}$$

- *Let $id_\Gamma : \Gamma \rightarrow V(\mho) \cup \mho$ defined for $c \in \Gamma$ such that $\bullet_c \rightarrow c$ by*

$$id_\Gamma(c) = \begin{cases} id_\Pi(n) & \text{iff } \bullet_c \rightarrow n \text{ with } n \in \Pi \\ id_\Gamma(c') & \text{iff } \bullet_c \rightarrow \bullet' \text{ with } \bullet' \rightarrow c' \\ sp_\Omega(\bullet_c) & \text{iff } \bullet_c \in \Omega' \end{cases}$$

Summarizing we can define the identity function $id : \Pi \cup \Gamma \cup \Omega \rightarrow \mho \cup V(\mho)$ and the spatial function $sp : \Pi \cup \Gamma \cup \Omega \rightarrow \mho \cup V(\mho) \cup N$ by:

$$id(s) = \begin{cases} id_\Pi(s) \text{ iff } s \in \Pi \\ id_\Gamma(s) \text{ iff } s \in \Gamma \\ id_\Omega(s) \text{ iff } s \in \Omega \end{cases} \quad sp(s) = \begin{cases} sp_\Pi(s) \text{ iff } s \in \Pi \\ sp_\Gamma(s) \text{ iff } s \in \Gamma \\ sp_\Omega(s) \text{ iff } s \in \Omega \end{cases}$$

Observe that while the range of id is $\mho \cup V(\mho)$, the range of sp is $\mho \cup V(\mho) \cup N$ (we consider here natural numbers as cardinals[4] so that no structure anomaly emerges as long as $N \subset \mho \cup V(\mho)$). Hereafter, for the sake of the presentation, we will still consider natural numbers and not cardinals.

We identify the sets U_A of urelements chosen for ambients, U_P of urelements chosen for atomical processes, and the set of sets of urelements O that contain all the addresses of the elements in Ω'.

We now define *labeled syntax tree* for a given syntax tree of an ambient process.

[4] Informally, we treat 0 as $\emptyset$, 1 as $\{\emptyset\}$, 2 as $\{\emptyset, \{\emptyset\}\}$, 3 as $\{\emptyset, \{\emptyset\}, \{\emptyset, \{\emptyset\}\}\}$ and so on.

Definition 7. *Let $S_P = (S, \rightarrow)$ be the syntax tree of the ambient process P. We call* the labeled syntax tree *of it the triplet $Sl_P = (S, \rightarrow, \phi)$ where ϕ is the function defined on the nodes of the syntax tree by*

$$\phi(s) = \langle id(s), sp(s) \rangle \text{ for all } s \in S.$$

Remark 1. It is obvious the central position of the function id in the previous definitions. For a particular ambient process, once we defined the function id, all the construction, up to the labeled syntax tree, can be done inductively on the structure of the ambient process. Because of this, our construction of the labeled syntax tree is unique up to the choice of urelements (i.e. of U_P and U_A).

Definition 8. *For a given labeled syntax tree $Sl = (S, \rightarrow, \phi)$ we define the functions:*

- *$ur : \Pi \cup \Omega' \rightarrow U_P \cup U_A \cup O$ by:*

$$ur(s) = \begin{cases} id(s) \text{ if } s \in \Pi \\ sp(s) \text{ if } s \in \Omega' \end{cases}$$

  This function associates to each node of the structure graph the set-theoretical identity defined by the labeled syntax tree
- *Let $e : U_P \cup U_A \cup O \rightarrow \mho \cup V(\mho)$ be the function defined by*

$$e(\nu) = sp(ur^{-1}(\nu))$$

  It associates to each ambient and compound process the set of addresses of its children.
- *$f : U_P \cup U_A \cup O \rightarrow \Lambda \cup \Pi$, where Λ is the set of names of ambients of Ambient Calculus, and Π is the set of atomical processes. For each $\nu \in U_P \cup U_A \subset \mho$, $f(\nu)$ is the name of the process with which ν is associated by id[5], and $f(\nu) = \langle 0, 0 \rangle$ if $\nu \in O$. By the function f each urelement (or set of urelements) used as identity will receive the name of the ambient or atomical process that il is pointing to (the sets receive the name $\langle 0, 0 \rangle$).*
- *$F : U_P \cup U_A \cup O \rightarrow \Gamma^*$ for each $\nu \in U_P \cup U_A \cup O$, $F(\nu) = \langle c_1, c_2, ...c_k \rangle$ where $c_i \in \Gamma$ such that $\forall i \in N$, $id(c_i) = \nu$, $sp(c_i) = i$ and $\nexists c_{k+1} \in \Gamma$ such that $id(c_{k+1}) = \nu$ and $sp(c_{k+1}) = k + 1$. In the case that, for ν we cannot find any such c_i, we define $F(\nu) = \langle \varepsilon, \varepsilon, ... \rangle$, ε being the null capability. We adopt the following enrichment of the relation of equality on capability chains $=_\Gamma$ defined by the next rules* [6]*:*

[5] informally we could say that, on $U_A \cup U_P$, we have $f = id^{-1}$, but this is not exact for the reason that id is an injective function while f is not. Because if we have two processes named P, then, for both, the value by f will be P, but, by id^{-1}, they point to different nodes in the syntax tree.

[6] these rules are allowed by the syntax of Ambient Calculus together with the rules of structural congruence over processes.

- $\langle c_1, c_2, c_3, ...c_n \rangle - \langle c_1 \rangle =_\Gamma \langle c_2, c_3, ...c_n \rangle$
- $\langle \varepsilon, c_1, ..., c_k \rangle =_\Gamma \langle c_1, c_2, ..., c_k, \varepsilon \rangle =_\Gamma \langle c_1, ..., c_t, \varepsilon, c_{t+1}, ...c_k \rangle =_\Gamma \langle c_1, c_2, ..., c_k \rangle$,
- $\langle \varepsilon, \varepsilon, ...\varepsilon \rangle =_\Gamma \emptyset$.

The function F associates with each of these the list of capabilities that exists in front of the process they point to.

Definition 9. *Let* $\mathcal{S} = (S, \rightarrow, \phi)$ *be a labeled syntax tree of the ambient process* P*. We will call the* canonical labeled syntax tree *associated with* P*, denoted by* $\mathcal{S}^+ = (S^+, \rightarrow_+, \phi_+)$*, the restriction of the labeled syntax tree to the set* $S^+ = \{n | \ n \in S, \ f(n) \neq 0$ *and* $F(n) \neq \langle \varepsilon, ...\varepsilon \rangle\}$*, where* 0 *is the null process and* ε *is the null capability.*

Further we will analyze only canonical labeled trees (by extension canonical processes), these being those who evolves during the ambient calculus computations, so are those who really matters for our purpose.

Other aspects concerning the definition of the labeled syntax tree for situations that involves the new name operator, the replication operator, or recursive processes can be found in [16]. Also we introduce an algebra of labeled trees in order to analyze their composition.

In [16] we proved that the function that associates to each ambient process the set $\langle U_P, U_A, O, e, f, F \rangle$ is generating a sound model for Ambient Calculus. Being this result we construct the logic as having these ordered sets as states. We say that a process satisfies a formula of our logic, if its ordered set (as state) satisfies it.

4 The Logic

The logic we construct is a branching propositional temporal logic, CTL^*[7]. The requirements for such a construction [14] are to organize a structure $\mathcal{M} = (S_0, \Sigma, \Re, \mathcal{L})$ where S_0 is the initial state of our model, Σ is the class of all possible states in our model, $\Re$ is the accessibility relation between states, $\Re \subseteq \Sigma \times \Sigma$, and $\mathcal{L} : \Sigma \rightarrow \mathcal{P}(\mathcal{A})$ is a function which associates to each state $S \in \Sigma$ a set of atomical propositions $\mathcal{L}(S) \subseteq \mathcal{P}(\mathcal{A})$ - the set of the atomical propositions true in the state S ($\mathcal{A}$ will be the class of atomical propositions and $\mathcal{P}$ the power-set operator).

We propose to use the ordered sets $S = \langle U_A, U_P, O, e, f, F \rangle$ as states in our logic. The choice of the initial state should depend on the purpose of our analysis. If we are interested in the future of an ambient calculus process P by itself, then its ordered set will be the initial state. But if P will interact with another process Q, or will become child of an ambient, or both like in $m[P|Q]$, then, even if we have a particular interest in P, the initial state should be the ordered set of $m[P|Q]$. For this purpose we defined computation operations over these ordered sets such that to be able, starting from the sets constructed for

[7] we choose CTL^* because is more expressive then CTL, but a CTL is possible as well

some initial processes to obtain the sets for other processes constructed in top of these, for more see [16].

The construction of Σ should be done in such a way to contain all the possible future states of the initial state. For this reason we take

$$\Sigma = \{S_i = \langle U_A^i, U_P^i, O^i, e_i, f_i, F_i\rangle \mid U_A^i = U_A^0,\ U_P^i = U_P^0,\ \mathit{and}\ O_i = O_0\}$$

where $S_0 = \langle U_A^0, U_P^0, O_0, e_0, f_0, F_0\rangle$ is the initial state. The intuition is that no matter how the process will evolve, it is not possible to appear in it new elements then those that already exist in the initial state[8].

Our main idea is to define the atomic propositions such that to express the basic equations that defines the spatial relations between parts of our process. So, we could define the set of atomical propositions as:

$$\mathcal{A} = \{xiny | x \in U_P \cup U_A \cup O \ \mathit{and}\ y \in U_A \cup O\}.$$

In our logic we want $xiny$ to be just an atomical proposition and x, y just letters. The cardinality of $\mathcal{A}$ is $card(U_P \cup U_A \cup O) \times card(U_A \cup O)$ which depends (polynomial) on the number of atomical processes and ambients in the ambient calculus process S_0.

Further, the interpretation function $\mathcal{L} : \Sigma \rightarrow \mathcal{P}(\mathcal{A})$ is defined by:

$$\mathcal{L}(S) = \{xiny \mid x \in e_y \text{ if } x \in U_P, \text{ or } e_x \in e_y \text{ if } x \in U_A \cup O\}$$

As it concerns the accessibility relation $\Re \subseteq \Sigma \times \Sigma$, following the previous intuition we could define it for two states S_0 and S_1, constructed for the processes P_0 and P_1, by $\langle S_0, S_1\rangle \in R$ iff $P_0 \rightarrow P_1$ (i.e. P_1 can be reached from P_0 in one step of ambient calculus reduction).

Further, we could introduce the syntax of the CTL* logic in the usual way [14]. We inductively define a class of state formulae (formulae which will be true or false of states) and a class of path[9] formulae (true or false of paths), starting from $\mathcal{A}$. We accept, as basic operators the logical operators $\wedge$ and $\neg$, the temporal operators X (*next time*) and $\cup$ (*until*) and the path quantifier E (*for some futures*). We will derive from them all the usual propositional logic operators, the temporal operators G (*always*) and F (*sometimes*) and the path quantifier A (*for all futures*).

4.1 Semantics

Now we define $\models$ inductively. We write $\mathcal{M}, S_0 \models p$ to mean that the state formula p is true at state S_0 in the model $\mathcal{M}$, and $\mathcal{M}, x \models p$ to mean that the path formula p is true for the fullpath x in the structure $\mathcal{M}$. The rules are:

---

[8] we include here also the situations where some ambients were dissolved by consuming, for example, *open* capability; we consider, in this case, that these ambients still exist in our process but they have an "empty position".

[9] A *fullpath* is an infinite sequence $S_0, S_1, ...$ of states such that $(S_i, S_{i+1}) \in \Re$ for all i. We use the convention that if $x = (S_0, S_1, ...)$ denotes a fullpath, then x^i denotes the suffix path $(S_i, S_{i+1}, S_{i+2}, ...)$.

$\mathcal{M}, S_0 \models P$ iff $P \in \mathcal{L}(S_0)$, where $P \in \mathcal{A}$
$\mathcal{M}, S_0 \models p \wedge q$ iff $\mathcal{M}, S_0 \models p$ and $\mathcal{M}, S_0 \models q$
$\mathcal{M}, S_0 \models \neg p$ iff it is not the case that $\mathcal{M}, S_0 \models p$
$\mathcal{M}, S_0 \models Ep$ iff $\exists$ fullpath $x = (S_0, S_1, ...)$ in $\mathcal{M}$ with $\mathcal{M}, x \models p$
$\mathcal{M}, S_0 \models Ap$ iff $\forall$ fullpath $x = (S_0, S_1, ...)$ in $\mathcal{M}$ with $\mathcal{M}, x \models p$
$\mathcal{M}, x \models p$ iff $\mathcal{M}, S_0 \models p$
$\mathcal{M}, x \models p \wedge q$ iff $\mathcal{M}, x \models p$ and $\mathcal{M}, x \models q$
$\mathcal{M}, x \models \neg p$ iff it is not the case that $\mathcal{M}, x \models p$
$\mathcal{M}, x \models p \cup q$ iff $\exists i \left(\mathcal{M}, x^i \models q \text{ and } \forall j \left(j < i \text{ implies } \mathcal{M}, x^j \models p\right)\right)$
$\mathcal{M}, x \models Xp$ iff $\mathcal{M}, x^1 \models p$

Definition 10. *A state formula p (resp. path formula p) is* valid *provided that for every structure $\mathcal{M}$ and every state S (resp. fullpath x) in $\mathcal{M}$ we have $\mathcal{M}, s \models p$ (resp. $\mathcal{M}, x \models p$). A state formula (resp. path formula) p is* satisfiable *provided that for some structure $\mathcal{M}$ and some states S (resp. fullpath x) in $\mathcal{M}$ we have $\mathcal{M}, S \models p$ (resp. $\mathcal{M}, x \models p$).*

4.2 Describing the State of a System

Consider the example of the interaction between the Virus and Macrophage discussed before. If the mathematical model chosen to describe the interaction is appropriate, then our system should have the property that, independently of the path of time that it will choose, always we will meet, in the future, the situation $n[Infect|Digest]$. Our logic allows us to formulate all these as a logical statement. We have:

$$u[k'[open\ k.k''[Infect]]|n[k[out\ n.in\ k'.in\ n.0]|open\ k'.open\ k''.Digest]] \quad (1)$$

For 1 we choose the urelements: α for u, β for n, o for 0, κ for k, κ' for k', κ'' for k'', p for $Infect$ and q for $Digest$ with $\alpha, \beta, \kappa, \kappa', \kappa'', p, q, o \in \mho$. So, $U_A = \{\alpha, \beta, \kappa, \kappa', \kappa''\}$, $U_P = \{q, p, o\}$, $O = \emptyset$; f is defined by: $f(\alpha) = u$, $f(\beta) = n$, $f(o) = 0$, $f(\kappa) = k$, $f(\kappa') = k'$, $f(\kappa'') = k''$, $f(q) = Infect$, $f(p) = Digest$ and e is defined by:

$$\begin{array}{lll}
e(\alpha) = \{e(\kappa'), e(\beta)\} \Longrightarrow \begin{cases} e(\kappa') \in e(\alpha) \\ e(\beta) \in e(\alpha) \end{cases} & \Longrightarrow \begin{cases} \kappa' in \alpha \text{ is true} \\ \beta in \alpha \text{ is true} \end{cases} \\
e(\kappa') = \{e(\kappa'')\} \Longrightarrow \{\ e(\kappa'') \in e(\kappa') & \Longrightarrow \{\ \kappa'' in \kappa' \text{ is true} \\
e(\beta) = \{e(\kappa), p\} \Longrightarrow \begin{cases} e(\kappa) \in e(\beta) \\ p \in e(\beta) \end{cases} & \Longrightarrow \begin{cases} \kappa in \beta \text{ is true} \\ p in \beta \text{ is true} \end{cases} \\
e(\kappa'') = \{q\} \Longrightarrow \{\ q \in e(\kappa'') & \Longrightarrow \{\ q in \kappa'' \text{ is true} \\
e(\kappa) = \{o\} \Longrightarrow \{\ o \in e(\kappa) & \Longrightarrow \{\ o in \kappa \text{ is true}
\end{array}$$

The property we are interested in could be expressed as

$$Macrophage|Virus \models AF(\beta in \alpha \bigwedge q in \beta \bigwedge p in \beta)$$

It says that in all time paths exists at least a reachable state for which n is a child of the master ambient $u = f(\alpha)$, $Infect = f(q)$ and $Digest = f(p)$ are children of the Macrophage ambient $n = f(\beta)$. Further, for checking the truth value of this statement, a model checker could be used. Proving that our logical formula is true it finally means that our mathematical model for describing our problem is a correct one. Vice versa, if is not valid, the model checker will give us a counter example that will show the conflict in our model.

4.3 Algorithms for the Accessibility Relation

The accessibility relation computation is based on analysis of the initial state structure and all its possible derivatives. What basically defines the possible evolutions (in time) are the prefixes of the processes involved. For every type of Ambient Calculus reduction (i.e. for each type of capability, c, and for communication) we construct an algorithm able to verify if the conditions of reduction are fulfilled (c-condition algorithm) and an algorithm which computes the final state of the system (c-reduction algorithm). These algorithms are then used within a more general procedure (the general algorithm) that handles the full structure of the initial state.

In order to perform the analysis it is useful to arrange the information in the initial state in two matrices. Consider the following example:

$$u\,[m\,[in\ n.P]\ \mid n\,[\ Q\ ]]$$

If we choose $f(\alpha) = u$, $f(\beta) = m$,$f(\gamma) = n$,$f(p) = P$, and $f(q) = Q$, then the two functions and the matrix are:

matrix T_1

T_1	α	β	γ	p	q
α	0	1	1	0	0
β	0	0	0	1	0
γ	0	0	0	0	1
p	0	0	0	0	0
q	0	0	0	0	0

matrix T_2

T_2	f	F
α	u	ε
β	m	ε
γ	n	ε
p	P	*in m*
q	Q	ε

Example (*)

The functions F and f are bundled into T_2 matrix, while the matrix T_1 has one line for each element of $U_P \cup U_A \cup O$, one column for each element of $U_A \cup O$, and is made by setting the entry of column x and row y to 1, if the proposition $xiny$ is true. All the empty entries are set to 0.

In what follows we present the general algorithm and the algorithms for *in*-capability only, the rest of the cases being similar.

General Algorithm. Assume that the initial state S_1 is described by T_1 and T_2 matrices. The first step in the algorithm is to pick the first column of the F-part in T_2 matrix. For the Example (*) it would be Row $= \{in\ m\}$, just one element.

Now specific c-condition algorithm checks the possibility of using reduction rules of the ambient calculus semantics, and if all the necessary conditions hold then the specific *c*-reduction part is performed to compute the next state (by updating T_1 and T_2 matrices). In the other case another capability might be chosen in the cycle until either *c*-reduction algorithm is finally performed or the *Row* set is empty. The algorithm computes exact one state on-forward. See **Algorithm1**.

Algorithm 1 General form of the accessibility algorithm

```
1: Row ⇐ {c | c is the first column of F-part of the T2 matrix }\{ε, output}
2: while Row ≠ ∅ do
3:    choose c ∈ Row
4:    c-condition
5:    if condition then
6:       c-reduction
7:       Row ⇐ ∅
8:    else
9:       Row ⇐ Row\{c}
10:   end if
11: end while
```

where *c* can be *In*, *Out*, *Open* or *Communication* in *c*-condition and *c*-reduction, which depends on chosen capability at the third line.

While the empty place ε is excluded from the set *Row* for the obvious reason, the *output* action is not accepted for avoiding overlapping actions with the accepted *input* action.

The notation $S_1 \models_{alg} S_2$ denotes that S_2 state is obtained from S_1 in one step using algorithm 1 instantiated with suitable *c*-condition and *c*-reduction parts. We can prove that the accessibility relation between states fulfill the condition:

$$S_1 \Re S_2 \text{ iff } S_1 \models_{alg} S_2.$$

In-Condition, *In*-Reduction Algorithms

$$n\,[in\ m.P \mid Q] \mid m\,[R] \longrightarrow m\,[n\,[P \mid Q] \mid R]$$

The representation of the initial state of the process is the following:

matrix T_1 matrix T_2

T_1	α	β	p	q	r	T_2	f	F
α	0	0	1	1	0	α	n	ε
β	0	0	0	0	1	β	m	ε
p	0	0	0	0	0	p	P	*in m*
q	0	0	0	0	0	q	Q	ε
r	0	0	0	0	0	q	R	ε

The *In*-condition and *In*-reduction algorithms implement the in-reduction rule of Ambient Calculus semantics.

The *In*-condition algorithm checks if there is an ambient with the same name as the one in-capability refer to (m in the particular case), at the same nested level as the *parent* process of the capability owner process; it checks also if there is no prefix in front of either ambient processes that will be involved in the reduction. If all the conditions hold then the *In*-reduction will be performed. It consists in updating T_1 and T_2 such that to represent the final state.

The *Out-*, *Open-* and *Communication-* condition/reduction algorithms differ from the above w.r.t. the Ambient Calculus reduction rules they describe. For the sake of space, we not discuss them here (for complete details, the reader is referred to [15]).

Algorithm 2 *In*-condition

$condition \Leftarrow false$
$UrBundle \Leftarrow f_{S_1}^{-1}(m)$
while $UrBundle \neq \emptyset$ **do**
  choose $\nu \in UrBundle$
  if $parent(parent(p)) = parent(\nu)$ AND
  $F_{S_1}(parent(p)) = \varepsilon$ AND
  $F_{S_1}(\nu) = \varepsilon$ AND
  $\forall \mu \in f_{S_1}^{-1}(\langle 0,0 \rangle), \nu \in \mu \Rightarrow F_{S_1}(\mu) = \varepsilon$ **then**
    $condition \Leftarrow true$
    $UrBundle \Leftarrow \emptyset$
  else
    $UrBundle \Leftarrow UrBundle \setminus \{\nu\}$
  end if
end while

Algorithm 3 *In*-Reduction

{update T_2}
$F_{S_2}(p) \Leftarrow F_{S_1}(p) - \langle in\ m \rangle$
$F_{S_2}(x) \Leftarrow F_{S_1}(x)$ for all $x \neq p$
$f_{S_2}(x) \Leftarrow f_{S_1}(x)$
if $f_{S_2}(p) = \langle 0,0 \rangle \wedge F_{S_2}(p) = \varepsilon$ **then**
  $\forall \mu \in p, F_{S_2}(\mu) \Leftarrow F_{S_2}(\mu) - \langle \star \rangle$
end if

{update T_1}
$\beta in \alpha \Leftarrow 0$
$\beta in \gamma \Leftarrow 1$
if $f_{S_2}(p) = \langle 0,0 \rangle \wedge F_{S_2}(p) = \varepsilon$ **then**
  $\forall \mu \in p, \mu inp \Leftarrow 0$
end if

5 Implementation Details

We present here the details of the implementation we developed for this logic in order to perform model checking analysis for Ambient Calculus. We use the NuSMV model checker for analyzing CTL* logic. Anyway, having the CTL* logic developed for Ambient Calculus, we can use for our purpose any model checker able to analysis temporal logics.

The implementation consists in the construction of a translator (in top of the algorithms presented before) that accepts as input a mobile ambient process and gives, as output, a model specification file for NuSMV model checker. Hereafter we sketch this construction.

The translator assigns to each atomical process or ambient (to each urelement), to each capability and to each ambient process name a natural number, and so it generates the constant definitions for the NuSMV model.

In order to adapt our approach to the requirements of the NuSMV software, we had to represent the matrices T_1 and T_2 by means of arrays.

For the matrix T_1, representing the urelements by natural numbers and using the function *parent* we obtain the representation in NuSMV model as follows:

matrix T_1

T_1	α	β	γ	δ
α	0	1	1	0
β	0	0	0	1
γ	0	0	0	0
δ	0	0	0	0

representation of T_1

parent[α]	parent[β]	parent[γ]	parent[δ]
$< no\_parent >$	α	α	β

In this case the translator converted the 4×4 matrix having 0 or 1 as entries into an array of 4 elements where each of them can have one of the values 0, 1, 2 or 3 (these values represent the identities of the processes and play the same role as the urelements).

For the representation of the matrix T_2, the translator generates the next arrays (functions):

cap2proc: $N_c \rightarrow N_p$
cap2order: $N_c \rightarrow N_c$
nextCap: $N_p \rightarrow N_c$
cap2name: $N_c \rightarrow N_n$
proc2name: $N_p \rightarrow N_n$
enabled: $N_p \rightarrow$ *boolean*

where N_c, N_p and N_n are integers used to identify, respectively, a capability (N_c), a process (N_p) or a name (N_n).

The array *cap2proc* stores the information that the capability with identity N_c is prefixing the process with identity N_p.

The array *cap2order* points out the order in which the capabilities prefixing the same process can be used for reductions. For example, *cap2order*[*in* γ] = *out* γ means that *out* γ might be used only after *in* γ was used.

The array *nextCap* associates to a process the leftmost capability that prefix it. For instance, $nextCap[\delta] = in\ \gamma$ means that the capability $in\ \gamma$ is enabled in the process δ.

The arrays *cap2name* and *proc2name* handle the storing information about names that are used in a process formula. For instance, $cap2name[in\ \gamma] = m$ express that $in\ \gamma$ can only be applied in the case of an ambient with the name m, while $cap2name[\gamma] = m$ is used to express the fact that γ was chosen to name a process with the name m.

The array *enabled* is used to block the action of some capabilities. For example, it is syntactically possible that the use of a capability to be conditioned by the use of another one which do not belong to the same process (so *cap2order* is not enough). This is the case for $c_1.(P|c_2.Q)$, where c_2 cannot be consumed before c_1, but this case can arise in presence of communications as well. So, $enabled[in\ \gamma] = 1$ allows the capability to be used while $enabled[in\ \gamma] = 0$ forbid the use of the capability.

Using the procedure described above the translator is able to encode the information behind each state of the system. Further it generates the model for the initial state and for the possible next states using the functions already presented. The initial state consists in an assignment of values for variables. Then using the general algorithm it computes the models of the possible next states.

Fairness constraints generated by the syntactical structure of the ambient process are defined by the translator in order to avoid the stuck of the system and to prevent the appearance of impossible paths.

Finally the translator converts the property we want to verify in a form consistent with the one of the system. In this way the interface with NuSMV is complete.

6 Conclusions

The logic we constructed in top of Ambient Calculus opens the perspective of using model checking algorithms (or software) developed for temporal logics in analyzing mobile computations and, in this way, to predict over the future of the systems (biological systems) described using the calculus.

Having the description of the states, together with the algorithms for accessibility relation, all we have to do for having model checking for mobile computations, is to use, further, the algorithms for model checking CTL* (or the tools already constructed for this purpose). Here we presented the possibility of using NuSMV model checker.

Our ongoing research makes us confident in the possibility to construct such a logic for other calculi used for describing biological systems, e.g. BioAmbients Calculus, or Brane Calculi. In such a way, we could move towards predictions about the future of (the structures of) biological systems that can be described using these calculi.

References

1. HyTech: The HYbrid TECHnology Tool. `http://www-cad.eecs.berkeley.edu/~tah/HyTech/`.
2. NuSMV: A new symbolic model checker. `http://nusmv.irst.itc.it/`.
3. Receptors directly activating trimetric g proteins. `http://courses.washington.edu/conj/gprotein/trimericgp.htm`.
4. The SMV system. `http://www-2.cs.cmu.edu/~modelcheck/smv.html`.
5. VIS homepage. `http://www-cad.eecs.berkeley.edu/~vis/`.
6. B. Alberts, A. Johnson, J. Lewis, M. Raff, K. Roberts, and P. Walter. *Molecular Biology of the Cell.* Garland Publishing, Inc., fourth edition, 2002.
7. A.Regev, E.M.Panina, W.Silverman, L.Cardelli, and E.Shapiro. Bioambients: An abstraction for biological compartments. `http://www.luca.demon.co.uk/`, to appear in Theoretical Computer Science, 2003.
8. J. Barwise and L. Moss. *Vicious Circles. On the Mathematics of Non-Wellfounded Phenomena.* CLSI Lecture Notes Number 60 Stanford: CSLI Publication, 1996.
9. L. Cardelli. Brane calculi. `http://www.luca.demon.co.uk/`.
10. L. Cardelli and L. Caires. A spatial logic for concurrency (part i). *Information and Computation, Vol.186/2, pages:194-235*, 2003.
11. L. Cardelli and A.D. Gordon. Ambient logic. `http://www.luca.demon.co.uk/` to appear in Mathematical Structures in Computer Science.
12. L. Cardelli and A.D. Gordon. Anytime, anywhere. modal logics for mobile ambients. *Proceedings of the 27th ACM Symposium on Principles of Programming Languages*, pages 365–377, 2000.
13. L. Cardelli and A.D. Gordon. Mobile ambients. *Theoretical Computer Science, Special Issue on Coordination, D. Le Metayer Editor*, pages 177–213, June 2000.
14. E. A. Emerson. Temporal and modal logic. *Handbook of Theoretical Computer Science*, B: Formal Models and Sematics:995–1072, 1990.
15. R. Mardare and C. Priami. Computing the accessibility relation for ambient calculus. Technical report, Dipartimento di Informatica e Tlc, University of Trento, 2003. Available at `http://www.dit.unitn.it` following the link Publications.
16. R. Mardare and C. Priami. A propositional branching temporal logic for the ambient calculus. Technical report, Dipartimento di Informatica e Tlc, University of Trento, 2003. Available at `http://www.dit.unitn.it` following the link Publications.

A The Construction of a Labeled Syntax Tree

We present further the construction of a labeled syntax tree. Consider the ambient calculus program:

$$m[open\ n.Q|s[out\ m.in\ m.n[open\ t.(\ out\ s.(open\ s.P|R)|K)]\ ]\ ]\ |n[P]. \quad (2)$$

As a general rule, we embed our program into a *master ambient*[10] (the master ambient will have a fresh name). Our program becomes:

$$u[m[open\ n.Q|s[out\ m.in\ m.n[open\ t.(\ out\ s.(open\ s.P|R)|K)]\ ]\ ]\ |n[P]] \quad (3)$$

[10] This is a technical trick that is not disturbing our analysis because of the rule (RedAmb): $P \to Q \Rightarrow n[P] \to n[Q]$, [13], but it helps to treat the processes as a whole from the spatial point of view.

The syntax tree of this process is in Figure A.

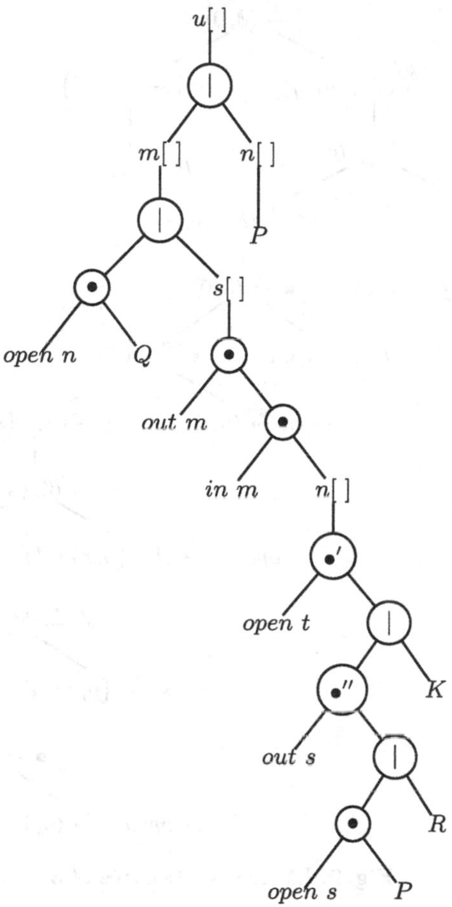

Fig. 1. Syntax tree of the process 3

For constructing the labeled syntax tree we will define ϕ. We define the identity function id as:

$id(u) = \alpha$, $id(m) = \beta$, $id(n) = \gamma$ (the child of u), $id(s) = \delta$, $id(n) = \mu$, $id(Q) = q$, $id(P) = p'$ (the child of that n which have γ as identity), $id(P) = p$ (the child of that n which have μ as identity), $id(R) = r$, $id(K) = k$, where $\{\alpha, \beta, \gamma, \delta, \mu, p, q, p', r, k\} \subset \mho$.

Observe that in our situation $\Omega' = \{\bullet', \bullet''\}$ (see Figure A). The space function sp for $\Pi \cup \Omega'$ will be defined starting from the values of id for atomic processes and following the definition of decoration:

$$sp(u) = \{sp(m), sp(n)\} \text{ (here } n \text{ is the child of } u\text{)}, \; sp(m) = \{sp(s), q\},$$
$$sp(n) = \{p'\} \text{ (the child of } u\text{)}, \; sp(s) = \{sp(n)\}, \; sp(n) = \{sp(\bullet')\},$$
$$sp(\bullet') = \{k, sp(\bullet'')\}, \; sp(\bullet'') = \{p, r\}.$$

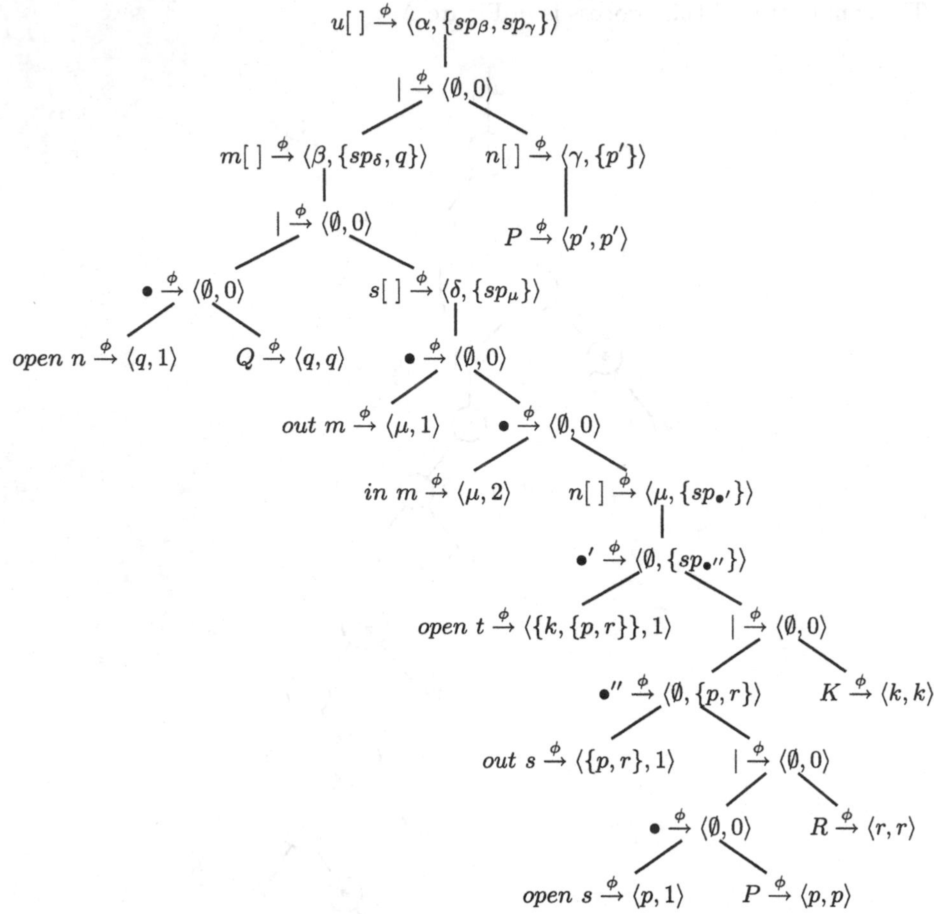

Fig. 2. Labeled syntax tree of 3

For capabilities the identity function have the values:

$$id(open\ n) = q,\ id(out\ m) = \mu,\ id(in\ m) = \mu,\ id(open\ t) = \{k, \{p, r\}\},$$
$$id(out\ s) = \{p, r\},\ id(open\ s) = p$$

and the spatial function:

$$sp(open\ n) = 1,\ sp(out\ m) = 1,\ sp(in\ m) = 2,\ sp(open\ t) = 1,\ sp(out\ s) = 1,$$
$$sp(open\ s) = 1$$

Concluding, the function ϕ will be defined as (we will denote $sp(x)$ by sp_x):

$\phi(u) = \langle \alpha, \{sp_m, sp_n\}\rangle$, $\phi(m) = \langle \beta, \{sp_s, q\}\rangle$,
$\phi(n) = \langle \gamma, \{p'\}\rangle$,(the child of u) $\phi(P) = \langle p', p'\rangle$(the child of n),
$\phi(open\ n) = \langle q, 1\rangle$, $\phi(Q) = \langle q, q\rangle$,
$\phi(s) = \langle \delta, \{sp_n\}\rangle$, $\phi(out\ m) = \langle \mu, 1\rangle$,
$\phi(in\ m) = \langle \mu, 2\rangle$, $\phi(n) = \langle \mu, sp_{\bullet'}\rangle$,
$\phi(\bullet') = \langle \emptyset, \{sp_{\bullet''}\}\rangle$, $\phi(open\ t) = \langle \{k, \{p, r\}\}, 1\rangle$,
$\phi(\bullet'') = \langle \emptyset, \{p, r\}\rangle$, $\phi(K) = \langle k, k\rangle$,
$\phi(out\ s) = \langle \{p, r\}, 1\rangle$, $\phi(R) = \langle r, r\rangle$,
$\phi(open\ s) = \langle p, 1\rangle$, $\phi(P) = \langle p, p\rangle$,
for all $\bullet \in \Omega \setminus \Omega'$, $\phi(\bullet) = \langle \emptyset, 0\rangle$, for all $| \in \Omega$, $\phi(|) = \langle \emptyset, 0\rangle$,

The labeled syntax tree is in Figure A.
We can define now the functions ur, e, f and F.

$$ur(u) = \alpha,\ ur(m) = \beta,\ ur(n) = \gamma \text{ (the child of } u),\ ur(s) = \delta,\ ur(n) = \mu,$$
$$ur(Q) = q,\ ur(P) = p' \text{ (the child of } n),\ ur(P) = p,\ ur(R) = r,\ ur(K) = k,$$
$$ur(\bullet') = \{k, \{p, r\}\},\ ur(\bullet'') = \{p, r\}$$

We can define now the function f:

$$f(\alpha) = u,\ f(\beta) = m,\ f(\gamma) = n,\ f(\delta) = s,\ f(\mu) = n,\ f(q) = Q,\ f(p) = P,$$
$$f(p') = P,\ f(r) = R,\ f(k) = K,\ f(\{p, r\}) = \langle 0, 0\rangle,\ f(\{k, \{p, r\}\}) = \langle 0, 0\rangle$$

We define, as before, $U_A = \{u \in \mho \mid f(u) \in \Lambda\}$ and $U_P = \{u \in \mho \mid f(u) \in \Pi\}$, which in our example became:
$U_P = \{p, q, r, k, p'\}$, $U_A = \{\alpha, \beta, \gamma, \delta, \mu\}$ and $O = \{\{k, \{p, r\}\}, \{p, r\}\}$.
The function e (as before, we denote $e(x)$ by e_x):

$$e_\alpha = \{e_\beta, e_\gamma\}\ ,\ e_\beta = \{e_\delta, q\},\ e_\gamma = \{p'\}\ ,\ e_\delta = \{e_\mu\},\ e_\mu = \{e_{\{k,\{p,r\}\}}\},$$
$$e_{\{k,\{p,r\}\}} = \{k, e_{\{p,r\}}\},\ e_{\{p,r\}} = \{p, r\}.$$

The function F:

$$F(\alpha) = \langle \varepsilon, \varepsilon, ...\rangle,\ F(\beta) = \langle \varepsilon, \varepsilon, ...\rangle,\ F(\gamma) = \langle \varepsilon, \varepsilon, ...\rangle\ F(\delta) = \langle \varepsilon, \varepsilon, ...\rangle$$
$$F(\mu) = \langle out\ m, in\ m, \varepsilon\rangle,\ F(q) = \langle open\ n, \varepsilon\rangle,\ F(p) = \langle \varepsilon, \varepsilon, ...\rangle,$$
$$F(p') = \langle open\ s, \varepsilon\rangle,\ F(r) = \langle \varepsilon, \varepsilon, ...\rangle,\ F(k) = \langle \varepsilon, \varepsilon, ...\rangle,\ F(\{p, r\}) = \langle out\ s, \varepsilon\rangle,$$
$$F(\{\{p, r\}, k\}) = \langle open\ t, \varepsilon\rangle.$$

Modeling the Molecular Network Controlling Adhesion Between Human Endothelial Cells: Inference and Simulation Using Constraint Logic Programming

Eric Fanchon[1], Fabien Corblin[1,2], Laurent Trilling[2], Bastien Hermant[1], and Danielle Gulino[1]

[1] LCM and LIM, Institut de Biologie Structurale Jean Pierre Ebel, CEA-CNRS-Université Joseph Fourier, 41, rue Jules Horowitz, 38027 Grenoble Cedex 1, France
eric.fanchon@ibs.fr

[2] IMAG-LSR, Université Joseph Fourier BP 53, 38041 Grenoble Cedex 9, France
corblinf@ufrima.imag.fr
laurent.trilling@imag.fr

Abstract. Cell-cell adhesion plays a critical role in the formation of tissues and organs. Adhesion between endothelial cells is also involved in the control of leukocyte migration across the endothelium of blood vessels. The most important players in this process are probably identified and the overall organization of the biochemical network can be drawn, but knowledge about connectivity is still incomplete, and the numerical values of kinetic parameters are unknown. This calls for qualitative modeling methods. Our aim in this paper is twofold: (i) to integrate in a unified model the biochemical network and the genetic circuitry. For this purpose we transform our system into a system of piecewise linear differential equations and then use Thomas theory of discrete networks. (ii) to show how constraints can be used to infer ranges of parameter values from observations and, with the same model, perform qualitative simulations.

1 Introduction - Modeling Objectives

With the development of high-throughtput projects the quantity of molecular level data is exploding. It is now clear that biology is entering a new era in which all these molecular components have to be assembled into a *system* in order to reach new levels of understanding.

In general terms, our goal is to formalize 'verbal models' or, stated differently, build a formal model from a word description of a biological phenomenon. This means in practice that the knowledge is incomplete, that most information is not precise but qualitative, and that we may have to deal with several hypotheses. In any case, we want to be able to exploit what we do have, even if it is qualitative information. In such a state of partial knowledge we view modeling as a tool

V. Danos and V. Schachter (Eds.): CMSB 2004, LNBI 3082, pp. 104–118, 2005.

to formalize different competing hypotheses and explore their consequences; to help in interpreting new data and use data to discriminate among competing models; to infer parameters and to devise maximally informative experiments. In short we look for rigorous methods to reason about models and data in the context of incomplete knowledge on complex systems.

Our first goal is the integration of biochemical reactions and genetic regulatory interactions in a single unified framework. It is possible in the case of genetic networks to describe the regulatory interactions by logical (or discrete) equations [22] without explicit reference to Ordinary Differential Equations (ODEs). The situation is different in the case of biochemical or signal transduction networks because the types of reaction are more diverse (phosphorylation, complexation, transport, *etc.*). So, although differential equations are not well suited to this knowledge level, it is nevertheless useful to describe the cellular process in term of differential equations (even with unknown parameters) and to transform them into a discrete model with the same logical structure. The modeling process can be summarized as follows: graph of biochemical reactions $\longrightarrow_{(1)}$ detailed ODEs $\longrightarrow_{(2)}$ simplified ODEs $\longrightarrow_{(3)}$ PLDEs $\longrightarrow_{(4)}$ discrete equations/interaction graph. Step (1) is straightforward. It relates two equivalent descriptions and can be automated [19]. Step (2) involves biochemical approximations to obtain a system of lower dimensionality based on sigmoids. In transformation (3) sigmoid functions are replaced by step functions. At this stage the system often reduces to piecewise linear differential equations (PLDEs). For step (4) R. Thomas and others [20, 21, 22, 13] have developpped a method to transform a special class of PLDE into discrete equations to perform a qualitative analysis of the dynamics.

The second goal is to show that such a formal description of the biological system at hand can be easily exploited via a Constraint Logic Programming (CLP) implementation. The advantages of the CLP approach are (i) that the implementation is expressed in a very similar way to the formal specification, thus guarantying the correctness of the implementation, (ii) that many different queries can be easily asked to this formal specification due to its logical form. For example, queries equivalent to simulation as well as inference of model parameters in a context of incomplete knowledge.

These principles are illustrated by the study of endothelial cell-cell adhesion. This case makes clear that the chemical reaction graph should not be confused with the interaction graph (as one might think from the study of genetic networks alone). Once a discrete model is at hand, we focus on the inference of parameter values satisfying some constraints about the existence of steady states and the existence of paths corresponding to the junction repairing process.

The organization of the paper is the following: after a short introduction to the biological phenomenon, we describe the pathway and its components. In the section 3 the modeling choices and approximations are explained. This allows us to transform the initial ODEs into a simpler PLDE system, with reduced dimensionality. The form of this system is slightly different from the one usually encountered in the field of genetic networks and we show how the theory of

Snoussi and Thomas [20] can be extended to account for this kind of piecewise linear system. We then perform an analysis following the lines of this theory. We explain briefly how the model is implemented in the CLP language Prolog IV [4] and give examples of queries.

2 Description of the Biological System

The phenomenon of cell-cell adhesion and its control by the cell is rather complex and our knowledge about it is far from complete.

Blood vessels are lined with a monolayer of endothelial cells that form a barrier between blood and underlying tissues. This monolayer, called the endothelium, plays a central role in regulating the recruitment of leukocytes at sites of injury or inflammation. It does this by detecting changes in both the flow and chemical composition of blood, which triggers the expression and/or release on the cell surface of a variety of mediators.

Junctions between endothelial cells (adherens junctions) rely on the interaction of Vascular Endothelium (VE) cadherin molecules that are specifically expressed at these junctions [14, 15]. VE-cadherin is directly involved in the maintenance of endothelium permeability [9, 12] and in the control of the traffic of leukocytes from blood toward inflamed tissues [10]. Thus, following close contact between leukocytes and endothelial cells, proteases previously stored within leukocytes are transported at the cell surface and locally cleave VE-cadherin. This results in the subsequent disruption of adherens junctions that open the way for leukocyte migration. Once leukocytes have gone through, the integrity of endothelium must be restored to avoid an excessive accumulation of leukocytes within inflamed tissues.

This restoration process is the focus of our study. We are able to grow in culture endothelial cells extracted from human umbilical cords. The cells grow to confluence and reproduce an endothelium on a 2-dimensional plate. The migration of leukocytes through the tissue and subsequent destruction of adherens junctions is simulated in the culture by anti-cadherin antibodies which destabilize the junctional complexes. The biochemical structure of the system is informally illustrated on Fig. 1. We give now a brief description of the different components.

Adherens junctions in endothelial cells are constituted of VE-cadherin hexamers [16]. These VE-cadherins are membrane proteins with a cytoplasmic tail and a multidomain extracellular part. An hypothesis is that three VE-cadherin molecules from cell 1 and three from cell 2 self-assemble to form antiparallel hexamers. These hexameric units are held together via intracellular partners of VE-cadherin.

β-catenin is one of these intracellular partners. It binds to the cytoplasmic part of VE-cadherin, and Fig. 1 displays its central role in adhesion. One hypothesis is that the VE-cadherin oligomer assemble first and then β-catenin binds to it. This is the hypothesis we incorporate in our model although this point is still controversial. β-catenin links cadherins to the actin cytoskeleton and has thus

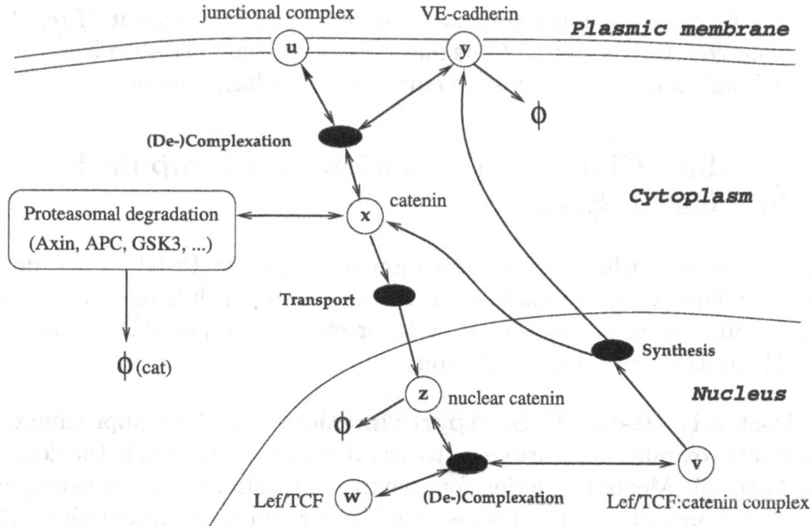

Fig. 1. Schematic and informal representation of the endothelial cell-cell adhesion system. This graph stresses the overall architecture of the system. Circles represent chemical species and black ovals chemical reactions. ϕ represents a degradation process. The molecules are distributed over three cellular locations: cytoplasm, nucleus and plasmic membrane. The concentration variables (x, y, z, u, v and w) are defined by the triples (concentration variables, chemical species label, verbal definition) as follows: (x, cat, cytoplasmic unphosphorylated β-catenin), (y, cad, monomeric cadherin in the membrane), (z, catnuc, β-catenin in the cell nucleus), (u, catcad3, complex of β-catenin with the cadherin trimer), (v, lefcat, complex of Lef/Tcf and β-catenin), (w, lef, Lef/Tcf transcription factor). Single-headed and double-headed arrows denote irreversible and reversible reactions, respectively

a structural role in the cell, but it is also able to act as a signalling molecule. Under certain conditions it can enter the nucleus where it complexes with members of the Tcf family of transcription factors, controlling gene expression. It is probable that this move to the nucleus is an active transport process (by opposition to passive diffusion) : a protein analogous to β-catenin, p120, is known to bind to motor proteins and to move along microtubules toward the cell nucleus (microtubules are molecular tubes made from a protein called tubulin). The real mechanism is poorly characterized and we will consider below two different ways of modeling it.

Cytoplasmic β-catenin is degraded by a complex called the proteasome. β-catenin is first complexed to APC and Axin (a scaffold protein), phosphorylated by two kinases acting in sequence (Ck1, Gsk3) and then released in the cytoplasm. The phosphorylated β-catenin is then tagged for destruction by the protasome. The levels of proteins like Ck1 and Gsk3 are probably regulated but there is currently no information available. We thus considered that these proteins are maintained at fixed levels and consequently that proteasomal degradation can be considered as a separate module with a single input. If regulatory interactions coupling proteins of this module to the system studied here are later identified, our model will have to be embedded in a larger model.

3 Modeling Choices - Derivation of a Simplified Differential System

We give here an outline of the derivation of simplified PLDEs from detailed 'chemical' ODEs. Our purpose is to obtain a system which is plausible from the biological point of view, knowing that other choices are possible in the current state of knowledge on cellular adhesion.

Pseudo-steady State (PSS) Approximation. The PSS approximation is the main approximation we are going to use. It is classically used in the derivation of the Michaelis-Menten equation for enzymatic reactions (see for example [17] for a detailed presentation) and of the equation for regulatory interactions. It is in fact potentially usable in cases where the first step is a reversible complexation-decomplexation reaction.

The classical mechanism of the Michaelis-Menten kinetics is:

$$\mathrm{S} + \mathrm{E} \rightleftharpoons^{k_1}_{k_{-1}} \mathrm{S:E} \rightarrow_{k_2} \mathrm{P} + \mathrm{E}$$

where S is the substrate (concentration s), E the enzyme (concentration e), S:E the transient complex (concentration c) and P the product (concentration p). The parameters k_1, k_{-1} and k_2 are the kinetic constants. The substrate binds reversibly (double arrow) to the enzyme E. When the substrate is bound, reaction occurs, product P is released and enzyme E is ready for a new event. This mechanism is represented by a system of four differential equations (one for each chemical species). The rate of variation of the concentration c is given by: $\dot{c} = k_1.s.e - k_{-1}.c - k_2.c$.

In most biological situations the concentration of substrate is much larger than that of enzyme. There is a short transient time during which c increases very fast and then reaches a steady level ($\dot{c} = 0$). This is the pseuso-steady state (PSS) hypothesis. In a closed system, substrate concentration decays slowly as reaction proceeds, and product concentration rises accordingly.

From $\dot{c} = 0$ one gets the algebraic equation $(k_{-1}+k_2).c = k_1.s.e$ in which time does not appear explicitly, only implicitly through variables s and e. When using the PSS hypothesis we neglect the transient phase and assume that the concentrations adjust *instantly* to the steady-state values after a perturbation. In other words the characteristic time of the transient phase is supposed to be short with respect to the kinetics of the perturbation.

The total quantity of enzyme is conserved: $e_0 = e + c$. From this and the above relation we get: $c = f(s) = \frac{e_0.s}{K+s}$ with $K = \frac{k_{-1}+k_2}{k_1}$.

The function $f(s)$ for $s \geq 0$ is a hyperbola branch. When cooperative binding occurs as in the case of allosteric enzymes (enzymes made of several sub-units and thus several binding sites), the function f has sigmoid shape. The rate of production of P is: $\dot{p} = k_2.c = k_2.f(s)$.

The parallel with a regulatory interaction is readily seen: regulatory protein R (the analog of substrate S) binds reversibly to promoter Pr (DNA sequence); When R is bound to Pr, gene transcription occurs. The kinetic constant k_p associated to this process represents the average number of 'polymerase start' events per time unit. The previous reasoning applies, with the difference that nothing is consumed here. This is in fact a particular case in which $k_2 = 0$ and thus the system is in quasi-*equilibrium* at all times. The rate of production by the regulated gene is then: $\dot{p} = k_\mathrm{p}.occ_\mathrm{Pr} = k_\mathrm{p}.f_\mathrm{p}(r)$ where r is the concentration of R and occ_Pr the promoter occupancy. Since regulatory proteins are often dimers, binding to the promoter is cooperative and function f_p has sigmoid shape in most cases. In this treatment concentration fluctuations are assumed to be negligible (no stochastic effects). We apply the PSS approximation in two instances below.

Reactions Taking Place in the Cell Nucleus. The rates of variation of the nuclear species (see Fig. 1) are given informally by:

$$\begin{aligned}\dot{z} &= [\text{transport}] + [\text{decomplexation lefcat}] - [\text{complexation lef \& catnuc}] \\ &\quad -[\text{z degradation}] \\ \dot{v} &= [\text{complexation lef \& catnuc}] - [\text{decomplexation lefcat}] \\ \dot{w} &= [\text{decomplexation lefcat}] - [\text{complexation lef \& catnuc}]\end{aligned}$$

The lefcat complex activates several genes among which the β-catenin and cadherin genes [9]. This is a typical situation where the PSS approximation can be used. First at the usual step of binding to DNA (binding of the lefcat complex to the regulatory site) and also at the step of complexation of lef and catnuc: $\dot{v} = -\dot{w} \approx 0$. The analysis made above can be reproduced with protein Lef/Tcf in the role of E and nuclear β-catenin in the role of S:

$$v = f_\mathrm{v}(z) = \frac{w_0.z}{K_\mathrm{v} + z} \tag{1}$$

where w_0 is the total quantity of Lef/Tcf in the nucleus, and K_v is the Michaelis-Menten constant of the enzyme.

The transport and degradation terms are given by:
$[\text{transport}] = k_\mathrm{t}.f_\mathrm{t}(x)$, $\quad [\text{z degradation}] = m_\mathrm{z}.z$

Reactions Taking Place in the Cytoplasm and Within or Near the Plasmic Membrane. Trimerization of cadherin and complexation of the trimer with β-catenin: we assume that these reactions are at quasi-equilibium (they are fast with respect to synthesis, degradation, transport, etc...). Consequently the global reaction cat + 3.cad $\rightleftharpoons$ catcad$_3$ gives the relation: $u = K_\mathrm{c}.x.y^3$, where K_c is the equilibrium constant.

Binding of β-catenin to cadherin may be cooperative, in which case the relation is different. This does not matter in our context, the important thing being that we have an algebraic relation $g(x, y, u) = 0$.

The remaining equations are:

$\dot{x}$ = [x synthesis] – [transport] – [x proteasomal degradation]
$\dot{y}$ = [y synthesis] – [y degradation]

The bracketed terms are given by: [x synthesis] $= k_{x0} + k_{xs}.\sigma^+(v, s_v)$
[x proteasomal degradation] $= (m_{x1} + m_{x2}.\sigma^+(x, s_{x2})).x$
[y synthesis] $= k_{y0} + k_{ys}.\sigma^+(v, s_v)$
[y degradation] $= m_y.y$

where $\sigma^+(a, s_a)$ is a positive sigmoid with an inflection point at s_a.

Proteasomal degradation differs from spontaneous degradation by the existence of two levels (m_{x1} and $m_{x1} + m_{x2}$). Note also that we include a basal level of expression of the β-catenin and cadherin genes (k_{x0} and k_{y0}).

All occurences of v are replaced by $f_v(z)$ (Eq. 1). The [x synthesis] term can be rewritten as follows:
[x synthesis] $= k_{x0} + k_{xs}.\sigma^+(f_v(z), s_v)$. Under the condition that the threshold on lefcat is less than the lefcat saturation value, the composition of the hyperbola branch with a sigmoid gives a slightly deformed sigmoid (with parameters different from those of the original sigmoid): [x synthesis] $= k_{x0} + k_{x1}.\sigma^+(z, s_{z1})$ where k_{x1} is a new constant defined from k_{xs} and the hyperbola parameters.

We are left with the following system :

$$\begin{cases} \dot{x} = k_{x0} + k_{x1}.\sigma^+(z, s_{z1}) - k_t.f_t(x) - (m_{x1} + m_{x2}.\sigma^+(x, s_{x2})).x \\ \dot{y} = k_{y0} + k_{y1}.\sigma^+(z, s_{z2}) - m_y.y \\ \dot{z} = k_t.f_t(x) - m_z.z \\ g(x, y, u) = 0 \end{cases}$$

Note that y does not influence other variables, and can be integrated once $z(t)$ is known. Thus we can work on the sub-system constituted by the two differential equations on x and z. From a given solution $(x(t), z(t))$, y and u can be calculated.

We consider two ways of *qualitatively* representing transport, called 'sigmoidal' and 'linear' for short:

– Linear transport: $f_t(x) = x$
– Sigmoidal transport : $f_t(x) = \sigma^+(x, s_{x1})$

In this last case the system we obtain is close to the classical form used in the field of genetic regulatory networks:

$$\dot{x_j} = h(x_1, \ldots, x_i, \ldots) - m_j.x_j \tag{2}$$

where h is a *sum* of products of (positive or negative) step functions depending on several variables x_i (possibly including x_j itself).

The noticeable differences between these equations and ours are twofold: (i) the coefficient in the x degradation term is not a constant but contains a step function, (ii) the complete system includes an algebraic equation. The three differential equations can be solved independently of the 4th equation since they do not involve u.

The version with a linear transport term introduces a third difference: the differential system is not diagonal any more since $\dot{z}$ involves z and x.

4 Discrete Models and Thomas-Snoussi Theory

L. Glass [7, 8], R. Thomas [22] and others have developped a logical description for genetic regulatory networks. It allows to analyse *qualitatively* the dynamics of such networks. As said above the logical equations are derived from PLDEs of the form (Eq. 2). The variables x_i can be viewed as the concentrations of the proteins produced by the genes. A state $\mathbf{x}$ of the system is defined by a vector $(x_1, \ldots, x_j, \ldots)$ and a path is a sequence of states.

In the asynchronous description a focal point $\mathbf{X}$ is associated to each state $\mathbf{x}$: $\mathbf{X} = f(\mathbf{x})$. The state $\mathbf{X}$ is the state toward which the system *tends*. If only one variable differs between $\mathbf{X}$ and $\mathbf{x}$, this variable changes value in the transition $t \to t + 1$ and state $\mathbf{X}$ is reached at the next step. If n variables differ between $\mathbf{X}$ and $\mathbf{x}$, n transitions are considered because by definition *only one* variable can change value in a transition. In such a description the system evolution is non-deterministic.

When logical equations are viewed as abstractions of PLDEs, it is easily seen that the probability for two or more variables to switch synchronously is marginal. The state space (concentration space) is divided into rectangular domains by the thresholds existing along each axis x_i: in each domain D_k the DE system is linear, and a focal point F_k can be associated to D_k: $F_k = \phi(D_k)$. Another important aspect is that the kinetic parameters are poorly known or totally unknown. Snoussi, Thomas and colleagues [20, 21, 22] have developped a method in which these parameters are discretized on the scale defined by the thresholds. This means that the location of the focal points is not precisely known. What is known is, for each domain D_k, which domain D_f the focal point F_k belongs to. From this mapping, the transitions between domains can be established and the *transition graph* built.

As mentioned at the end of section 3, the differential equations we obtain do not have exactly the form of those used to describe genetic networks. We have shown (to be published) that PLDEs with piecewise linear degradation terms can be transformed to the form used by Thomas and colleagues. It is thus possible to represent the logical structure of the system by an equivalent interaction graph and perform an analysis in term of circuits [20, 21, 22]. It is also in principle possible to extend the theory to PLDEs with off-diagonal terms, but the transformation becomes complicated, and implies the introduction of many discrete $\mathsf{K}_{\mathrm{a},i}$ parameters (see next section for the definition). In the following, we present only the discrete equations obtained on our specific case.

5 Model Implementation in Constraint Logical Programming

CLP is a programming technique based on a declarative approach. A program is constituted of a set of predicates stating what is known to hold true on the system.

The construction of a program proceeds in two steps:

1. Statement of a logical specification. In our case this step corresponds to the statement of predicates defining the formalism used (asynchronous networks with discrete parameters) and those defining the model itself. The model is represented a set of constraints corresponding to the discrete equations.
2. Statement of queries. Some of the parameters involved in the queries can be constrained while others are left unconstrained. In our context the parameters are typically the discretized kinetic parameters and sequences of states of the biological system. Examples are shown below.

In the case of a query with known kinetic parameters and unknown sequence of states the program returns the paths compatible with the parameters. This corresponds to a qualitative simulation. In the case of a query with known sequence of states and unknown parameters, the program returns the parameters compatible with the paths, or in other words an inference on the kinetic parameters (reverse-engineering). Mixed requests can also be formulated when part of paths have been observed and parameters are partially characterized. The reversibility property is one of the most powerful and interesting in this context. The same model description can be used for simulation, inference and mixed requests. Finally, constraints allow an efficient representation of *sets of* models. It is the addition of new constraints which allows to reduce the set of possible models.

5.1 The Two Models

The real parameters used in the models are the following:

$$K_{\mathrm{x},0} = \frac{k_{\mathrm{x}0}}{m_{\mathrm{x}1}} \quad K_{\mathrm{x},3} = \frac{k_{\mathrm{x}0}}{m_{\mathrm{x}1}+m_{\mathrm{x}2}} \quad K_{\mathrm{x},6} = \frac{k_{\mathrm{x}0}}{m_{\mathrm{x}1}+k_{\mathrm{t}}} \quad K_{\mathrm{x},9} = \frac{k_{\mathrm{x}0}}{m_{\mathrm{x}1}+m_{\mathrm{x}2}+k_{\mathrm{t}}}$$

$$K_{\mathrm{x},1} = \frac{k_{\mathrm{x}1}}{m_{\mathrm{x}1}} \quad K_{\mathrm{x},4} = \frac{k_{\mathrm{x}1}}{m_{\mathrm{x}1}+m_{\mathrm{x}2}} \quad K_{\mathrm{x},7} = \frac{k_{\mathrm{x}1}}{m_{\mathrm{x}1}+k_{\mathrm{t}}} \quad K_{\mathrm{x},10} = \frac{k_{\mathrm{x}1}}{m_{\mathrm{x}1}+m_{\mathrm{x}2}+k_{\mathrm{t}}}$$

$$K_{\mathrm{x},2} = \frac{k_{\mathrm{t}}}{m_{\mathrm{x}1}} \quad K_{\mathrm{x},5} = \frac{k_{\mathrm{t}}}{m_{\mathrm{x}1}+m_{\mathrm{x}2}} \quad K_{\mathrm{x},8} = \frac{k_{\mathrm{t}}}{m_{\mathrm{x}1}+k_{\mathrm{t}}} \quad K_{\mathrm{x},11} = \frac{k_{\mathrm{t}}}{m_{\mathrm{x}1}+m_{\mathrm{x}2}+k_{\mathrm{t}}}$$

$$K_{\mathrm{z}} = \frac{k_{\mathrm{t}}}{m_{\mathrm{z}}}$$

At this point, we make an additional approximation by replacing sigmoids (σ) by step functions ($\mathfrak{s}$).

Model A : Model with Linear Transport. From the PLDEs it is possible to derive the following equations:

$$X = \mathfrak{s}^-(x, s_{x2}).[K_{x,6} + K_{x,7}.\mathfrak{s}^+(z, s_{z1})] + \mathfrak{s}^+(x, s_{x2}).[K_{x,9} + K_{x,10}.\mathfrak{s}^+(z, s_{z1})]$$
$$Z = K_z.X$$

giving the coordinates (X,Z) of the focal point associated to the point (x,z).

Now, following Thomas [22], we define discretization operators. If for example, a real variable a has two thresholds ($s_{\mathrm{Sup}}, s_{\mathrm{Inf}}$ with $s_{\mathrm{Sup}} > s_{\mathrm{Inf}}$) it is abstracted in a discrete variable $\mathsf{a} = d(a)$ as follows:

$\mathsf{a} = 0 \Leftrightarrow a < s_{\mathrm{Inf}}$
$\mathsf{a} = 1 \Leftrightarrow s_{\mathrm{Inf}} < a < s_{\mathrm{Sup}}$
$\mathsf{a} = 2 \Leftrightarrow s_{\mathrm{Sup}} < a$

We introduce the following discrete variables:

$$\mathsf{X} = d_x(X),\ \mathsf{Z} = d_z(Z),\ \mathsf{x} = d_x(x),\ \mathsf{z} = d_z(z).$$

There are three thresholds in this model: s_{x2}, s_{z1}, s_{z2} and consequently two threshold orders have to be examined: $s_{z1} < s_{z2}$ (model A1) and $s_{z2} < s_{z1}$ (model A2). The parameterized discrete models A1 and A2 are defined in Table 1. Note that, even though s_{z2} does not appear in these two equations, it must be taken into account because y depends on this threshold.

Table 1. Tables for models A1 and A2

x z	X	Z	X	Z
0 0	$\mathsf{K}_{x,6}$	$\mathsf{K}_{z,6}$	$\mathsf{K}_{x,6}$	$\mathsf{K}_{z,6}$
0 1	$\mathsf{K}_{x,6+7}$	$\mathsf{K}_{z,6+7}$	$\mathsf{K}_{x,6}$	$\mathsf{K}_{z,6}$
0 2	$\mathsf{K}_{x,6+7}$	$\mathsf{K}_{z,6+7}$	$\mathsf{K}_{x,6+7}$	$\mathsf{K}_{z,6+7}$
1 0	$\mathsf{K}_{x,9}$	$\mathsf{K}_{z,9}$	$\mathsf{K}_{x,9}$	$\mathsf{K}_{z,9}$
1 1	$\mathsf{K}_{x,9+10}$	$\mathsf{K}_{z,9+10}$	$\mathsf{K}_{x,9}$	$\mathsf{K}_{z,9}$
1 2	$\mathsf{K}_{x,9+10}$	$\mathsf{K}_{z,9+10}$	$\mathsf{K}_{x,9+10}$	$\mathsf{K}_{z,9+10}$

There are eight discrete parameters $\mathsf{K}_{x,i}$ and $\mathsf{K}_{z,i}$ for these two models. These parameters $\mathsf{K}_{a,i}$ are defined as follows: $\mathsf{K}_{a,i} = d_a(K_{a,i})$, $\mathsf{K}_{a,i+j} = d_x(K_{a,i} + K_{a,j})$. From this definition one can deduce the constraints: $\mathsf{K}_{x,6} \leq \mathsf{K}_{x,6+7}$, $\mathsf{K}_{x,9} \leq \mathsf{K}_{x,6}$, $\mathsf{K}_{x,9+10} \leq \mathsf{K}_{x,6+7}$, $\mathsf{K}_{x,9} \leq \mathsf{K}_{x,9+10}$, $\mathsf{K}_{z,6} \leq \mathsf{K}_{z,6+7}$, $\mathsf{K}_{z,9} \leq \mathsf{K}_{z,6}$, $\mathsf{K}_{z,9+10} \leq \mathsf{K}_{z,6+7}$, $\mathsf{K}_{z,9} \leq \mathsf{K}_{z,9+10}$.

Due to the above inequalities, if $\mathsf{K}_{z,6+7} < 2$ then $\mathsf{K}_{z,6} < 2$, $\mathsf{K}_{z,9} < 2$ and $\mathsf{K}_{z,9+10} < 2$. This means that Z cannot be equal to 2 (see Table 1) and thus no state with $\mathsf{z} = 2$ is reachable. We make the biological hypothesis that our system can reach states in which both β-catenin and cadherin are produced, which means that z is above both s_{z1} and s_{z2} ($\mathsf{z} = 2$), so that we have necessarily $\mathsf{K}_{z,6+7} = 2$. For each case (A1, A2), 84 parameter sets are compatible with the above constraints.

Model B : Model with Sigmoidal Transport. From the PLDEs it is possible to derive the following equations:

$$X = \mathfrak{s}^-(x, s_{x2}).[(K_{x,0} - K_{x,2}) + K_{x,1}.\mathfrak{s}^+(z, s_{z1}) + K_{x,2}.\mathfrak{s}^-(x, s_{x1})]$$
$$+\mathfrak{s}^+(x, s_{x2}).[(K_{x,3} - K_{x,5}) + K_{x,4}.\mathfrak{s}^+(z, s_{z1}) + K_{x,5}.\mathfrak{s}^-(x, s_{x1})]$$
$$Z = K_z.\mathfrak{s}^+(x, s_{x1})$$

There are four thresholds in this model: s_{x1},s_{x2}, s_{z1}, s_{z2} and consequently four pairs of threshold orders have to be examined: $s_{x1} < s_{x2}$ and $s_{z1} < s_{z2}$ (model B1), $s_{x1} < s_{x2}$ and $s_{z2} < s_{z1}$ (model B2), $s_{x2} < s_{x1}$ and $s_{z1} < s_{z2}$ (model B3), $s_{x2} < s_{x1}$ and $s_{z2} < s_{z1}$ (model B4). The parameterized discrete models B1, B2, B3 and B4 are defined in Table 2.

Table 2. Tables for models B1, B2, B3 and B4

x z	X	Z	X	Z	X	Z	X	Z
0 0	$K_{x,0}$	0	$K_{x,0}$	0	$K_{x,0}$	0	$K_{x,0}$	0
0 1	$K_{x,0+1}$	0	$K_{x,0}$	0	$K_{x,0+1}$	0	$K_{x,0}$	0
0 2	$K_{x,0+1}$	0	$K_{x,0+1}$	0	$K_{x,0+1}$	0	$K_{x,0+1}$	0
1 0	$K_{x,0-2}$	K_z	$K_{x,0-2}$	K_z	$K_{x,3}$	0	$K_{x,3}$	0
1 1	$K_{x,0+1-2}$	K_z	$K_{x,0-2}$	K_z	$K_{x,3+4}$	0	$K_{x,3}$	0
1 2	$K_{x,0+1-2}$	K_z	$K_{x,0+1-2}$	K_z	$K_{x,3+4}$	0	$K_{x,3+4}$	0
2 0	$K_{x,3-5}$	K_z	$K_{x,3-5}$	K_z	$K_{x,3-5}$	K_z	$K_{x,3-5}$	K_z
2 1	$K_{x,3+4-5}$	K_z	$K_{x,3-5}$	K_z	$K_{x,3+4-5}$	K_z	$K_{x,3-5}$	K_z
2 2	$K_{x,3+4-5}$	K_z	$K_{x,3+4-5}$	K_z	$K_{x,3+4-5}$	K_z	$K_{x,3+4-5}$	K_z

There are nine discrete parameters $K_{x,i}$ and K_z for these four models, which are constrained by their definitions: $K_{x,0-2} \leq K_{x,0}$, $K_{x,0} \leq K_{x,0+1}$, $K_{x,0-2} \leq K_{x,0+1-2}$, $K_{x,0+1-2} \leq K_{x,0+1}$, $K_{x,3-5} \leq K_{x,3}$, $K_{x,3} \leq K_{x,3+4}$, $K_{x,3-5} \leq K_{x,3+4-5}$, $K_{x,3+4-5} \leq K_{x,3+4}$, $K_{x,3} \leq K_{x,0}$, $K_{x,3+4} \leq K_{x,0+1}$, $K_{x,3-5} \leq K_{x,0-2}$, $K_{x,3+4-5} \leq K_{x,0+1-2}$, $K_z = 2$. The last constraint is due to the biological hypothesis, as stated above.

5.2 Organization of the CLP Program

Now a few words about the overall organization of the CLP program. A general predicate defines the asynchronous multivalued framework. The predicate `multivalued_async_model(Model,Path)` is true if `Path` is a possible path of the model `Model`. A path is a list of states. A model is defined by the type of transport and the threshold order. It is represented by a transition table (see Tables 1 and 2) and a set of constraints between the model parameters which are deduced from the definition of the model. A general constraint defined in `multivalued_async_model` is for example that a variable value changes by unit steps in a transition (it cannot jump from 0 to 2 or from 2 to 0).

Due to the fact that we obtain discrete equations from the PLDEs, it is possible to use the constraint solver on intervals of Prolog IV. In our case, variables have finite domains. As an example, consider a case where a model Model is known, defined by a table T and parameters P. Then a typical query concerning the existence of steady states is the following:

```
Path = [S,S] ,
Model = [T,P] ,
multivalued_async_model(Model,Path) .
```

5.3 Example of Queries and Results

Questions about Steady States. In the case of sigmoidal transport (four models: B1, B2, B3, B4), we ask the list of all possible steady states (whatever the threshold orders). We find four states: (0, 0), (1, 0), (1, 2), (2, 2). An additional query allows to see that there exists no set $\{\mathsf{K}_{a,i}\}$ having more than two steady states. We obtain the same result, i.e. no more than two stable states, in the case of linear transport (two models: A1 and A2).

The next query is: What are the sets of $\mathsf{K}_{a,i}$ values having just two steady states? For model B3 for instance, there are five solutions: one having steady states (0,0) and (2,2) and four having (1,0) and (2,2). In the case {(0,0),(2,2)}, we interpret state (0,0) as the normal state in which β-catenin is present in the cytoplasm and in the nucleus at low levels. In all cases, (2,2) is a pathological state of constant over-expression.

Question About the Perturbation. From this point, one would like to know in case B with the two steady states (0,0) and (2,2) whether it is possible to take into account the repairing of the junctions. More precisely, when junctions are destabilized by antibodies, cadherins are destroyed and β-catenins are released in the cytoplasm. In our modeling the perturbation is represented by setting the system in a state with larger values of x. It implies that (2,2) cannot be considered as a normal state of the cell (because x cannot be augmented). The resulting query checks that it exists a path beginning in state (1,0) and containing a state where $\mathsf{z} = 1$ (the nuclear β-catenin concentration is observed increasing [9]) which finishes in state (0,0) (the normal steady state is reached). It appears that only model B2 is acceptable.

Also, one would like to know the parameter sets in case B for a unique steady state (excluding (2,2)), which takes into account the repairing of junctions. The resulting query checks that it exists a path beginning with the perturbed state and containing a state where z has been increased which finishes in the steady state. Model B3 admits 29 parameter sets having exactly one steady state. The added constraint on the existence of a restoration path eliminates 14 sets.

6 Conclusion and Perspectives

Using the PSS approximation we have been able to represent our biochemical system by PLDEs very similar to the ones used for genetic networks. This allowed us to describe in a unified way regulatory interactions and biochemical reactions. This kind of analysis is not general and must be performed on each specific case.

Then, using an exploratory strategy and refining progressively the queries, we were able to deduce interesting properties having biological interpretation. We focused on stable states and models exhibiting 'return to stable state' paths and it appeared that such constraints reduce notably the number of possible parameter sets. We were not interested in oscillating behaviors, but it is course possible to express constraint-based queries about cylic paths.

Related Work. Constraint Satisfaction Problem (CSP) technology is used by V. Devloo [6] to discover efficiently the steady-state of completely instantiated asynchronous models. It is mentionned also that this approach *could* be used, as we do in this paper, to induce parameters from system behaviors.

Chabrier and Fages [2] describe a model-checking approach in CLP [5] (with linear arithmetic constraints) to check properties of qualitative or quantitative systems expressed in Computation Tree Logic (CTL): this study considers that the biological system is known.

Peres and Comet [18] pursue a similar goal as ours (inference of a Thomas network) by using model-checking and CTL, but without using constraints. They generate 27 model instances and check the validity of a CTL formula on each of them. This allow them to reduce to 14 possibilities. An approach similar to ours was advocated by J. Cohen for the case of gene regulatory networks [3].

Bockmayr and Courtois [1] use HCC (Hybrid Concurrent Constraint) which can tackle more general differential equations. But this technology does not appear to possess the full capabilities of constraints, namely to infer pre-conditions from post-conditions. For the purpose of taking into account more general differential equations. We are thinking to rather use an approach similar to the one proposed by Hickey [11].

It is worth noting that the CLP technology as used in this paper, provide not only constraint solvers (in particular CSP solvers) but also a very flexible and powerful way to express queries via logical formulas. The queries mentioned in section 5 are typical of this aspect as each of them introduce new constraints to be added to the constraints defined by the predicate `multivalued_async_model`.

Perspectives. DNA chip experiments as well as proteomic experiments are being done in our lab to identify new players (genes and proteins). This will extend the molecular network but will also bring new data and thus new constraints. Other biological hypotheses should be included concerning for example the adherens junction assembly process. Taking into account combinations of such hypotheses, we will generate a 'model space'. It will then be even more important to have formal reasoning tools to discriminate these models.

Also we intend to study carefully the language of the interesting queries which can be answered efficiently. The present network is tractable, but it could be a different matter as complexity increases.

References

1. A. Bockmayer and A. Courtois. Using Hybrid Concurrent Constraint Programming to Model Dynamic Biological Systems. In Proceedings of the 18th International Conference on Logic Programming, LNCS 2401, Springer, 85–99 (2002).
2. N. Chabrier and F. Fages. Symbolic Model Checking of Biochemical Networks. Computational Methods in Systems Biology. In Computational Methods in System Biology 2003, C. Priami (ed.), LNCS 2602, Springer, 149–162 (2003).
3. J. Cohen. Approaches for simulating and modeling cell regulation : search for a unified view using constraints. In Linköping Electronic Articles in Computer and Information Science, 3, no7 (2001).
4. A. Colmerauer. Prolog – Constraints Inside, Manuel de Prolog, PROLOGIA, Case 919, 13288 Marseille cedex 09, France (1996).
5. G. Delzanno and A. Podelski. Model Checking in CLP. In Proceeding of the 5th International Conference TACAS'99, Springer, LNCS 1579, 223–239 (1999).
6. V. Devloo, P. Hansen and M. Labbé. Identification of All Steady States in Large Biological Systems by Logical Analysis. Bulletin of Mathematical Biology 65, 1025–1051 (2003).
7. L. Glass, S. A. Kauffman. Co-operative components, spatial localization and oscillatory cellular dynamics. J. Theor. Biol. 34, 219–237 (1972).
8. L. Glass, S. A. Kauffman. The logical analysis of continuous, non-linear biochemical control networks. J. Theor. Biol. 39, 103–129 (1973).
9. D. Gulino, E. Delachanal, E. Concord, Y. Genoux, B. Morand, M. O. Valiron, E. Sulpice, R. Scaife, M. Alemany and T. Vernet. Alteration of Endothelial Cell Monolayer Integrity Triggers Resynthesis of Vascular Endothelium Cadherin. The Journal of Biological Chemistry 273, 29786–29793 (1998).
10. B. Hermant, S. Bibert, E. Concord, B. Dublet, M. Weidenhaupt, T. Vernet and D. Gulino-Debrac. Identification of Proteases Involved in the Proteolysis of Vascular Endothelium Cadherin during Neutrophil Transmigration. The Journal of Biological Chemistry 278, 14002–14012 (2003).
11. T. J. Hickey and D. K. Wittenberg. Rigorous Modeling of Hybrid Systems using Interval Arithmetic Constraints. In Technical Report CS-03-241, Computer Science Departement, Brandeis University (2003).
12. P. L. Hordijk, E. Anthony, F. P. Mul, R. Rientsma, L. C. Oomen and D. Roos. Vascular-Endothelial-Cadherin Modulates Endothelial Monolayer Permeability. J. Cell Sci. 112, 1915–1923 (1999).
13. H. de Jong, J.-L. Gouzé, C. Hernandez, M. Page, T. Sari, and J. Geiselmann. Hybrid modeling and simulation of genetic regulatory networks: A qualitative approach. In Hybrid Systems : Computation and Control (HSCC 2003), A. Pnueli and O. Maler (ed.), LNCS 2623, Springer, 267–282 (2003).
14. M. G. Lampugnani, M. Corada, L. Caveda, F. Breviario, O. Ayalon, B. Geiger and E. Dejana. The molecular organization of endothelial cell to cell junctions: differential association of plakoglobin, β-catenin, and α-catenin with vascular endothelial cadherin (VE-cadherin). J. Cell Biol. 129, 203–217 (1995).
15. M. G. Lampugnani, M. Resnati, M. Raiteri, R. Pigott, A. Pisacane, G. Houen, L. P. Ruco and E. Dejana. A novel endothelial-specific membrane protein is a marker of cell-cell contacts. J. Cell Biol. 118, 1511–1522 (1992).
16. P. Legrand, S. Bibert, M. Jaquinod, C. Ebel, E. Hewatt, F. Vincent, C. Vanbelle, E. Concord, T. Vernet and D. Gulino. Self-assembly of the vascular endothelial cadherin ectodomain in a Ca^{2+}-dependent hexameric structure. Journal of Biological Chemistry 276, 3581–3588 (2001).

17. J. D. Murray, Mathematical Biology, Springer-Verlag (1989).
18. S. Peres and J. P. Comet. Contribution of Computational Tree Logic to Biological Regulatory Networks: Example from Pseudomonas Aeruginosa. In CMSB 2003, C. Priami (ed.), LNCS 2602, 47–56 (2003).
19. B. E. Shapiro, A. Levchenko and E. Mjolsness. Automatic model generation for signal transduction with applications to MAPK pathway. In Foundations of Systems Biology, H. Kitano (ed.), MIT Press (2002).
20. E. H. Snoussi and R. Thomas. Logical Identification of All Steady States : The Concept of Feedback Loop Characteristic States. Bulletin of Mathematical Biology 55, 973–991 (1993).
21. D. Thieffry, M. Colet and R. Thomas. Formalisation of Regulatory Networks : a Logical Method and Its Automation. Math. Modelling and Sci. Computing 2, 144–151 (1993).
22. R. Thomas and M. Kaufman. Multistationarity, the Basis of Cell Differentiation and Memory. II. Logical Analysis of Regulatory Networks in Term of Feedback Circuits. Chaos, 11, 180–195 (2001).

Modelling Metabolic Pathways Using Stochastic Logic Programs-Based Ensemble Methods

Huma Lodhi and Stephen Muggleton

Department of Computing, Imperial College,
London SW7 2BZ, UK
{hml, shm}@doc.ic.ac.uk

Abstract. In this paper we present a methodology to estimate rates of enzymatic reactions in metabolic pathways. Our methodology is based on applying stochastic logic learning in ensemble learning. Stochastic logic programs provide an efficient representation for metabolic pathways and ensemble methods give state-of-the-art performance and are useful for drawing biological inferences. We construct ensembles by manipulating the data and driving randomness into a learning algorithm. We applied failure adjusted maximization as a base learning algorithm. The proposed ensemble methods are applied to estimate the rate of reactions in metabolic pathways of Saccharomyces cerevisiae. The results show that our methodology is very useful and it is effective to apply SLPs-based ensembles for complex tasks such as modelling of metabolic pathways.

1 Introduction

Metabolic pathways can be viewed as series of enzyme-catalysed reactions where product of one reaction becomes substrate for the next reaction. These pathways can be branched and interconnected via shared substrates. Quantitative analysis of enzymatic reactions is very important in biomedical applications, biotechnology and drug design. Estimation of reaction rates of enzymes in metabolic pathways is a key problem in quantitative analysis and modelling of metabolism. Behaviour of enzymes in metabolic pathways can be studied using Michaelis-Menten (MM) framework that is very useful in biological kinetics and pharmacokinetics. However information required for MM equation is not easily available and the application of MM equation is not free from problems [1]. In this paper we present a methodology that applies stochastic logic learning in ensemble learning to calculate enzymatic reaction rates.

Ensemble methods are state-of-the-art learning algorithms to solving prediction problems. The underlying aim of ensemble methods is to construct a highly accurate predictor. These methods accomplish this aim by constructing a series of predictors (models). The final model performs the estimation task by aggregating the estimations of individual models. Bagging [2] and boosting [3] are the most popular examples of these methods. Experimental results [4, 5, 6] have demonstrated ensemble methods' ability to generate highly accurate predictors

V. Danos and V. Schachter (Eds.): CMSB 2004, LNBI 3082, pp. 119–133, 2005.

and their surprising contradiction of Ockham's razor that give preference to simple hypotheses over complex ones. In order to understand this phenomenon ensembles have been analysed as they relate to the margin theory [7, 8]. Ensemble methods have also been analysed in terms of bias (measure of the goodness of the average predictor's approximation of the target function) and variance (a measure of the diversity among the base learning algorithm's guess) [5]. Bagging is a variance reduction technique and is very effective for unstable learning methods. It has also been shown that bias and variance can be expressed in terms of margin and margin can be expressed in terms of bias and variance [9]. In this paper we focus on bagging that is particularly useful for unstable predictors and possess characteristics such as parallelization. Parallelization is particularly important for stochastic logic learning where run time can be high depending on the complexity of a problem. These properties make bagging an ideal method to combine with SLPs.

SLPs [10] are generalisations of Hidden Markov Models (HMMs) [11] and Stochastic Context Free Grammars (SCFGs) [12]. They were viewed as a compact approach to representing a probabilistic preference function to provide as a parameter to ILP algorithms. HMMs and SCFGs have been extremely successful in sequence-oriented applications in natural language and bioinformatics. They provide a compact representation of a probability distribution over sequences. This contrasts with Bayesian networks, which represent conditional independences between a set of propositions. It is natural to think of HMMs and SCFGs as representing probabilities over objects in the domain (Halpern's type 1 approach from [13]) and Bayesian networks as representing probabilities over possible worlds (Halpern's type 2 approach). Learning of SLPs can be viewed as learning of parameter estimation or structural learning. In this paper we focus on the parameter estimation task, as we want to study the performance of SLPs-based ensembles to obtain the rate information of reactions in metabolic pathways. In order to learn the parameters over SLPs we employed failure adjusted maximisation (FAM) as a base learner.

The combination of boosting with Inductive Logic Programming (ILP) has been pioneered by Quinlan [14]. Recently Dutra et al. [15] have investigated bagging in Inductive Logic Programming. However ensemble methods have never been applied in conjunction with SLPs. This is the first combination of ensemble learning with SLP learning.

We adapt bagging in SLPs to perform rate estimation task in metabolic pathways. We also present another method, ranbag, to construct an ensemble by driving randomness into a learning algorithm. Bagging and ranbag obtain predictors from FAM. The final estimation is obtained by computing the average of the outputs of all the base predictors. We evaluate SLPs-based bagging and ranbag to modelling metabolic pathway of Saccharomyces cerevisiae. The results show that it is useful to apply ensembles for learning SLPs to modelling metabolic pathways.

The paper is organised as follows. A brief overview of metabolic pathways has been given in Sect. 2. Section 3 explains SLPs and FAM. In Sect. 4 we have described bagging and ranbag. Section 5 explains experimental results.

2 Metabolic Pathways

Genomic data is now being obtained on an industrial scale. Complete drafts of the human genome were published during 2001 [16, 17]. Projects are under way to sequence the genomes of the mouse, rat, zebra fish and puffer fishes *T. nigoviridis* and *Takifugu rubripes.* The focus of genome research is moving to the problem of identifying the biological functions of genes. An application of inductive logic programming to functional genomics is described in [18].

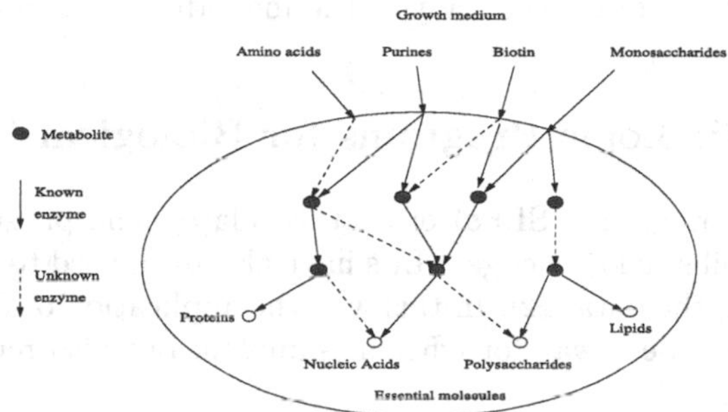

Fig. 1. Illustration of Cell with metabolic network involved in converting growth media into molecules essential for life

Figure 1 illustrates the way in which a network of metabolic reactions within a cell convert a growth medium (input compounds) into molecules, which are essential for life (output compounds). Each intermediate compound (associated with nodes of the graph) is known as a metabolite and each metabolic reaction is mediated by an enzyme (arcs in the graph). Presently not all enzymes involved in metabolic reactions are known. Online databases such as KEGG[1], WIT[2] and BRENDA[3] describe relationships between tens of thousands of enzymes. Measuring the rates of catalysed reactions can identify enzymes. Michaelis-Menten (MM) framework can be used to estimate reaction rates of enzymes. In order to compute these rates, using MM equation, information can be obtained from online database discussed above. These databases may not contain all the required information. Alternatively enzyme kinetics can be studied using learning techniques such as SLPs-based bagging and ranbag. SLPs provide an ideal representation for such data as rates can be viewed as probabilities, which can capture rates as proportions. In this way, the rates of enzymatic reactions can be estimated by computing the parameters over SLPs. In other words SLPs-based ensemble methods provide an efficient way to study enzyme kinetics.

Metabolic pathways have been studied using mathematical models that comprise sets of ordinary differential equations (ODEs). ODEs model the dynamic

[1] http://www.genome.ad.jp/kegg/
[2] http://wit.mcs.anl.gov/WIT2/
[3] http://www.brenda.uni-koeln.de/

proprieties of enzyme-catalysed reactions. There exist simulation tools for ODE cellular modelling. Gepasi [19] and DBsolve [20] are example of such simulation software. In order to handle non-deterministic characteristics of biological system, stochastic techniques such as Next Reaction Method [21] have been developed. In order to apply learning methods such as bagging and ranbag to ODE models or stochastic models, we need base learner that can learn these models. Once a sequence of ODE models or stochastic models is obtained from base learner, bagging-based methods can be applied to generate the final model. The final model that is a combination of individual ODE models or stochastic models can have a higher accuracy than any of the individual base models.

3 Stochastic Logic Programs for Biological Domains

Stochastic logic programs (SLPs) extend standard logic programs in order to represent probabilistic knowledge. SLPs have also been used to provide distribution for sampling the data [22]. In this way the application of SLPs are twofold, they not only provide a way for efficient sampling but also represents complex uncertain knowledge.

Syntax for SLPs: An SLP is a definite labelled logic program. In an SLP all or some of the clauses are associated with probability labels (parameters) and is known as pure SLP or impure SLP respectively. An SLP is said to be normalised if label of the clauses with same predicate symbol in the head sum to one and unnormalised otherwise. Formally an SLP S is a set of labelled definite clauses $p : C$ where $p \in [0, 1]$ is a probability label or parameter and C is a range-restricted definite clause. In this way an SLP provides an efficient representation to model metabolic pathways, where the set of clauses can describe enzymes and probability labels depict rates of reactions in metabolic pathways. Figure 2 shows the syntax of SLPs.

SLP	eg.
Labelled Definite Clause	0.3: like(X,Y) ← pet(Y,X)
Labelled Program (impure)	0.3: proteinfold1(X) ← ..
	0.3: proteinfold2(X) ←..
Labelled Program (normalised)	0.5: coin(head) ←
	0.5: coin(tail) ←

Fig. 2. Syntax for SLPs

Semantics for SLPs: The semantics for SLPs is illustrated in Fig. 3. SLPs have a distributional semantics, that is one, which assigns a probability distribution to the atoms of each predicate in the Herbrand base of the underlying (unlabelled) logic program. An interpretation M is a model of an SLP S if all the atoms a have a probability assigned by M which is at least the sum of the probabilities of derivations of a with respect to S.

SLP	eg.
Distributional Interpretation	0.3: p(a) .. 0.4: q(a) ..
Distributional Model	0.3: p(a) .. 0.3: q(a) ..
$P \models Q$	0.3: q(a), 1.0: p(X) ← q(X) ⊨ 0.3: p(a)

Fig. 3. Semantics for SLPs

SLP	eg.
SSLD derivation	{0.3: q(a), 1.0: p(X) ← q(X) } refutes goal 0.3: ← p(a)

Fig. 4. Proof for SLPs

Proof for SLPs: Figure 4 shows proofs for SLPs. Probabilities are assigned to atoms according to an SLD-resolution strategy which employs a stochastic selection rule. Derivations can be viewed as Markov chains in which each stochastic selection is made randomly and independently. Thus the probability of deriving any particular atom a is the sum of products of the probability labels on the derivations of a.

Failure Adjusted Maximization (FAM)- An Example of Learning Methods for SLPs
Expectation Maximization (EM) [23] is a well-known maximum likelihood parameter estimation technique. EM is an iterative algorithm that performs the parameter estimation task from incomplete data. Failure adjusted maximization (FAM) [24] uses EM algorithm to compute maximum likelihood estimates for pure, normalised SLPs. In order to apply EM algorithm for parameter estimation task a complete dataset of atoms has its natural representation as incomplete dataset of atoms. A set of atoms yielding the refutation from a complete dataset of derivations makes an incomplete dataset of atoms.

Given a logic program and a set of initial (prior) parameters FAM computes the maximum likelihood estimates in a two step (expectation step and maximization step) iterative learning process. In the expectation step FAM computes the weighted contribution of the clause to deriving a data point and weighted contribution of the clause due to failed derivation. In the maximization step the contribution of the clause is maximised. The value associated with each clause is normalised and becomes an input for the next iteration of FAM. In this way, at each iteration FAM improves the current estimates of the parameters. This process is repeated till convergence. In this paper we applied FAM as a base learner to compute the reaction rate of metabolic pathways.

4 Bagging and Variants

Ensemble methods such as bagging and ranbag work by repeatedly calling a base learner to produce a series of predictors. The final predictor is a combination of individual predictors and generally has a higher accuracy than any of the individual base learners. Although this higher accuracy is due to uncorrelated errors among the base predictors, the following factors also contribute to the success of the ensemble methods. 1) The base learner is too simple (weak) to generate a hypothesis with low error. A predictor that is a combination of these hypotheses can have high accuracy. 2) The base learner is unstable such as decision trees, neural network, an inductive logic programming algorithm and a learning algorithm for SLPs. For an unstable base learner a small change in the learning set significantly affects the generated predictor. 3) The base learning algorithm suffers from some problems (employing search strategies that are not good enough to select a good hypothesis) that can be overcome using a combination of predictors generated by these learning algorithms.

We can view the learning process of bagging and ranbag as comprising two stages. In the first stage base predictors are generated and in the second stage these predictors are combined.

Bootstrap Aggregating (Bagging): We now describe how we have adapted bagging for learning the parameters over SLPs to modelling metabolic pathways. Bagging is based on the idea of resampling and combining. In order to obtain a predictor from the base learner, bagging provides the base learner with bootstrap replicates [25] of the learning set. A bootstrap replicate is constructed by randomly drawing, with replacement, n instances from the learning set of size n. These instances are drawn according to a uniform distribution that is kept on the learning set. The bootstrap replicate may not contain all of the instances from the original learning set and some instances may occur many times. On average bootstrap replicate contain 63.2% of the distinct instances in the learning set.

Require:
Learning Set: $L = \{x_1, , \ldots, x_n\}$ where $x_i \in X$.
A base learner that takes an underlying logic program representing enzymes in a metabolic pathway and a set of prior parameters that gives an initial guess of the rates of enzyme-catalysed reactions.
for $t = 1$ to T do
/* Generate bootstrap sample L^B from a learning set L */
/* Call the base learner with underlying logic program LP and prior parameters P_0. Set the prior parameters according to uniform distribution */
$h_t = BL(L^B, LP, P_0)$
end for
/* The bagged estimation is */
$h_{bag} = \frac{1}{T}\sum_{t=1}^{T} h(\hat{P})_i$

Fig. 5. Bagging for rates estimation in metabolic pathways

Pseudocode for bagging is given in Fig. 5. As input, bagging requires a learning set L of instances of the form $L = \{x_1, \ldots, x_n\}$. The instances are generated independently and identically according to the probability distribution D. In our setting, the learning set contains all the information that is required to measure the reaction rates of enzymes. In order to estimate these rates NMR or mass spectrometric data can be used as learning set. Alternatively an SLP representing metabolic pathway can be used to generate a learning set.

As described, bagging calls learning algorithm BL for T number of times. In order to perform rates estimation a learning algorithm such as FAM is specified. FAM is provided with a bootstrap sample L^B, an underlying logic program and a set of prior parameters (initial guess). Note that the prior parameters are set according to a uniform distribution. For a particular bootstrap sample the estimated parameters (predictions) are denoted by $h(\hat{P})$. This process of drawing a bootstrap sample and obtaining predictions is repeated for T times. The bagged prediction is obtained by computing the average of the outputs of all the base predictors. The bagged estimation for ith parameter is $h_{bag} = 1/T \sum_{t=1}^{T} h(\hat{P})_i$.

Random Prior Aggregating (Ranbag): In order to find out the reaction rates of enzymes in a metabolic pathway we introduce, ranbag, a variant of bagging. Ranbag performs the rates estimation task by driving randomness into FAM. In this ways ranbag is based on the idea of combining a set of diverse predictors. In order to obtain these predictors the prior parameters of FAM are set randomly. The obtained base predictors can be substantially diverse as FAM depends on the selection of prior parameters. Pseudocode for ranbag is given in Fig. 6. As described, ranbag requires a learning set L and a base learner that takes a set of prior parameters and underlying logic program. Let the learning set L represents the data containing the information required to calculate reactions rates of enzymes and the logic program LP is a set of clauses where each clause represents an enzyme. Let the prior parameters provide the initial guess for reaction rates. Ranbag calls learning algorithm BL such as FAM, for T number of iterations. At each iteration FAM is provided with same learning set L, and underlying logic program but prior parameters P_t are set randomly. The estimated parameters (predictions) are denoted by $h(\hat{P})$. This process of

Require:
Learning Set: $L = \{x_1, , \ldots, x_n\}$ where $x_i \in X$.
A base learner that takes an underlying logic program and a set of prior parameters.
for $t = 1$ to T do
/* Set the prior parameters P_t according to random distribution.
/* Call the base learner with underlying logic program LP and prior parameters P_t.
$h_t = BL(L^B, LP, P_t)$
end for
/* The final estimation is */
$h_{ranbag} = \frac{1}{T} \sum_{t=1}^{T} h(\hat{P})_i$

Fig. 6. Ranbag for the estimation of the rates of enzyme-catalysed reactions

setting the prior randomly and obtaining predictors is repeated for T times. The final estimation is obtained by computing the average of the outputs of all the base predictors. The final estimation for ith parameter is $h_{ranbag} = 1/T \sum_{t=1}^{T} h(\hat{P})_i$.

5 Experimental Analysis

In this section, we describe a series of extensive and systematic experiments. We empirically evaluated bagging and ranbag to calculate the rates of enzyme-catalysed reactions in metabolic pathways.

Datasets. We applied bagging and ranbag to modelling aromatics amino acid pathway of Saccharomyces cerevisiae (baker's yeast, brewer's yeast) [26]. We compared modelling performance of bagging with ranbag's performance. Figure 7 shows the aromatic amino acid pathway of yeast. SLPs naturally represent metabolic pathways as they can capture the rate information by way of probabilities (parameters over SLPs). We used the implementation of FAM available at[4]. The metabolic pathways have been represented by an SLP comprising of 21 stochastic clauses. Each clause of SLP provides the probabilistic information about the occurrence and non-occurrence of a reaction. In this way an SLP represents a metabolic pathway and tell the reactions' rates. Furthermore, modelling performance of bagging and ranbag has also been evaluated where a branch has been added in the metabolic networks. A branching metabolic networks is obtained by adding a branch in the same metabolic pathway (shown in Fig. 7). This phenomenon provides us with a new SLP.

As discussed, SLPs provide an efficient way for sampling the data, we generated the data using SLPs for both the chain and branching metabolic pathways. In our experimental setting the two SLPs represent chain and branching scenarios and two datasets are generated using these SLPs. These dataset hereafter are referred to as Chain dataset (non-branching metabolic network) and Branch dataset (branching metabolic network).

Experimental Methodology. Bagging and ranbag obtain base predictors from an SLP learning algorithm, FAM. The coordinates described below can control the performance of FAM and ensembles.

Convergence Criteria: FAM allows specifying the convergence criterion. We set it the log likelihood.

Prior: This corresponds to prior (initial) parameters of FAM. The prior parameters can be set randomly or uniformly. We set the prior parameters of FAM according to a uniform distribution for bagging. The prior has been set according to random distribution for ranbag.

Stopping Criterion: In bagging and ranbag we specify the number of base models T. We set the number of models T to 100.

---

[4] http://www-users.cs.york.ac.uk/~nicos/sware/

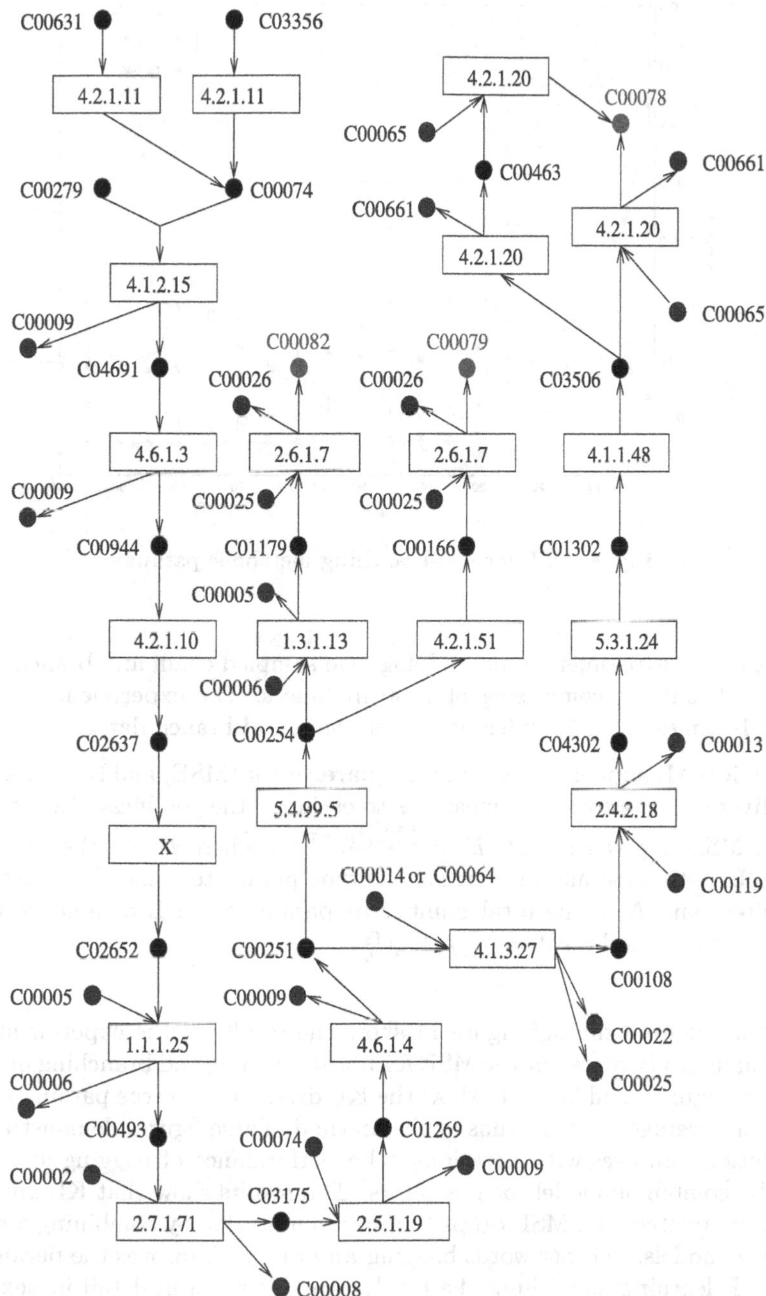

Fig. 7. The aromatic amino acid pathway of yeast. A chemical reaction is represented by a rectangle with its adjacent circles where rectangles represent enzymes and circles represent metabolites. In this figure metabolites are labelled by their KEGG accession numbers and enzymes by the EC number

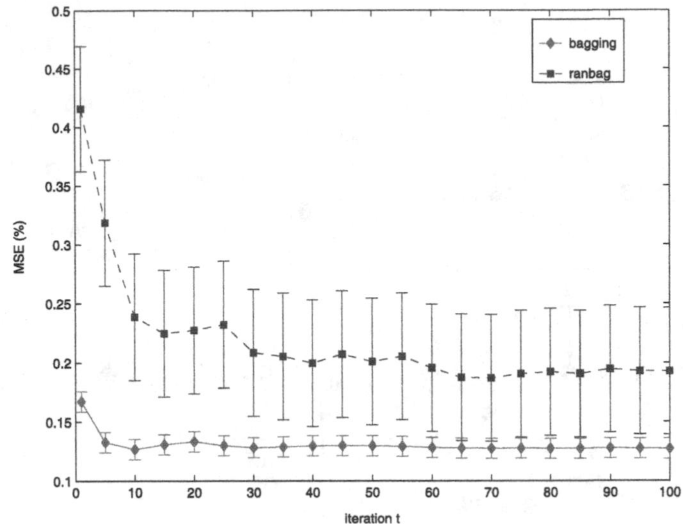

Fig. 8. MSE for non-branching metabolic pathway

As part of our experimental methodology we sampled chain and branch dataset where each dataset comprising of 1000 instances. The experiments were performed 10 times using 10 different sets of chain and branch data.

Evaluation Measure: We used mean squared error (MSE) and Kullback-Leiber (KL) divergence performance measures to estimate the goodness of bagging and ranbag. MSE is given by, $MSE = \frac{\sum_{i=1}^{N}(p_i - \hat{p_i})^2}{N}$, where p_i are the parameters need to be estimated and are termed the true parameters and $\hat{p_i}$ are estimated parameters and N is the total number of parameters. KL divergence to true parameters is given by $KL = \sum_i p_i \log(\frac{p_i}{\hat{p_i}})$

Results. Figure 8 through figure 11 show the results of the experiments. Figure 8 and figure 10 represent the MSE for non-branching and branching metabolic pathway. Figure 9 and figure 11 show the KL divergence to true parameters. The results are averaged over 10 runs of the method. These figures demonstrate how the solution improves with iterations. The performance of bagging and ranbag varies by combining models or predictors. The results show that KL divergence to true parameters and MSE drops to a minimum value by combining a number of diverse models. In other words bagging and ranbag improves the performance of an SLP learning algorithm. The results are described in detail in next paragraphs.

We first consider the non-branching metabolic pathway (chain dataset). The average error for a single model is 0.17% for bagging and 0.29% for ranbag. The results show that both bagging and ranbag improves the performance. Average error for bagging is 0.12% and average error for ranbag is 0.14%. The curves

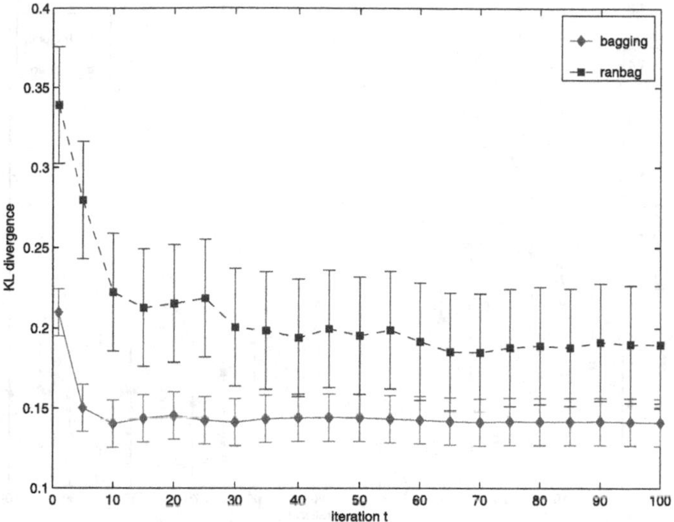

Fig. 9. KL divergence for non-branching metabolic pathway

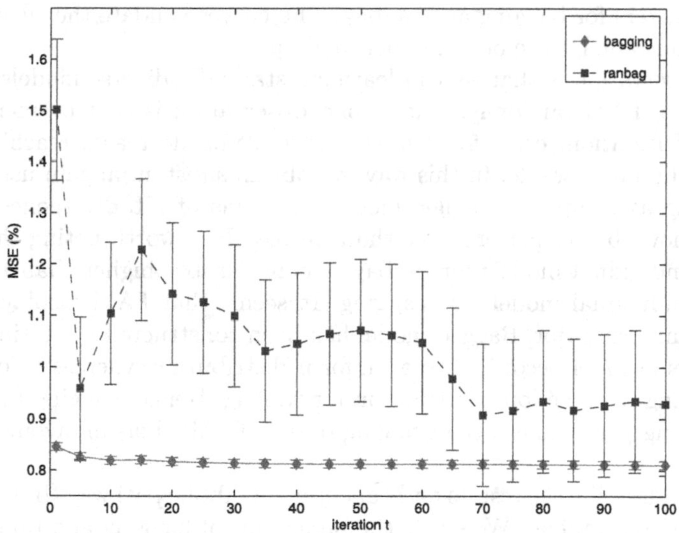

Fig. 10. MSE for branching metabolic pathway

show that both bagging and ranbag achieve substantial improvements. Bagging obtains the improvements within first 15 iterations and ranbag achieves the maximum gain by combining 70 models. The curves also show that after some initial iterations there is no significant improvement in the performance of bagging but ranbag does not show this phenomenon. Figure 9 tells the KL divergence to

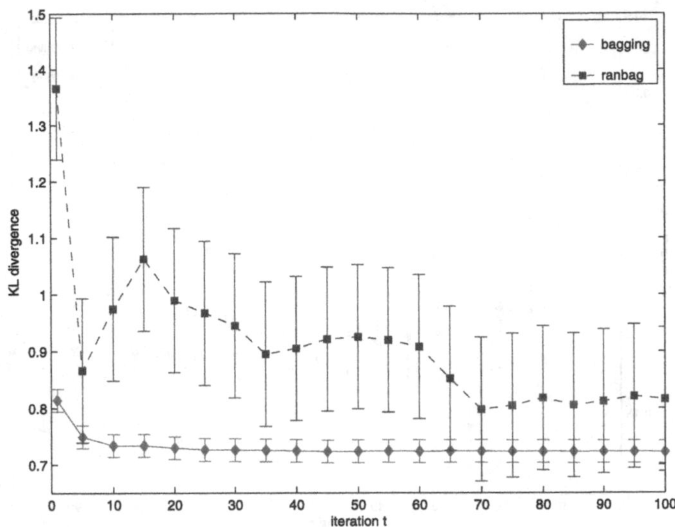

Fig. 11. KL divergence for branching metabolic pathway

true parameters for bagging and ranbag. The curves validate the effectiveness of bagging and ranbag to modelling metabolic pathways.

Our observation is that we can learn substantially diverse models by setting the priors of FAM randomly. Our second observation is that by increasing the number of iteration, error for ranbag and bagging decreases reaching a minimum and becomes stable. In this way, we obtain substantial gain using bagging and ranbag as compare to single model. In terms of KL divergence and MSE bagging shows better performance than ranbag. It is worth noting that average error for individual models for ranbag is considerably higher than the average error for individual models for bagging. It seems that FAM displays a bias in favour of uniform prior. Bagged model has been constructed by setting the prior parameters of FAM according to a uniform distribution whereas prior has been set according to a random distribution for ranbag. Hence, bagging achieves better modelling performance than ranbag due to FAM's bias in favour of uniform prior.

We now describe the results for branching metabolic pathway (branch dataset) for bagging and ranbag. We study the behaviour of bagging and ranbag in conjunction with FAM in a scenario where an alternative path has been added to a pathway. Figure 10 shows MSE for bagging and ranbag. The average MSE for single model for bagging is 0.85% and 1.5% for ranbag. Average error for bagging is 0.81% that is obtained by combining 25 models. Ranbag obtains minimum average MSE of 0.9%. The results show that bagging improves the performance of a single model but the improvement is not substantial. Success of bagging-based methods depends on the fit of the style of the function with the particular data and base learner. It seems that bagging is unable to achieve substantial improvement in performance due to underlying SLP where dataset has also been generated using SLP. The average MSE for ranbag is substantially better than the average MSE for a single model for ranbag. However average MSE for ranbag is not better than the average MSE for single model for bagging. As discussed in the preceding paragraph FAM's bias in favour of uniform prior seems a cause of the occurrence of this phenomenon. Furthermore, underlying SLP can also account for this phenomenon. Ranbag's performance can be improved by selecting individual models on the basis of some statistical test. Figure 11 represents KL divergence to true parameters for bagging and ranbag. Average KL divergence for single model is 0.81% for bagging and 1.4% for ranbag. Bagging and ranbag substantially minimises KL divergence to true parameters. Average KL divergence for bagging is 0.72% and 0.79% for ranbag.

The results demonstrate the effectiveness of ensemble methods. Sometimes we obtain only a modest gain applying bagging in conjunction with FAM. This happens especially for MSE. The parameter learning process of FAM is influenced by the underlying SLP. It seems that FAM's ability to learn good starting model due to underlying SLP, and FAM's bias in favour of uniform prior account for the modest gain for particular dataset. In summary it is effective and efficient to apply bagging and ranbag to modelling metabolic pathways.

6 Conclusion

Bagging is a useful learning technique especially for unstable predictors. This paper presents a novel study of SLPs-based ensemble methods and addresses an important problem of modelling metabolic pathways. We have focused on the parameter estimation task over SLPs as reaction rates of enzymes in metabolic pathways can be computed by estimating the parameters over SLPs. We have shown how bagging can be adapted to perform maximum likelihood parameter estimation task. We have also shown that an effective ensemble can be constructed by driving randomness into an SLP learning algorithm. The empirical results demonstrate the efficacy of these methods and show that bagging and ranbag obtain substantial gain in performance. In terms of KL divergence and MSE these techniques show sizeable improvements in performance.

Deterministic models (such as ODE models) and stochastic models are popular for cellular modelling. We believe that application of bagging-based methods to ODE models and stochastic models will be effective and efficient. We are looking at the ways to develop base learners to learn ODE models and stochastic models and to apply them in conjunction with bagging-based methods.

SLPs provide an efficient representation for metabolic pathways where each clause of an SLP contains probabilistic information about enzyme-catalysed reactions. One of the important issues to be addressed in our future work is to learn really unknown parameters. In order to capture enzymatic reaction rates each clause of an SLP can be augmented by incorporating temporal information. One way to capture temporal dimension in stochastic logic framework is to add an expression specifying time as an argument in each clause of an SLP.

Acknowledgements

The authors would like to acknowledge the support of the DTI Beacon project "Metalog - Integrated Machine Learning of Metabolic Networks Applied to Predictive Toxicology", Grant Reference QCBB/C/012/00003. Thanks to Nicos Angelopoulos for technical help. Thanks to 2 anonymous reviewers for useful comments.

References

1. Dugglery, R.G., Clarke, R.B.: Experimental design for estimating the parameters of the Michaelis-menten equation from progress curves of enzyme-catalysed reactions. Biochim. Biophys. Acta **1080** (1991) 231–236
2. Breiman, L.: Bagging predictors. Machine Learning **24** (1996) 123–140
3. Schapire, R.E.: A brief introduction to boosting. In: Proceedings of the Sixteenth International Conference on Artificial Intelligence. (1999) 1401–1406
4. Dietterich, T.G.: An experimental comparison of three methods for constructing ensembles of decision trees: Bagging, boosting, and randomisation. Machine Learning **40** (2000) 139–157
5. Bauer, E., Kohavi, R.: An empirical comparison of voting classification algorithm: bagging, boosting and variants. Machine Learning **36** (1999) 105–142
6. Lodhi, H., Karakoulas, G., Shawe-Taylor, J.: Boosting strategy for classification. Intelligent Data Analysis **6** (2002) 149–174
7. Schapire, R.E., Freund, Y., Barlett, P., Lee, W.S.: Boosting the margin: A new explanation for the effectiveness of voting methods. The Annals of Statistics **5** (1998) 1651–1686
8. Lodhi, H., Karakoulas, G., Shawe-Taylor, J.: Boosting the margin distribution. In Leung, K.S., Chan, L.W., Meng, H., eds.: Proceedings of the Second International Conference on Intelligent Data Engineering and Automated Learning (IDEAL 2000), Springer Verlag (2000) 54–59
9. Domingos, P.: A unified bias-variance decomposition for zero-one and squared loss. In: Seventeenth National Conference on Artificial Intelligence, AAAI (2000) 564–569
10. Muggleton, S.H.: Stochastic logic programs. In de Raedt, L., ed.: Advances in Inductive Logic Programming. IOS Press (1996) 254–264
11. Rabiner, L.R.: A tutorial on hidden markov models and selected applications in speech recognition. Proceedings of the IEEE **77** (1989) 257–286
12. Lari, K., Young, S.J.: The estimation of stochastic context-free grammars using the inside-outside algorithm. *Computer Speech and Language* **4** (1990) 35–56
13. Halpern, J.Y.: An analysis of first-order logics of probability. *Artificial Intelligence* **46** (1990) 311–350
14. Quinlan, J.R.: Boosting first-order learning. In Arikawa, S., Sharma, A., eds.: Proceedings of the 7th International Workshop on Algorithmic Learning Theory. Volume 1160 of LNAI., Berlin, Springer (1996) 143–155
15. Dutra, I.C., Page, D., Shavilk, J.: An emperical evaluation of bagging in inductive logic programming. In: Proceedings of the International Conference on Inductive Logic Programming. (2002)
16. Venter, J.C., Adams, M.D., et al, E.W.M.: The sequence of human genome. Science **291** (2001) 1304–1351

17. Consortium, I.H.G.S.: Initial sequencing and analysis of the huma genome. Nature **409** (860–921)
18. Bryant, C.H., Muggleton, S.H., Oliver, S.G., Kell, D.B., Reiser, P., King, R.D.: Combining inductive logic programming, active learning and robotics to discover the function of genes. Electronic Transactions in Artificial Intelligence **6-B1** (2001) 1–36
19. Gepasi, M.P.: A software package for modelling the dynamics, steady states and control of biochemical and other systems. Comput. Appl. Biosci **9** (1993) 563–571
20. Goryanin, I., Hodgman, T.C., Selkov, E.: Mathematical simulation and analysis of cellular metabolism and regulation. Bioinformatics **15** (1999) 749–758
21. Gibson, M.A.: Computational methods for stochastic biological systems. PhD thesis, California Institute of Technology (2000)
22. Muggleton, S.H.: Learning from positive data. Machine Learning (2001)
23. Dempster, A.P., Laird, N.M., Rubin, D.B.: Maximum likelihood from incomplete data via the em algorithm. J. Royal statistical Society Series B **39** (1977) 1–38
24. Cussens, J.: Parameter estimation in stochastic logic programs. Machine Learning **44** (2001) 245–271
25. Efron, B., Tibshirani, R.: An introduction to bootstrap. Chapman and Hall (1993)
26. Angelopoulos, N., Muggleton, S.: Machine learning metabolic pathway descriptions using a probabilistic relational representation. Electronic Transactions in Artificial Intelligence **6** (2002)

Projective Brane Calculus

Vincent Danos[1,*] and Sylvain Pradalier[2]

[1] Université Paris 7 & CNRS
[2] ENS Cachan

Abstract. A refinement of Cardelli's *brane calculus* [1] is introduced where membrane actions are directed. This modification brings the language closer to biological membranes and also obtains a symmetric set of membrane interactions. An associated structural congruence, termed the *projective* equivalence, is defined and shown to be preserved under all possible system evolutions. Comparable notions of projective equivalence can be developed in other hierarchical process calculi and might be of interest in other applications.

1 Introduction

An estimated third of the proteins in eukaryotic cells are membrane proteins. Fusions and fissions of membranes happen often in cells, so often in fact that the outer plasmic membrane renews itself completely in half an hour! Both figures give a sense of how intense membrane-bound trafficking is, and of how important a component of cell computing it is.

Biological membrane interactions seem worth a formal investigation, in the larger perspective of laying out formal languages to describe, simulate, combine and otherwise analyze biological systems. Living organisms are complicated organizations and while engineering is clearly needed to understand them further, it seems a formal approach might also prove useful in defining suitable levels of abstractions, workable notions of properties and observations, and appropriate tools for constructing, investigating, and refining models.

Languages specifically dealing with protein-protein and protein-DNA interactions, are already well under way [2, 3, 4, 5, 6, 7, 8]. Comparatively few are addressing membrane interactions. The idea of *membrane computing* was introduced in the lively subject of natural computing by Paŭn [9], mostly in the tradition of automata and formal language theory. Another language of membranes and compartments, *bio-ambients*, following the tradition of process algebras, was proposed by Regev *et al.* [10], adressing description and simulation issues, and evolved later into the *brane calculus* proposed recently by Cardelli [1].

The brane calculus stands on its own as an elegant formalism, and does well in abstracting and describing the intricate biological processes involving membranes. Membranes are represented as nested multisets of actions, representing

* *Corresponding author*: Équipe PPS, Université Paris 7 Denis Diderot, Case 7014, 2 Place Jussieu 75251 PARIS Cedex 05, `Vincent.Danos@pps.jussieu.fr`

V. Danos and V. Schachter (Eds.): CMSB 2004, LNBI 3082, pp. 134–148, 2005.

the fusion and fission capabilities of the branes they sit *on*. Note that this in contrast with the original Ambient calculus, from which the brane calculus is derived, where capabilities sit *in* the ambients they control. These fission and fusion capabilities are meant to abstract actual proteins, or protein complexes, inserted in a membrane and defining the membrane potential interactions with various signals and other membranes. Membrane interactions are preserving the nesting parity, a principle which Cardelli refers to as *bitonality*, and which one observes in biological systems.

The purpose of the present paper is to incorporate another comparable principle, that of *projective invariance*. Specifically, we work out a revision of the brane calculus, as suggested by Cardelli [11], which consists in replacing actions with *directed actions*. Introducing directed actions takes the language a little step further down in the details, by actually telling whether an action, that is an interaction capability, is looking inwards or outwards. While still clearly an abstraction, the concept of directed actions meshes with what is known of membrane interactions at the molecular level.

In eukaryotic cells, new membranes patches are mainly produced from the second membrane surrounding the nucleus, called ER (Endoplasmic Reticulum). Some of them are dispatched to the outer membrane of the cell, called the plasmic membrane, with specific sugar decorations that control the correct interaction of the cell with its environment. These lipids bearing sugar signatures are first grown on the outer membrane of ER, and only then flipped by a protein known as a *flippase*, so that they face the inside of ER [12]. Once this is done, a vesicle breaks off from ER and bubbles up to the surrounding plasmic brane, with the glycolipid facing the inside of the brane. When finally the vesicle fuses with the outer membrane, its own membrane is flipped inside-out, and the glycolipid faces the outside, as it should to be functional.

A similar case, is the production of transmembrane proteins designed to be receptors on the outer membrane. These proteins are inserted in ER during their translation from RNA, with their specific receptor domain facing inward, so that when they are dispatched to their final destination, they ultimately face the outer solution.

Both examples, show that biological membrane interactions operate under the strong constraint that molecular implementations of actions are directed. It seems therefore natural to enrich the original brane-calculus in order to represent them, obtaining a more accurate description of biological membranes. One could fear that the resulting calculus could become complicated, but it turns out to be structurally simpler. Indeed, the constraint generated by the fact that actions are directed reflects into the calculus as a new invariance principle, which we call *projective invariance*.

The contribution of this paper is to define properly this modified brane calculus, called the *projective brane calculus*, and to give a rigorous definition and proof of projective invariance. A few key events in protein trafficking, such as fusions, sorting, tagging and routing are then described. All events are integrated in an example of a retrovirus infecting a cell, using the cell synthesis chain for its own replication, and then escaping.

Outline. The paper is organized as follows. First, an informal comparison between directed and undirected actions serves both as a motivation and a way to introduce the principle of projective invariance. Then the refined syntax is given, together with an algebraic notation of *bitonal trees* representing projective equivalence in a concrete way. With this alternative notation projective invariance is easily proved. Finally, an example is detailed and at the same time various practical ways to enhance the syntax are discussed.

Acknowledgements. This paper benefitted greatly from discussions with Luca Cardelli.

2 Why Directed Actions Are Better

Anticipating somewhat on our definitions of the next section, we discuss briefly why refining ordinary actions into directed ones. This discussion is also meant to be an introduction to the syntax.

2.1 An Example System

Consider a solution $\mathcal{S}$ with one big membrane containing two smaller ones, and suppose further the two smaller ones may fuse (a capability represented below by the actions $\mathfrak{f}_2$), and the first may also fuse with the outer one (a capability represented below by the actions $\mathfrak{f}_1$). Our notation for this situation will be:

$$\mathcal{S} := \langle \_; \mathfrak{f}_1 \rangle \llparenthesis \langle \mathfrak{f}_1, \mathfrak{f}_2; \_ \rangle \llparenthesis P \rrparenthesis, \langle \mathfrak{f}_2; \_ \rangle \llparenthesis Q \rrparenthesis \rrparenthesis \tag{1}$$

where P and Q represent the respective contents of the two inner membranes. Since we are dealing with directed actions, the notation is introducing a distinction between outward and inward actions. We use a place-value notation for this. For instance $\langle \mathfrak{f}_1, \mathfrak{f}_2; \_ \rangle \llparenthesis P \rrparenthesis$ represents a membrane containing P with the fusion actions $\mathfrak{f}_1$ and $\mathfrak{f}_2$ both pointing outwards, because they stand on the left of the semi-column. All actions are pointing outwards in $\mathcal{S}$, except for the $\mathfrak{f}_1$ borne by the outer membrane, which is pointing inwards.

Our two pairs of fusions $\mathfrak{f}_1$ and $\mathfrak{f}_2$ can cause either an horizontal fusion (mate) or a vertical one (exo) as long as they are "facing" each other. The system $\mathcal{S}$ can therefore evolve in two ways depending on whether the horizontal fusion or the vertical one happens first:

$$\begin{array}{l}
\langle _; \mathfrak{f}_1 \rangle \llparenthesis \langle \mathfrak{f}_1, \mathfrak{f}_2; _ \rangle \llparenthesis P \rrparenthesis, \langle \mathfrak{f}_2; _ \rangle \llparenthesis Q \rrparenthesis \rrparenthesis \xrightarrow{\mathfrak{f}_2} \langle _; \mathfrak{f}_1 \rangle \llparenthesis \langle \mathfrak{f}_1; _ \rangle \llparenthesis P, Q \rrparenthesis \rrparenthesis \xrightarrow{\mathfrak{f}_1} P, Q, \langle _; _ \rangle \llparenthesis \rrparenthesis \\
\langle _; \mathfrak{f}_1 \rangle \llparenthesis \langle \mathfrak{f}_1, \mathfrak{f}_2; _ \rangle \llparenthesis P \rrparenthesis, \langle \mathfrak{f}_2; _ \rangle \llparenthesis Q \rrparenthesis \rrparenthesis \xrightarrow{\mathfrak{f}_1} P, \langle _; \mathfrak{f}_2 \rangle \llparenthesis \langle \mathfrak{f}_2; _ \rangle \llparenthesis Q \rrparenthesis \rrparenthesis \xrightarrow{\mathfrak{f}_2} P, Q, \langle _; _ \rangle \llparenthesis \rrparenthesis
\end{array}$$

The first step of the second row is the interesting bit. The action $\mathfrak{f}_2$ that was outwards has been *reversed* and is now pointing inwards. This is the formal counterpart of the membrane reversion happening upon vertical membrane fusion (see the flippase example discussed in the introduction). The $\mathfrak{f}_1$ step reverses directions just as the real thing does. As a consequence both $\mathfrak{f}_2$ actions are still face to face, and, as said, still able to interact. No matter in which order the fusions are happening, they converge to the same end solution.

2.2 Same Example Undirected

The situation is different in the undirected case, obtained by forgetting the distinction between inward and outward pointing actions. The same example cast in the original undirected brane calculus reads now as follows:[1]

$$\mathcal{T} := \langle \mathfrak{f}_1 \rangle (\!| \langle \mathfrak{f}_1, \mathfrak{f}_2' \rangle (\!| P |\!), \langle \mathfrak{f}_2' \rangle (\!| Q |\!) |\!) \tag{2}$$

where $\mathfrak{f}_1$ might cause a vertical fusion, while $\mathfrak{f}_2'$ might cause an horizontal one. Note that since actions are now undirected, there is no longer a notion of two actions facing each other. Therefore, one has to let the action itself tell whether it is meant to trigger a vertical fusion or an horizontal one. Thus, the system evolves as follows:

$$\langle \mathfrak{f}_1 \rangle (\!| \langle \mathfrak{f}_1, \mathfrak{f}_2' \rangle (\!| P |\!), \langle \mathfrak{f}_2' \rangle (\!| Q |\!) |\!) \xrightarrow{\mathfrak{f}_2'} \langle \mathfrak{f}_1 \rangle (\!| \langle \mathfrak{f}_1 \rangle (\!| P, Q |\!) |\!) \xrightarrow{\mathfrak{f}_1} P, Q, \langle \text{-} \rangle (\!| |\!)$$
$$\langle \mathfrak{f}_1 \rangle (\!| \langle \mathfrak{f}_1, \mathfrak{f}_2' \rangle (\!| P |\!), \langle \mathfrak{f}_2' \rangle (\!| Q |\!) |\!) \xrightarrow{\mathfrak{f}_1} P, \langle \mathfrak{f}_2' \rangle (\!| \langle \mathfrak{f}_2' \rangle (\!| Q |\!) |\!) \not\xrightarrow{\mathfrak{f}_2'} P, Q, \langle \text{-} \rangle (\!| |\!)$$

To close the square in a way similar to the directed case above, one would have either to have the first interaction, as a side-effect, change both the action $\mathfrak{f}_2'$ in the outer part (easy) and the inner one in the membrane enclosing Q, which seems absurd, since this membrane doesn't take part in the interaction; or to suppose further that whatever can cause mate, can also cause exo, which, in the absence of directions, seems too lax a control to be expressive or even plausible.

Directed actions solve the square naturally. Now the important point is that not having this square is a violation of what we call *projective invariance*, which postulates that the nature of membrane interactions is such that the physics underlying the interaction can't tell the top from the bottom.

To see why, let us return at the example, and suppose the region (also called the lumen) containing the two membranes enclosing P and Q is projected at infinity. Under this change of viewpoint, what once was the outer space, becomes a bounded region and the corresponding term becomes ($\mathfrak{f}_1$ is inverted purposefully):

$$\mathcal{S}' := \langle \mathfrak{f}_1; \text{-} \rangle (\!| |\!), \langle \mathfrak{f}_1, \mathfrak{f}_2; \text{-} \rangle (\!| P |\!), \langle \mathfrak{f}_2; \text{-} \rangle (\!| Q |\!), \tag{3}$$

and from *that* new point of view, the two fusions are of the mate sort, and they clearly commute. Therefore, if one follows the principle, $\mathcal{S}$ and $\mathcal{S}'$ which only

[1] In the original calculus, "perps" breaking the symmetry between a fusion action and a fusion co-action are used. They are not shown here.

differ by a change of the point of view, or the top-level of the solution, should behave the same, which in this particular case means that $\mathcal{S}$ has to close the square explained above.

2.3 Summary

To summarize this discussion, we may say that the original calculus was designed in order to respect bitonality invariance, and that the present revision is designed to further incorporate projective invariance.

As a final note before to proceed to the definitions, we may observe that both invariance constraints can be derived from Cardelli's fundamental remark that the double layer structure of biological membranes essentially enforces a single local membrane operation of *switching* [13]. There could well be a way of presenting the projective brane calculus as an abstract rendition of a more detailed calculus, incorporating double layer membrane switching as a unique primitive operation, and satisfying both invariance principles.

3 Membrane Interactions

It remains to give a proper definition and prove a general form of the projective invariance investigated above.

3.1 Actions, Branes and Solutions

The syntax given in Table 1 defines *actions* and *solutions*. The set of actions will be denoted by $\mathcal{A}$. Basic actions can be of three sorts: fusions, wraps and bubbles, and actions can be combined freely by product and prefix to construct a membrane (abbreviated to brane). Solutions are products of solutions nested in membranes.

All comma-separated expressions, branes or solutions, are to be understood up to associativity and commutativity.

So far this language only concerns membranes and their interactions capabilities and doesn't include proteins and complexes, nor any other molecules. Following Cardelli, a somewhat richer language of actions will be used in the example at the end. In particular, we will introduce molecules and actions to bind and release those, it will also be convenient to use "banged" or inexhaustible actions written $!\sigma$, as well as a mechanism of recruitment to specify how actions are allocated to the sub-branes when a brane is divided. But this is for later, as none of what interests us for now is depending on these choices.

3.2 Membrane Interactions

Interactions are sorted in three groups: bubbles, fusions and wraps. Both bubbles and wraps are dividing a membrane. This is the reason why the associated actions have as arguments the actions dispatched to the membrane which they create. Fusions are fusing two membranes in one and don't need such an allocation. To ease reading action indices are not represented, nor are prefixes.

Table 1. Actions, Branes and Solutions

$\alpha :=$		**actions**
	$\mathfrak{f}_n$	fusion
	$\mathfrak{b}_n\langle\sigma;\sigma\rangle$	bubble
	$\mathfrak{w}_n\langle\sigma;\sigma\rangle$	wrap
	$\mathfrak{w}_n^\perp$	co-wrap
$\sigma :=$		**branes**
	$\_$	empty brane
	σ,σ	product
	$\alpha.\langle\sigma;\sigma\rangle$	prefix
$P :=$		**solutions**
	$\_$	empty solution
	$\langle\sigma;\sigma\rangle(\!|P|\!)$	membrane
	P,P	product

Bubbles: Drip & Pino

$$\langle\mathfrak{b}\langle\sigma';\tau'\rangle,\sigma;\tau\rangle(\!|P|\!) \xrightarrow{drip} \langle\sigma';\tau'\rangle(\!|\_|\!),\langle\sigma;\tau\rangle(\!|P|\!)$$
$$\langle\sigma;\tau,\mathfrak{b}\langle\sigma';\tau'\rangle\rangle(\!|P|\!) \xrightarrow{pino} \langle\sigma;\tau\rangle(\!|\langle\sigma';\tau'\rangle(\!|\_|\!),P|\!)$$

A bubble action divides the brane it is sitting on, creating a new brane, which is either outside or inside the original one, depending on whether the action points outwards (drip) or inwards (pino).

Fusions: Mate & Exo

$$\langle\mathfrak{f},\sigma;\tau\rangle(\!|P|\!),\langle\mathfrak{f},\sigma';\tau'\rangle(\!|Q|\!) \xrightarrow{mate} \langle\sigma,\sigma';\tau,\tau'\rangle(\!|P,Q|\!)$$
$$\langle\sigma;\tau,\mathfrak{f}\rangle(\!|\langle\mathfrak{f},\sigma';\tau'\rangle(\!|P|\!),Q|\!) \xrightarrow{exo} \langle\sigma,\tau';\tau,\sigma'\rangle(\!|P,Q|\!)$$

A pair of fusion actions facing each other induces a fusion which is either vertical (exo) or horizontal (mate) depending on where the two actions are sitting.

Wraps: Phago, Bud & Swap

$$\langle\mathfrak{w}^\perp,\sigma;\tau\rangle(\!|P|\!),\langle\mathfrak{w}\langle\sigma'';\tau''\rangle,\sigma';\tau'\rangle(\!|Q|\!) \xrightarrow{phago} \langle\sigma';\tau'\rangle(\!|\langle\sigma'';\tau''\rangle(\!|\langle\sigma;\tau\rangle(\!|P|\!)|\!),Q|\!)$$
$$\langle\sigma';\tau',\mathfrak{w}\langle\sigma'';\tau''\rangle\rangle(\!|\langle\mathfrak{w}^\perp,\sigma;\tau\rangle(\!|P|\!),Q|\!) \xrightarrow{bud} \langle\sigma'';\tau''\rangle(\!|\langle\sigma;\tau\rangle(\!|P|\!)|\!),\langle\sigma';\tau'\rangle(\!|Q|\!)$$
$$\langle\sigma;\mathfrak{w}^\perp,\tau\rangle(\!|\langle\mathfrak{w}\langle\sigma'';\tau''\rangle,\sigma';\tau'\rangle(\!|P|\!),Q|\!) \xrightarrow{swap} \langle\sigma;\tau\rangle(\!|\langle\tau'';\sigma''\rangle(\!|P,\langle\tau';\sigma'\rangle(\!|Q|\!)|\!)|\!)$$

A pair of a wrap and a co-wrap actions facing each other induces a division of the brane carrying the wrap action, which encloses the brane carrying the co-wrap action. The wrapped brane is either outside the wrapping brane (phago), or inside (bud). These two steps are similar to respectively pino and drip, except that the other brane (carrying the co-action) is wrapped in the process.

Table 2. Bubble and Fusions symmetries

$$\begin{array}{ccc}
\langle \mathfrak{b}\langle \sigma'; \tau' \rangle, \sigma; \tau \rangle \llparenthesis P \rrparenthesis, Q & \xrightarrow{drip} & \langle \sigma'; \tau' \rangle \llparenthesis _ \rrparenthesis, \langle \sigma; \tau \rangle \llparenthesis P \rrparenthesis, Q \\
\updownarrow \sim & & \updownarrow \sim \\
P, \langle \tau; \sigma, \mathfrak{b}\langle \sigma'; \tau' \rangle \rangle \llparenthesis Q \rrparenthesis & \xrightarrow{pino} & P, \langle \tau; \sigma \rangle \llparenthesis \langle \sigma'; \tau' \rangle \llparenthesis _ \rrparenthesis, Q \rrparenthesis
\end{array}$$

$$\begin{array}{ccc}
\langle \mathfrak{f}, \sigma; \tau \rangle \llparenthesis P \rrparenthesis, \langle \mathfrak{f}, \sigma'; \tau' \rangle \llparenthesis Q \rrparenthesis, R & \xrightarrow{mate} & \langle \sigma, \sigma'; \tau, \tau' \rangle \llparenthesis P, Q \rrparenthesis, R \\
\updownarrow \sim & & \updownarrow \sim \\
P, \langle \tau; \mathfrak{f}, \sigma \rangle \llparenthesis \langle \mathfrak{f}, \sigma'; \tau' \rangle \llparenthesis Q \rrparenthesis, R \rrparenthesis & \xrightarrow{exo} & P, Q, \langle \tau, \tau'; \sigma, \sigma' \rangle \llparenthesis R \rrparenthesis
\end{array}$$

However the swap interaction is different, and doesn't look like corresponding to any actual biological transformation. What it really is, is a phago step, but seen from the point of view of P, that is to say from the inside. It is there to close our set of interactions in the case when P is taken as the region at infinity.

Closing these basic interactions under any context results in a transition system over solutions, written $\rightarrow$ and called thereafter *reduction*.

3.3 Projective Equivalence

Projective equivalence, written $\sim$, is the least equivalence relation such that:

$$P, \langle \sigma; \tau \rangle \llparenthesis Q \rrparenthesis \sim \langle \tau; \sigma \rangle \llparenthesis P \rrparenthesis, Q \tag{4}$$

One has to be careful about two things here:

- σ and τ are exchanged above, because inward actions become outward actions and conversely;
- changing one's point of view is a global transformation and therefore not compatible with product:

$$\begin{aligned} R, P, \langle \sigma; \tau \rangle \llparenthesis Q \rrparenthesis &\sim \langle \tau; \sigma \rangle \llparenthesis R, P \rrparenthesis, Q \\ &\not\sim R, \langle \tau; \sigma \rangle \llparenthesis P \rrparenthesis, Q \end{aligned}$$

Proposition 1. *Reduction respects projective equivalence, that is to say, if $P \sim Q$ and $P \rightarrow P'$, then there exists a Q', such that $P' \sim Q'$ and $Q \rightarrow Q'$.*

This can be proved by induction on the number of steps of $\sim$ applied to go from P to Q. Key lemmas are given by Tables 2 and 3 and express the internal symmetry relating interactions of a same sort. However, a more perspicuous argument can be given if one changes notation. This new notation is developed in the next section. Of course it is more abstract and doesn't have the direct biological interpretation that the first syntax supports. It is nevertheless very convenient and in some sense, as we shall see, equivalent.

Table 3. Wrap symmetries

$$
\begin{array}{ccc}
P, \langle \tau; \mathfrak{w}^{\perp}, \sigma \rangle (\!| \langle \mathfrak{w} \langle \sigma''; \tau'' \rangle, \sigma'; \tau' \rangle (\!| Q |\!), R |\!) & \xrightarrow{swap} & P, \langle \tau; \sigma \rangle (\!| \langle \tau''; \sigma'' \rangle (\!| Q, \langle \tau'; \sigma' \rangle (\!| R |\!) |\!) |\!) \\
\updownarrow \sim & & \updownarrow \sim \\
\langle \mathfrak{w}^{\perp}, \sigma; \tau \rangle (\!| P |\!), \langle \mathfrak{w} \langle \sigma''; \tau'' \rangle, \sigma'; \tau' \rangle (\!| Q |\!), R & \xrightarrow{phago} & \langle \sigma'; \tau' \rangle (\!| \langle \sigma''; \tau'' \rangle (\!| \langle \sigma; \tau \rangle (\!| P |\!) |\!), Q |\!), R \\
\updownarrow \sim & & \updownarrow \sim \\
\langle \tau'; \mathfrak{w} \langle \sigma''; \tau'' \rangle, \sigma' \rangle (\!| \langle \mathfrak{w}^{\perp}, \sigma; \tau \rangle (\!| P |\!), R |\!), Q & \xrightarrow{bud} & \langle \sigma''; \tau'' \rangle (\!| \langle \sigma; \tau \rangle (\!| P |\!) |\!), Q, \langle \tau'; \sigma' \rangle (\!| R |\!)
\end{array}
$$

4 Bitonal Tree Representation

A graph is a pair (V, E), where V is a set of vertices (or nodes), and $E \subseteq V \times V$ is a set of edges. Such graphs are undirected, and have at most one edge between any two given vertices.

A *bitonal graph* consists of the following data:

- an undirected graph $G = (V, E)$;
- a labeling function λ mapping E to $\mathcal{A} \times \mathcal{A}$;
- a 2-colouring of G, that is a map κ from V to $\{0, 1\}$, such that $(x, y) \in E \Rightarrow \kappa(x) \neq \kappa(y)$.

Being 2-colourable G can only have even length cycles. When G is acyclic and connected one says G is a bitonal tree. A *pointed bitonal tree* $(x, V, E, \lambda, \kappa)$ is a bitonal tree together with a distinguished vertex $x \in V$.

There is an obvious map $g[\_]$ from solutions to pointed bitonal trees. Given a solution $\langle \sigma_1; \tau_1 \rangle (\!| P_1 |\!), \ldots, \langle \sigma_n; \tau_n \rangle (\!| P_n |\!)$ one represents it as:

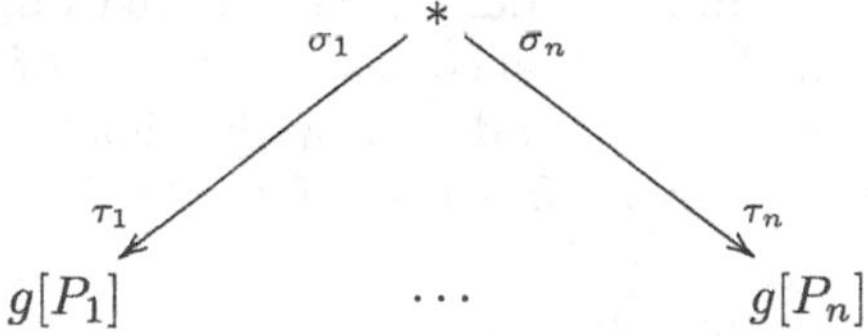

where the distinguished vertex is denoted $*$. Up to associativity and commutativity on the side of solutions, this map is a bijection. In the vocabulary of graph theory, $g[P]$ is the dual graph of P, its nodes are regions (or lumens) and its edges represent region adjacency. Changing one's "point of view" amounts simply in this dual representation to changing the distinguished vertex in the dual world of bitonal trees.

Proposition 2. *Let P and P' be two solutions, then $P \sim P'$ if and only if $g[P]$ and $g[P']$ differ only in their distinguished vertex.*

This is easily seen, and is what we wanted. Bitonal trees are giving a concrete representation of projective equivalence classes. Furthermore, membrane reductions can be defined directly on this representation.

4.1 Tree Interactions

Tree interactions are presented in Table 4. Even numbered vertices corresponding to one colour and odd numbered ones to the other. One sees that for each type of membrane interaction, there remains only one interaction in the world of trees. It is easy to verify that:

Table 4. Rewriting bitonal trees

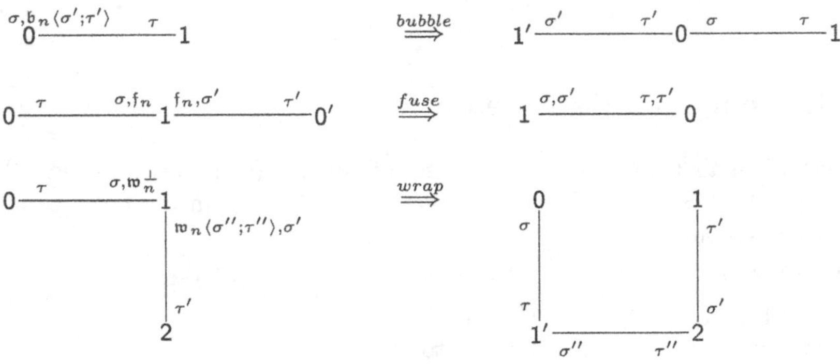

- these transformations preserve bitonality,
- that they are definable with general 2-colourable graphs,[2]
- that they preserve acyclicity and connectivity, and
- most importantly, that once one chooses a particular point of view and therefore an orientation of the tree, they correspond exactly to the membrane interactions of the preceding section.

To illustrate the last point, consider the case of the wrap rewriting. Depending on where the tree is rooted, wrap will correspond to one of phago, swap and bud. Specifically, if in the partial order induced by the choice of a root, $1 > 0$ and 2, then it corresponds to a phago, if $2 > 1$ and 0 then it is a bud, and if $0 > 1$ and 2, then it is a swap.

Proposition 1 follows now easily.

5 Enriching the Language

Next, we would like to test on a concrete and reasonably large example how well our modified brane language does in expressing biological interactions involving

[2] Fuse and wrap only apply when the two left hand side edges are distinct and then even nodes must be distinct as well, since multi arcs are not allowed.

membranes. Yet, for the example to be interesting we need to flesh in a bit our spartan syntax and adjust some notations. This is what we do in this section.

5.1 Recruitment

The first natural modification, already suggested by Cardelli, is to have membrane dividing actions (wraps and bubbles) explicitly recruit the actions allocated to the new membrane. This is particularly useful when there might be very few copies of some action of interest.

For instance a pino allocating actions σ to the inner bubble becomes:

$$\langle \sigma; \mathfrak{b}\langle \_; \sigma\rangle\rangle (\!| P |\!) \xrightarrow{pino} \langle \_; \_\rangle (\!| \langle \_; \sigma\rangle (\!| \_ |\!), P |\!)$$

The recruited actions σ are consumed and no longer part of the outer membrane they were taken from.

Adding a recruitment mechanism has the side effect to make even clearer which membranes are turned upside down and when. Pino above has to flip σ because the inner part of the new membrane corresponds to the outer part of the former membrane it was pinched off. This is not new and we knew it from the exo case. The same considerations applies to phago, but not to the "direct" interactions drip, mate and bud where the branes orientation are unchanged.

In the example we will use it specifically to describe the trip of the action corresponding to the *escape receptor* that the virus uses to exit the cell once it is properly replicated.

5.2 Protein-Membrane Interactions

As natural as our place-value syntactic convention for actions (having outward action and inward action separated by a semi-column) may be, it is not so readable in examples. So to ease reading, and starting from now, actions will have an explicit superscript indicating whether they are directed inward or outward. For instance an outward fusion action will be written $\mathfrak{f}^e$, while an inward fusion action will be written $\mathfrak{f}^i$.

The other modification that we need is a mechanism to control the insertion in membranes of various molecules such as proteins. Introducing the corresponding two actions, B and R, we keep Cardelli's elegant *bind-and-release* rules [1]:

$$\begin{array}{lcl}
m, \langle \mathsf{B}^e(m).\sigma, \tau\rangle (\!| P |\!) & \xrightarrow{Bind(out)} & \langle \sigma, \tau\rangle (\!| P |\!) \\
\langle \sigma; \mathsf{B}^i(m).\tau\rangle (\!| m, P |\!) & \xrightarrow{Bind(in)} & \langle \sigma, \sigma'\rangle (\!| P |\!) \\
\langle \mathsf{R}^e(m).\sigma, \sigma'\rangle (\!| P |\!) & \xrightarrow{Release(out)} & m, \langle \sigma, \sigma'\rangle (\!| P |\!) \\
\langle \mathsf{R}^i(m).\sigma, \sigma'\rangle (\!| P |\!) & \xrightarrow{Release(in)} & \langle \sigma, \sigma'\rangle (\!| m, P |\!)
\end{array}$$

We used the superscript notation to distinguish whether an action is directed inward or outward. In this simple model, inserted molecules are not considered to be directed.

5.3 Routing Molecules

The bind and release constructions can be combined to route molecules in the solution. A basic combination of actions is the following (in superscript notation):

$$\mathsf{B}^i(m).\mathfrak{b}^e\langle \mathfrak{f}^e_{dest}, \mathsf{R}^i(m)\rangle$$

the binding mechanism is looking inwards for an m, then pinches off a bubble with m inserted in the bubble, a bubble which is then sent to its destination, that is any membrane offering complementary fuse action.

A typical evolution would be then:

$$\begin{array}{ll}
\langle \mathsf{B}^i(m).\mathfrak{b}^e\langle \mathfrak{f}^e_{dest}, \mathsf{R}^i(m)\rangle\rangle(\!|m|\!), \langle \mathfrak{f}^e_{dest}\rangle(\!|_|\!) & \xrightarrow{Bind(in)} \\
\langle \mathfrak{b}^e\langle \mathfrak{f}^e_{dest}, \mathsf{R}^i(m)\rangle\rangle(\!|_|\!), \langle \mathfrak{f}^e_{dest}\rangle(\!|_|\!) & \xrightarrow{drip} \\
\langle_\rangle(\!|_|\!), \langle \mathfrak{f}^e_{dest}, \mathsf{R}^i(m)\rangle(\!|_|\!), \langle \mathfrak{f}^e_{dest}\rangle(\!|_|\!) & \xrightarrow{Release(in)} \\
\langle_\rangle(\!|_|\!), \langle \mathfrak{f}^e_{dest}\rangle(\!|m|\!), \langle \mathfrak{f}^e_{dest}\rangle(\!|_|\!) & \xrightarrow{mate} \\
\langle_\rangle(\!|_|\!), \langle\rangle(\!|m|\!) &
\end{array}$$

The way the macro is written above, m can be freed from the bubble membrane either before or after it has fused with the destination. Other combinations are possible by prefixing. For instance if one insists that m is freed in the bubble before the latter is fused, then one writes $\mathsf{B}^i(m).\mathfrak{b}^e\langle \mathsf{R}^i(m).\mathfrak{f}^e_{dest}\rangle$.

Bubble trafficking in the example will include actions with their sorts chosen so as to remind of their destination, *e.g.*, $\mathfrak{f}_G$ will denote fuse actions recognized by the Golgi apparatus.

6 A Virus Invasion Formalized

As said, our example will be the *modus operandi* of a virus. The system below is pictured roughly after the influenza virus, a retrovirus that carries genetic information as RNA and contains a retrotranscriptase, written DNAp below, converting it back to DNA so that all the virus components can be synthesized by the host cell.

One thing to keep track of, in relation to our choice of directed actions, is that the action corresponding to the escape receptor prefixes an action of the form $\mathsf{B}(\mathsf{C})$, meant to bind the final virus complex C, and help him escape. It therefore has to point ultimately inwards. Since the patch of membrane hosting it is fused with the outer membrane by an exo, that particular action has to point outwards by the time it is synthesized.

6.1 Description

The virus presents itself to the outer plasmic membrane of the cell and fools the cell in engulfing it as in the ordinary endocytotic pathway. Properly wrapped the virus is transported to a close organelle termed the early endosome, from which it escapes to deliver its cargo in the cytosol. The cargo, consists in genetic material that codes for the capsid (envelope), and various other proteins needed to reconstruct the virus. Among these proteins there is a receptor that is going to allow the virus, when reconstructed in the cell (possibly in many copies) to escape to the outer space.

We first define all the organelles involved in this process together with the virus itself:

$$\begin{array}{ll}
\mathsf{Virus} & := \langle \mathfrak{w}^e, \mathfrak{f}^e_{exo} \rangle (\!| \mathsf{C} |\!) \\
\mathsf{Cell} & := \langle !\mathfrak{w}^e(\mathfrak{f}^e_{endo}), !\mathfrak{f}^e_{endo}, !\mathfrak{f}^i_{out} \rangle (\!| \mathsf{Nucleus}, \mathsf{Endosome}, \mathsf{ER}, \mathsf{Golgi} |\!) \\
\mathsf{Endosome} & := \left\langle \mathfrak{f}^e_{endo}, \mathfrak{f}^i_{exo} \right\rangle (\!| _ |\!) \\
\mathsf{Nucleus} & := \langle !\mathsf{B}^e(\mathsf{RNA}_{vir}).\mathsf{R}^i(\mathsf{RNA}_{vir}), !\mathsf{B}^e(\mathsf{DNAp}).\mathsf{R}^i(\mathsf{DNAp}), \\
& \quad\; !\mathsf{B}^i(\mathsf{RNA}_{vir}).\mathsf{R}^e(\mathsf{RNA}_{vir}) \rangle (\!| \mathsf{RNAp} |\!) \\
\mathsf{ER} & := \langle !\mathsf{B}^e(\mathsf{RNA}_{vir}).(\mathsf{R}^e(\mathsf{RNA}_{vir}), \mathsf{R}^e(\mathsf{capsid}), \\
& \quad\; \mathsf{R}^e(\mathsf{DNAp}), send_receptor) \rangle (\!| _ |\!) \\
\mathsf{Golgi} & := \langle !\mathfrak{f}^e_G, !\mathfrak{b}^e \langle \mathfrak{f}^e_{out}, action_receptor \rangle \rangle (\!| _ |\!)
\end{array}$$

where we used the '!' to represent inexhaustible actions, and the actions:

$$\begin{array}{ll}
send_receptor & := \mathfrak{b}^e \langle \mathfrak{f}^e_G, action_receptor \rangle \\
action_receptor & := \mathsf{B}^e(\mathsf{C}).\mathfrak{b}^i \left\langle \mathfrak{w}^e, \mathfrak{f}^e_{exo}, \mathsf{R}^i(\mathsf{C}) \right\rangle
\end{array}$$

The actions sitting on the cell brane (the plasmic membrane) represent the cell endocytotic and exocytotic pathways, while the actions on top of the nucleus brane represent the workings of the nuclear pore complexes forming the only passageways between the nuclear lumen, where transcription take place, and the cytosolic one, where translation takes place. Translation is here modeled indirectly at the level of ER.

One also has the respective reactions of decomplexation, retrotranscription and forward transcription:

$$\begin{array}{ll}
\mathsf{C} & \leftrightarrow \mathsf{capsid} + \mathsf{RNA}_{vir} + \mathsf{DNAp} \\
\mathsf{RNA}_{vir} + \mathsf{DNAp} & \rightarrow \mathsf{DNA}_{vir} + \mathsf{DNAp} \\
\mathsf{DNA}_{vir} + \mathsf{RNAp} & \rightarrow \mathsf{DNA}_{vir} + \mathsf{RNAp} + \mathsf{RNA}_{vir}
\end{array}$$

We assume the first reaction to be reversible and suppose also that the last two reactions are happening in the nucleus.

6.2 Running the System

With the system definition in place, we can play our model. To keep a manageable notation we haven't systematically represented actions recruited and consumed during bubble formation, and only relevant actions, membranes and molecules are represented at each step.

$$\begin{array}{ll}
\mathsf{Virus},\ \mathsf{Cell} \rightarrow_{phago} & \langle !\mathfrak{w}^e(\mathfrak{f}^e_{endo}), !\mathfrak{f}^i_{out} \rangle (\!| \langle \mathfrak{f}^e_{endo} \rangle (\!| \langle \mathfrak{f}^e_{exo} \rangle (\!| \mathsf{C} |\!) |\!), \mathsf{Endosome}, \ldots |\!) \\
\rightarrow_{mate} & \langle \ldots \rangle (\!| \langle \mathfrak{f}^i_{exo} \rangle (\!| \langle \mathfrak{f}^e_{exo} \rangle (\!| \mathsf{C} |\!) |\!), \ldots |\!) \\
\rightarrow_{exo} & \langle \ldots \rangle (\!| \mathsf{C}, \mathsf{Nucleus}, \ldots |\!)
\end{array}$$

The virus just broke in and delivered the cargo in the cytosol. We don't write the outer membrane anymore. The C complex dissociates and after transfer to the nucleus, backward and forward transcription take place.

$$\begin{array}{ll}
\rightarrow_{decomplexation} & \mathsf{capsid}, \mathsf{RNA}_{vir}, \mathsf{DNAp}, \mathsf{Nucleus}, \ldots \\
\rightarrow_{Bind^2, Release^2} & \langle \ldots \rangle (\!| \mathsf{RNA}_{vir}, \mathsf{DNAp}, \mathsf{RNAp} |\!), \ldots \\
\rightarrow_{retrotranscription} & \langle \ldots \rangle (\!| \mathsf{DNA}_{vir}, \mathsf{RNAp} |\!), \ldots \\
\rightarrow_{transcription} & \langle \ldots \rangle (\!| \mathsf{RNA}_{vir}, \mathsf{DNA}_{vir}, \mathsf{RNAp} |\!), \ldots \\
\rightarrow_{Bind, Release} & \mathsf{Nucleus}, \mathsf{RNA}_{vir}, \mathsf{ER}, \ldots
\end{array}$$

The viral RNA has been duplicated and sent out again from the nucleus. It then binds to ER where it is translated. Products of this translation, capsid, DNAp, and RNA_{vir} are released in the cytosol, while the receptor, modeled here as an action, remains inserted in ER:

$$\begin{array}{l}
\rightarrow_{Drip} \mathsf{ER}, \langle \mathfrak{f}^e_G, action_receptor \rangle (\!| _ |\!), \mathsf{Golgi}, \ldots \\
\rightarrow_{Mate} \langle action_receptor, !\mathfrak{b}^e \langle \mathfrak{f}^e_{out}, action_receptor \rangle, \ldots \rangle (\!| _ |\!), \ldots
\end{array}$$

The drip action needs to recruit a copy of the receptor. Now that one is around, it can be triggered. We write again the plasmic outer membrane:

$$\begin{array}{l}
\rightarrow_{Drip} \langle \mathfrak{f}^i_{out}, \ldots \rangle (\!| \mathsf{Golgi}, \langle \mathfrak{f}^e_{out}, action_receptor \rangle (\!| _ |\!), \ldots |\!) \\
\rightarrow_{Exo} \langle \overline{action_receptor}, \ldots \rangle (\!| \mathsf{Golgi}, \ldots |\!)
\end{array}$$

with the following notation for the flipped action:

$$\overline{action\_receptor} := \mathsf{B}^i(\mathsf{C}).\mathfrak{b}^e \langle \mathfrak{w}^e, \mathfrak{f}^e_{exo}, \mathsf{R}^i(\mathsf{C}) \rangle$$

Observe that the suffix is flipped too in the exo. So now the receptor is bound to the outer membrane, and facing inwards as it should, waiting for the viral complex to escape.

Meanwhile the other components, capsid, DNAp and RNA_{vir}, have assembled again in the cytosol, and may now exit the cell with the help of the receptor:

$$\begin{array}{ll}
\rightarrow_{complexation} & \langle \mathsf{B}^i(\mathsf{C}).\mathfrak{b}^e \langle \mathfrak{w}^e, \mathfrak{f}^e_{exo}, \mathsf{R}^i(\mathsf{C}) \rangle, \ldots \rangle (\!| \mathsf{C}, \ldots |\!) \\
\rightarrow_{Bind} & \langle \mathfrak{b}^e \langle \mathfrak{w}^e, \mathfrak{f}^e_{exo}, \mathsf{R}^i(\mathsf{C}) \rangle, \ldots \rangle (\!| \ldots |\!) \\
\rightarrow_{Drip} & \mathsf{Virus}, \mathsf{Cell}
\end{array}$$

and the overall system is back to its original state, except that the cell contains now many copies of the virus components and is actively manufacturing more.

7 Conclusion

Cardelli's original brane calculus was designed to express biological interactions involving membranes. By design, its syntax reflected the bitonality invariant. We have presented here a refined brane calculus introducing directed actions, where the original calculus only used undirected ones. Thus our variant also incorporates a new notion of projective invariance. Together with the standard syntax, we gave an alternative syntax of bitonal trees, which reflects directly the new invariant, and where the number of primitive interactions downs to three: fusion, wrap and bubble. At the same time this revision of the original calculus is more convincing biologically in that it gives richer means to express how proteins are inserted in the membrane. The viral example developed and specifically the way the virus "escape" receptor is handled makes a good case for this refined language of actions.

While not completely understood, some observable membrane interactions within a cell are beginning to be explained at the protein level by today's molecular biology. Internal routing mechanisms, involving proteins such as t-snares and v-snares, are believed to be a pretty good account of how speficic routing is handled in the cell. Membrane fission, as in endocytosis, is also described in some cases at the level of proteins, for instance in clathrin-coated vesicle formation [14–pp.514–516]. It therefore seems tempting to pursue the effort and explore whether one can actually reduce membrane actions, which here were taken to be abstract atomic properties of a given membrane, to the properties of proteins (and glycolipids) actually inserted in the membrane. In other words one would like to merge a protein calculus such as the κ-calculus [15], and a brane calculus, in a bigger and more comprehensive language to be able to handle additional detail when additional detail is known. This is one interesting subject for further investigation.

Another question which might be worth exploring is that of endowing the brane-calculus with a quantitative operational semantics. If one thinks again about the virus infection scenario, a non-deterministic semantics is clearly not enough to describe what is happening. Whether an infection will succceed or not hinges on the kinetics of the various processes involved, and if one is willing to analyse this, it is important to incorporate quantitative aspects in the semantics. Using Gillespie's algorithm [16, 17], a non-deterministic transition system can be made to generate a continuous-time stochastic process over the system state space. This was done already in the calculus of bio-ambients [10], and it should be easy to adapt to the present framework. However, the challenge is not only to have a quantitative semantics, but also to reproduce some known regulation phenomena, and some more work has to be done here to get some interesting examples.

Finally, as said at the end of the second section, there is also the question of whether there is a more detailed calculus based on the single local operation of membrane switching described by Cardelli [13], from which one could reconstruct the projective brane calculus.

References

1. Luca Cardelli. Brane calculi. In *Proceedings of BIO-CONCUR'03, Marseille, France*, volume ? of *Electronic Notes in Theoretical Computer Science*. Elsevier, 2003. To appear.

2. Aviv Regev, William Silverman, and Ehud Shapiro. Representation and simulation of biochemical processes using the π-calculus process algebra. In R. B. Altman, A. K. Dunker, L. Hunter, and T. E. Klein, editors, *Pacific Symposium on Biocomputing*, volume 6, pages 459–470, Singapore, 2001. World Scientific Press.
3. Corrado Priami, Aviv Regev, Ehud Shapiro, and William Silverman. Application of a stochastic name-passing calculus to representation and simulation of molecular processes. *Information Processing Letters*, 2001.
4. Aviv Regev and Ehud Shapiro. Cells as computation. *Nature*, 419, September 2002.
5. Vincent Danos and Cosimo Laneve. Core formal molecular biology. In *Proceedings of the 12th European Symposium on Programming (ESOP'03, Warsaw, Poland)*, volume 2618 of *LNCS*, pages 302–318. Springer, April 2003.
6. Vincent Danos and Cosimo Laneve. Graphs for formal molecular biology. In *Proceedings of the First International Workshop on Computational Methods in Systems Biology (CMSB'03, Rovereto, Italy)*, volume 2602 of *LNCS*, pages 34–46. Springer, February 2003.
7. Vincent Danos and Jean Krivine. Formal molecular biology done in CCS. In *Proceedings of BIO-CONCUR'03, Marseille, France*, volume ? of *Electronic Notes in Theoretical Computer Science*. Elsevier, 2003. To appear.
8. Marc Chiaverini and Vincent Danos. A core modeling language for the working molecular biologist. In *Proceedings of CMSB'03*, volume 2602 of *LNCS*, page 166. Springer, 2003.
9. Gh. Paun. *Membrane Computing. An Introduction.* Springer-Verlag, Berlin, 2002.
10. Aviv Regev, Ekaterina M. Panina, William Silverman, Luca Cardelli, and Ehud Shapiro. Bioambients: An abstraction for biological compartments. *Theoretical Computer Science*, 2003. To Appear.
11. Luca Cardelli. Brane calculi (slides). Slides., 2003.
12. John W. Kimball. http://users.rcn.com/jkimball.ma.ultranet/BiologyPages/Kimball's Biology Pages. Online biology textbook, 2003.
13. Luca Cardelli. Bitonal membrane systems. Draft, 2003.
14. Bruce Alberts et al. *Essential Cell Biology.* International Series on Computer Science. Garland Science, New-York, 2004.
15. Vincent Danos and Cosimo Laneve. Formal molecular biology. *Theoretical Computer Science*, 325(1):69–110, September 2004.
16. Daniel T. Gillespie. A general method for numerically simulating the stochastic time evolution of coupled chemical reactions. *J. Comp. Phys.*, 22:403–434, 1976.
17. Daniel T. Gillespie. Exact stochastic simulation of coupled chemical reactions. *J. Phys. Chem*, 81:2340–2361, 1977.

Residual Bootstrapping and Median Filtering for Robust Estimation of Gene Networks from Microarray Data

Seiya Imoto[1,*], Tomoyuki Higuchi[2,*], SunYong Kim[1], Euna Jeong[1], and Satoru Miyano[1]

[1] Human Genome Center, Institute of Medical Science, University of Tokyo, 4-6-1 Shirokanedai, Minato-ku, Tokyo, 108-8639, Japan
{imoto, sunk, eajeong, miyano}@ims.u-tokyo.ac.jp
[2] Institute of Statistical Mathematics, 4-6-7, Minami-Azabu, Minato-ku, Tokyo, 106-8569, Japan
higuchi@ism.ac.jp

Abstract. We propose a robust estimation method of gene networks based on microarray gene expression data. It is well-known that microarray data contain a large amount of noise and some outliers that interrupt the estimation of accurate gene networks. In addition, some relationships between genes are nonlinear, and linear models thus are not enough for capturing such a complex structure. In this paper, we utilize the moving boxcel median filter and the residual bootstrap for constructing a Bayesian network in order to attain robust estimation of gene networks. We conduct Monte Carlo simulations to examine the properties of the proposed method. We also analyze *Saccharomyces cerevisiae* cell cycle data as a real data example.

1 Introduction

In recent years, estimation of gene networks based on microarray gene expression data has received considerable attention and the use of various computational methods, such as Boolean networks [1], differential equations [2, 3] and Bayesian networks [5, 6, 9, 10, 16], have been proposed in bioinformatics. In the estimation of gene networks based on microarray data, we need to consider two issues: One is how to capture the nonlinear relationships between genes. Due to the nonlinearity, the methods based on linear transformation of the data cannot be guaranteed to give sufficient results. The other problem arises from outliers included in microarray data. The outliers sometimes inhibit correct relationships or lead to spurious correlations between genes.

To estimate gene networks from microarray data, Bayesian networks [12] provide a probabilistic framework that is suitable for extracting effective information from high-dimensional noisy data. Unlike Bayesian network models based

*These authors contributed equally to this work.

V. Danos and V. Schachter (Eds.): CMSB 2004, LNBI 3082, pp. 149–160, 2005.

on linear regression, the discrete Bayesian networks [5, 6, 16] can capture nonlinear relationships between genes. However, since microarray data take continuous variables, the discretization possibly leads to information loss. Furthermore, the threshold values and the number of categories for discretization are parameters that should be optimized. To avoid the discretization and capture the nonlinearity, Imoto *et al.* [9, 10] proposed a Bayesian network and nonparametric regression model for estimating gene networks. However, a problem that still remains to be solved is how we treat the effect of outliers. Since nonparametric regression model employed in a Bayesian network is based on the Gaussian distribution, the outliers in microarray data sometimes affect the resulting networks. Therefore, development of statistical methods that can handle outliers and nonlinearity appropriately is considered as an important problem.

In this paper, we propose the use of moving boxcel median filter and residual bootstrap [4] for constructing Bayesian networks aimed at robust estimation of gene networks. By using the moving boxcel median filter, we can reduce the effect of outliers and can estimate nonlinear relationships between genes suitably. Residual bootstrap virtually realizes a model for measurement noise included in microarray data and gives a stable estimation of gene networks. In Section 2.1 we give the explanation of Bayesian networks. Since microarray data contain measurement noise, we introduce a "virtual sample method" to realize a measurement noise model in Section 2.2. The moving boxcel median filter and the residual bootstrap are introduced in Section 2.3 and 2.4, respectively. A greedy hill-climbing algorithm for choosing the optimal graph from candidates is introduced in Section 2.5. We conduct Monte Carlo simulations to show the effectiveness of the proposed method in Section 3.1. In Section 3.2, we analyze *Saccharomyces cerevisiae* cell cycle data collected by Spellman *et al.* [19] as a real data example.

2 Proposed Method

2.1 Bayesian Networks

In the context of Bayesian networks, we consider the directed acyclic graph (DAG) encoding the Markov assumption between nodes, i.e. a graph contains no cyclic regulations and a node depends only on its direct parents. In the Bayesian network models, a gene is regarded as a random variable and shown as a node. Under above assumptions, we can decompose the joint probability of all genes into the product of the conditional probabilities as

$$P(X_1, ..., X_p) = P(X_1|parent(X_1)) \times \cdots \times P(X_p|parent(X_p)), \quad (1)$$

where X_j $(j = 1, ..., p)$ is a random variable and corresponds to the jth gene, denoted by gene$_j$, and $parent(X_j)$ is a random variable vector of the direct parents of gene$_j$. For example, if gene$_2$ and gene$_3$ are the direct parents of gene$_1$ in the DAG, G, we then have $parent(X_1) = (X_2, X_3)^T$. Therefore, an essential problem for constructing a Bayesian network is the computation of each conditional probability $P(X_j|parent(X_j))$.

The computation of $P(X_j|parent(X_j))$ is essentially the same as the regression problem. In general, a regression model can capture the relationship between X_j and $parent(X_j)$ as

$$X_j = h_j(parent(X_j)) + \varepsilon_j, \tag{2}$$

where ε_j is noise satisfying $E[\varepsilon_j] = 0$ and $V(\varepsilon_j) = \sigma_j^2$, and $h_j(parent(X_j))$ is a function that describes the structure between X_j and $parent(X_j)$. Imoto *et al.* [9, 10] gave $h_j(parent(X_j))$ as the additive form

$$h_j(parent(X_j)) = h_{j1}(parent(X_j)_1) + \cdots + h_{jq_j}(parent(X_j)_{q_j}), \tag{3}$$

where $parent(X_j)_k$ $(k = 1, ..., q_j)$ is the kth parent of gene$_j$, and $h_{jk}(x)$ is a smooth function from R to R. For the noise, Imoto *et al.* [10] assumed the heterogeneous error variances for reducing the effect of outliers in microarray data. In the next section, we introduce the moving boxcel median filter to achieve more robustness in the estimation of the relationships between genes against outliers.

2.2 Virtual Samples

Suppose that we have p genes' expression data observed by n microarrays. That is, $\boldsymbol{x}_i = (x_{i1}, ..., x_{ij}, ..., x_{ip})^T$ is a p-dimensional gene expression vector from ith microarray and $\boldsymbol{X} = (\boldsymbol{x}_1, ..., \boldsymbol{x}_n)^T$ is an $n \times p$ gene expression matrix whose (i, j)th element, x_{ij}, is an expression value of gene$_j$ of ith microarray. Since microarray data contain various noise including measurement noise, we can decompose

$$\boldsymbol{x}_i = \boldsymbol{x}_i^{internal} + \boldsymbol{\eta}_i, \tag{4}$$

where $\boldsymbol{\eta}_i$ is a p-dimensional noise vector. By using $\boldsymbol{x}_i^{internal}$ and density functions instead of probability measure in (1), the purpose is to express a Bayesian network model as

$$f(\boldsymbol{x}_i^{internal}) = \prod_{j=1}^{p} f_j(x_{ij}^{internal}|\boldsymbol{p}_{ij}^{internal}),$$

where $\boldsymbol{p}_{ij}$ is a parent gene vector of gene$_j$ of ith microarray, i.e. if $parent(X_1) = (X_2, X_3)^T$, we have $\boldsymbol{p}_{i1} = (x_{i2}, x_{i3})^T$, and $\boldsymbol{p}_{ij}^{internal}$ is defined by the same as (4). In previous works, Bayesian network models are expressed as $f(\boldsymbol{x}_i) = \prod_{j=1}^p f_j(x_{ij}|\boldsymbol{p}_{ij})$. Therefore, one of the differences between the proposed method and previous works [9, 10] is in (4). However, gene expressions are observed by a few (typically one) microarrays for an experimental condition. Therefore, it is difficult to separate the true signal, $\boldsymbol{x}_i^{internal}$, from the noise like (4). As an alternative approach to realize the model (4), we make virtual observations for each gene as follows: We generate M virtual samples, $\{x_{ij}^{*1}, ..., x_{ij}^{*M}\}$, for each observation, x_{ij}, from the following system

$$x_{ij}^{*m} = x_{ij} + \varepsilon_{ij}^{*m}, \quad m = 1, ..., M, \tag{5}$$

where ε_{ij}^{*m} depends independently and normally on mean 0 and variance σ_{jo}^2. For the setting of σ_{jo}^2, we set $\sigma_{jo}^2 = \alpha \sum_i (x_{ij} - \sum_{i'} x_{i'j}/n)^2/n$ with $\alpha = 0.2$ empirically. However, it is often the case that the setting σ_{jo}^2 is inappropriate. We then update x_{ij}^{*m}'s by using the residual bootstrap described in Section 2.4. By using the system (5), we can model the measurement noise included in x_{ij} virtually. In standard regression methods, the measurement noise is usually ignored. On the other hand, the proposed method allows the measurement noise and estimate the relationship between variables suitably.

2.3 Moving Boxcel Median Filter

For constructing $h_j(\boldsymbol{x})$ in (2), we apply the moving boxcel median filter to $((\boldsymbol{p}_{ij}^{*m})^T, x_{ij})$ for $i = 1, ..., n$; $m = 0, ..., M$, where $\boldsymbol{p}_{ij}^{*0} = \boldsymbol{p}_{ij}$. To explain the moving boxcel median filter, we consider a simple example that gene$_1$ has one parent gene, gene$_2$. The moving boxcel median filter estimate ($X_1 = \hat{h}_1(X_2)$) is obtained as follows: First, we compute

$$\Delta x_j = \frac{\max\limits_{i=1,...,n}(x_{ij}) - \min\limits_{i=1,...,n}(x_{ij})}{2\lambda},$$

for $j = 2$. Here λ is a constant and we set 20 as an appropriate value in the later section. The estimated value at a, i.e. $\hat{h}_1(a)$, is the median of the observations of x_{i1} whose parent observations x_{i2}^* are included in the interval $I_{\Delta x_2}(a) = [a - \Delta x_2, a + \Delta x_2]$, where $a \in Val(X_2) = [\min_i(x_{i2}), \max_i(x_{i2})]$. The moving boxcel median filter curve is thus obtained by moving the interval $[a - \Delta x_2, a + \Delta x_2]$ from $a = \min_i(x_{i2})$ to $a = \max_i(x_{i2})$. Figure 1 shows an example of the moving boxcel median filter estimate. Figure 1 (a) is the scatterplot of the expression data of gene$_2$ and gene$_1$, which are generated numerically. Figure 1 (b) is the scatterplot of $\{(x_{i2}^{*m}, x_{i1}^*)|i = 1, ..., n; m = 0, ..., M\}$ with $x_{i2}^{*0} = x_{i2}$. We set $\alpha = 0.2$, $\lambda = 20$ and $M = 10$, and generate virtual samples. In Figure 1 (b), the moving boxcel median filter estimate at $X_2 = a$ is given as the median of the data in the shadowed area. In Figure 1 (c), the true relationship between gene$_1$ and gene$_2$ is shown as the dotted curve and the moving boxcel median filter estimate is the solid curve. It is clear that the moving boxcel median filter can reduce the effect of the outliers (in right-bottom) and construct a suitable relationship, which is close to the true one.

The moving boxcel median filter can be easily extended for more than two parent genes cases. If gene$_1$ has three parents, gene$_2$, gene$_3$ and gene$_4$, the interval defined above is extended as

$$I_{\Delta x_2}(a) \otimes I_{\Delta x_3}(b) \otimes I_{\Delta x_4}(c)$$

with $a \in Val(X_2)$, $b \in Val(X_3)$ and $c \in Val(X_4)$. Note that, by using the moving boxcel median filter, we do not need to assume the additive form (3) as a regressor and can model the relationship between genes by using more general form given in (2).

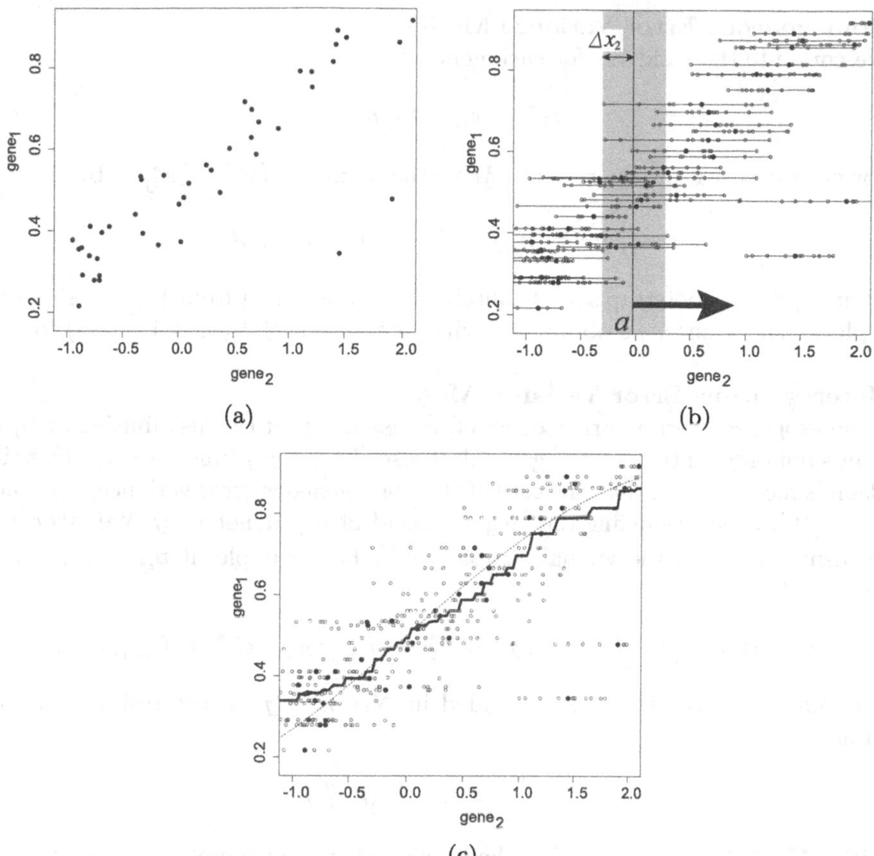

Fig. 1. Example of the moving boxcel median filter. (a) Scatterplot of $\{(x_{i2}, x_{i1})|i = 1, ..., n\}$. (b) Scatterplot of $\{(x_{i2}^{*m}, x_{i1}^{*m})|i = 1, ..., n; m = 0, ..., M\}$. (c) Solid curve: moving boxcel median filter estimate, Dotted curve: true curve

2.4 Residual Bootstrap

Since microarray data contain various noise including measurement noise, we model this fact by the model defined by (4) and (5). However, in the system (5), we determined the volume of the error variance σ_{j0}^2 empirically. Therefore, it is often the case that σ_{j0}^2 is not appropriate. Although the value of σ_{j0}^2 can be set as a somewhat appropriate value by using some information about the true relationships or resulting networks, it is clear that this approach has a limitation in practice. To solve this problem, we use the residual bootstrap method and recreate M virtual samples for each observation.

Suppose that parent genes of each gene are temporarily obtained and the moving boxcel median filter estimate for each relationship is computed, the procedure of the residual bootstrap can be expressed as follows:

Homogeneous Error Variance Model
We compute the residuals for each gene as

$$\varepsilon_{ij}^{*} = x_{ij} - \hat{h}_j(\boldsymbol{p}_{ij}).$$

For each observation, we recreate M virtual samples $\{x_{ij}^{*1}, ..., x_{ij}^{*M}\}$ by

$$x_{ij}^{*m} = \hat{h}_j(\boldsymbol{p}_{ij}) + \varepsilon_{ij}^{*m}, \quad m = 1, ..., M,$$

where ε_{ij}^{*m} is a bootstrap sample obtained by resampling from $\{\varepsilon_{1j}^{*}, ..., \varepsilon_{nj}^{*}\}$ with replacement. Hence, we obtain new virtual samples x_{ij}^{*m} $(m = 1, ..., M)$ for x_{ij}.

Heterogeneous Error Variance Model
In heterogeneous error variance model, we assume that the distribution of ε_{ij} depends not only on the index i (gene), but also the index j (microarray). This situation is more natural than the case of the homogeneous error variance, e.g. Imoto *et al.* [10]. First, we define the neighborhood of $\boldsymbol{p}_{ij}$, denoted by $Neighbor(\boldsymbol{p}_{ij})$, by using Δx_j and the virtual samples $\boldsymbol{p}_{ij}^{*m}$. For example, if $\boldsymbol{p}_{ij} = (x_{i2}, x_{i3})^T$, then we have

$$Neighbor(\boldsymbol{p}_{ij}) = \{(x_{i'2}^{*m'}, x_{i'3}^{*m'}) | x_{i'2}^{*m'} \in I_{\Delta x_2}(x_{i2}), x_{i'3}^{*m'} \in I_{\Delta x_3}(x_{i3})\}.$$

By using the virtual samples included in $Neighbor(\boldsymbol{p}_{ij})$, the residuals are obtained as

$$\varepsilon_{i'j}^{*m'} = x_{i'j} - \hat{h}_j(\boldsymbol{p}_{i'j}^{*m'})$$

with $\boldsymbol{p}_{i'j}^{*m'} \in Neighbor(\boldsymbol{p}_{ij})$. We then make M virtual samples for x_{ij} as

$$x_{ij}^{*m} = \hat{h}_j(\boldsymbol{p}_{ij}) + \varepsilon_{ij}^{*m}, \quad m = 1, ..., M,$$

where ε_{ij}^{*m} is a bootstrap sample from $\{\varepsilon_{i'j}^{*m'}\}_{i',m'}$.

After updating the virtual samples, we will fit the moving boxcel median filter described in the previous section to the microarray data and the updated virtual samples. We repeat this iteration until the stable estimate is obtained. Note that the moving boxcel median filter and the residual bootstrap can be applied when we set the parents of each gene. In the next section, we describe the selection of the graph structure and show an algorithm for estimating gene networks by using the moving boxcel median filter and the residual bootstrap method.

2.5 Graph Selection

In the estimation of gene networks from gene expression data, an essential problem is the choice of the optimal graph structure that gives the best approximation of the system underlying the data. From a statistical view point, this problem can be considered as a statistical model selection problem. For the graph selection problem, we use the residual sum of squares as a criterion for choosing the optimal graph structure

$$\sigma_G^2 = \frac{1}{p}\sum_{j=1}^{p}\hat{\sigma}_j^2, \tag{6}$$

where $\hat{\sigma}_j^2$ is defined by

$$\hat{\sigma}_j^2 = \begin{cases} \sigma_{j_0}^2/\alpha & \text{for top genes,} \\ \frac{1}{n}\sum_{i=1}^{n}\{x_{ij} - \hat{h}_j(\boldsymbol{p}_{ij})\}^2 & \text{otherwise.} \end{cases}$$

The optimal graph $\hat{G}$ is obtained as the minimizer of σ_G^2. Note that the criterion (6) can evaluate graphs that are obtained by the same λ, M and α.

When we focus on a small gene networks, the optimal graph structure can be obtained by using a suitable learning algorithm, e.g. Ott *et al.* [15]. However, for large gene networks, we use a greedy hill-climbing algorithm for learning graph structures. Our greedy hill-climbing algorithm can be written as follows:

Initial Step

Step1. For all genes $\boldsymbol{x}_{(j)} = (x_{1j}, ..., x_{nj})^T$, create M virtual samples

$$\boldsymbol{x}_{(j)}^* = (x_{1j}^{*1}, ..., x_{nj}^{*1}, ..., x_{1j}^{*m}, ..., x_{nj}^{*m}, ..., x_{1j}^{*M}, ..., x_{nj}^{*M})^T,$$

where $x_{ij}^{*m} \sim N(x_{ij}, \sigma_{j_0}^2)$.

Step2. For each pair $\{(x_{ik}^{*m}, x_{ij}) | i = 1, ..., n; m = 0, ..., M\}$, apply the moving boxcel median filter and take 10 best genes in terms of $\hat{\sigma}_j^2$ as candidate parents of gene$_j$. We denote 10 candidate parents of gene$_j$ as "pa$_{jk}$" for $k = 1, ..., 10$.

Learning Step

Step3. For each gene$_j$ $(j = 1, ..., p)$:

Step3-1. For each candidate parent pa$_{jk}$ $(k = 1, ..., 10)$:

Step3-1-(a). Test one of the following operations, apply moving median filter, and calculate $\hat{\sigma}_{j,test}^2$.

- If pa$_{jk}$ $\to$ gene$_j$ does not exist, add this edge.
- If pa$_{jk}$ $\to$ gene$_j$ exists, remove this edge.
- If pa$_{jk}$ $\leftarrow$ gene$_j$ exists, reverse this edge.

Step3-1-(b). If $\hat{\sigma}_{j,test}^2 < \hat{\sigma}_j^2$, apply the operation in Step3-1(a) and set $\hat{\sigma}_j^2 = \hat{\sigma}_{j,test}^2$. Otherwise, no operation is conducted.

Step3-2. Update virtual samples for gene$_j$.

Step4. Repeat Step3 until σ_G^2 converges.

In the learning step of the above algorithm, the resulting network depends on the learning order of genes. Therefore, we permute the learning order and take the best network out of 10 networks as the optimal one.

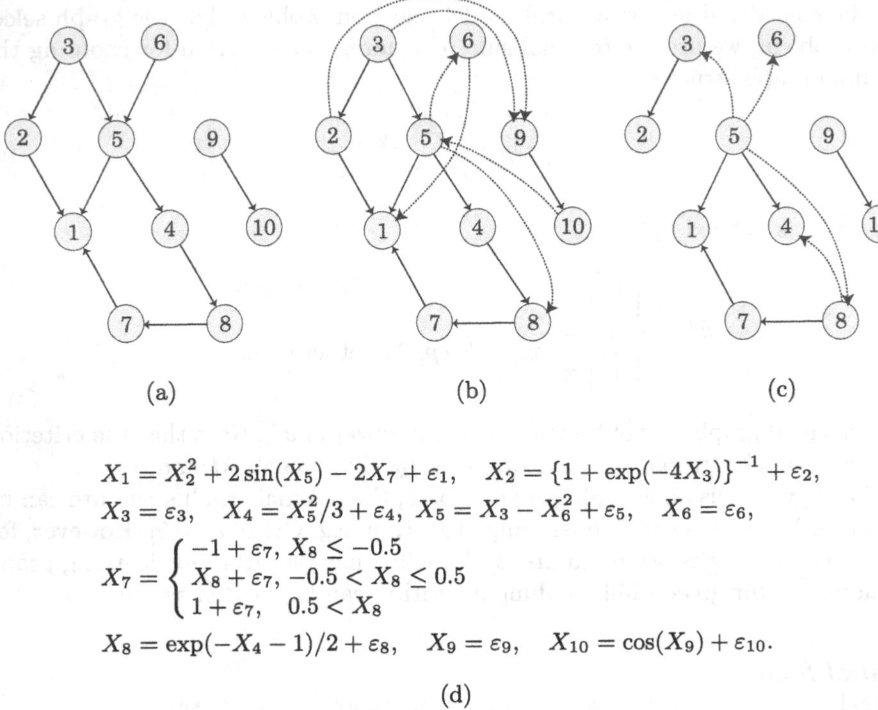

$$X_1 = X_2^2 + 2\sin(X_5) - 2X_7 + \varepsilon_1, \quad X_2 = \{1 + \exp(-4X_3)\}^{-1} + \varepsilon_2,$$
$$X_3 = \varepsilon_3, \quad X_4 = X_5^2/3 + \varepsilon_4, \; X_5 = X_3 - X_6^2 + \varepsilon_5, \quad X_6 = \varepsilon_6,$$
$$X_7 = \begin{cases} -1 + \varepsilon_7, \; X_8 \leq -0.5 \\ X_8 + \varepsilon_7, \; -0.5 < X_8 \leq 0.5 \\ 1 + \varepsilon_7, \quad 0.5 < X_8 \end{cases}$$
$$X_8 = \exp(-X_4 - 1)/2 + \varepsilon_8, \quad X_9 = \varepsilon_9, \quad X_{10} = \cos(X_9) + \varepsilon_{10}.$$

(d)

Fig. 2. True model and estimated networks of Monte Carlo simulations: (a) True network. (b) Estimated network by the previous method [9]. (c) Estimated network by the proposed method ($n = 100$, $\alpha = 0.2$, $\lambda = 20$ and $M = 100$). (d) Functions between nodes

3 Computational Experiments

3.1 Monte Carlo Simulations

We conduct Monte Carlo simulations to examine the properties of the proposed method by comparing with the previous method [9]. The simulated microarray data were generated from the artificial network of Figure 2 (a) with the functional structures between nodes shown in Figure 2 (d). The observations of the child variable are generated after transforming the observations of the parent variables to mean 0 and variance 1. After generating the data from the true system and making a matrix $\boldsymbol{X}' = (\boldsymbol{x}'_1, ..., \boldsymbol{x}'_n)^T$, we then add the noise corresponding to the measurement noise by

$$\boldsymbol{x}_i = \boldsymbol{x}'_i + \boldsymbol{\eta}_i, \quad i = 1, ..., n,$$

where $\boldsymbol{\eta}_i = (\eta_{i1}, ..., \eta_{ip})^T$ is a p-dimensional noise vector and η_{ij} depends on the mixture normal distribution of the form

$$\eta_{ij} \sim (1 - \kappa)N(0, \{Rng(x_{ij})/20\}^2) + \kappa N(0, \{Rng(x_{ij})/10\}^2).$$

Here we set $\kappa = 0.05$ and $Rng(x_{ij}) = \max_i(x_{ij}) - \min_i(x_{ij})$. Hence a microarray data matrix we used is defined by $\boldsymbol{X} = (\boldsymbol{x}_1, ..., \boldsymbol{x}_n)^T$. A network was rebuilt from simulated data consisting of $n = 50$ or $n = 100$ observations, which corresponds to 50 or 100 microarrays.

Figure 2 (b) and (c) are typical examples of the estimated networks for $n = 100$. We tried various settings of α, M and λ and set $\alpha = 0.2$, $M = 100$ and $\lambda = 20$ as appropriate values. In Figure 2 (b) and (c), the solid edges are correctly estimated edges and the dotted edges are the falsely estimated edges by the previous method [9] and the proposed method. It is clear that by adding measurement noise and some outliers, the previous method estimated some spurious relations that are false positives. On the other hand, by comparing with the previous method, the proposed method can reduce the number of false positives. We observe that a shortcoming of the proposed method from this simulation that the proposed method sometimes estimates edges that are inverse direction. However, if we consider the estimated network as an undirected graph, the sensitivity (the number of correctly estimated edges divided by the number of the estimated edges) of the proposed method is much higher than that of the previous method.

In the result of Monte Carlo simulations, it is shown that our method is robust to the noise which is independent of fluctuations of the gene network. Various instrumental and observation noises can be properly removed together with constructing Bayesian networks for estimating gene networks, resulting in giving robust and reliable estimates.

3.2 Real Data Example

In the real data example, we use *Saccharomyces cerevisiae* cell cycle data collected by Spellman *et al.* [19] and focus on 52 genes. These genes made a subnetwork estimated by Imoto *et al.* [9]. Figure 3 is the resulting network obtained by the proposed method with the same α, M and λ in the Monte Carlo simulations in the previous section.

The results of the real data example can be summarized as follows: In budding yeast, *Saccharomyces cerevisiae*, the homeodomain protein, *YOX1*, is a repressor that restricts early cell cycle boxes (ECB)-mediated transcription to the M/G1 phase of the cell cycle [17]. As a transcription factor, *YOX1* binds nearly 30 genes, including *CLN2* [7], that are important for DNA synthesis and repair. An ECB element (*CLN3*) activates SBF and MBF, two late G1-specific transcription complexes. Both SBF and MBF cause a burst of transcription of the late G1 cyclins, *CLN1* and *CLN2*, and other genes required for S phase (B cyclins), *CLB5* and *CLB6*, respectively. The two late G1 cyclins, *CLN1* and *CLN2*, have important effects for progression into S phase, i.e. an increase in the level of *CLB5,6-CDC* kinase activity sufficient to permit initiation of DNA replication [8]. In the resulting network shown in Figure 3, *YOX1* connects directly to *CLN2* and *CLN2* connects to *CLN1* and *CLB6*, while the resulting network of Imoto *et al.* [9] contains connections from *CLN2* to *CLN1* through *YOX1*. In comparison with Imoto *et al.* [9], our network can reflect functional correlations

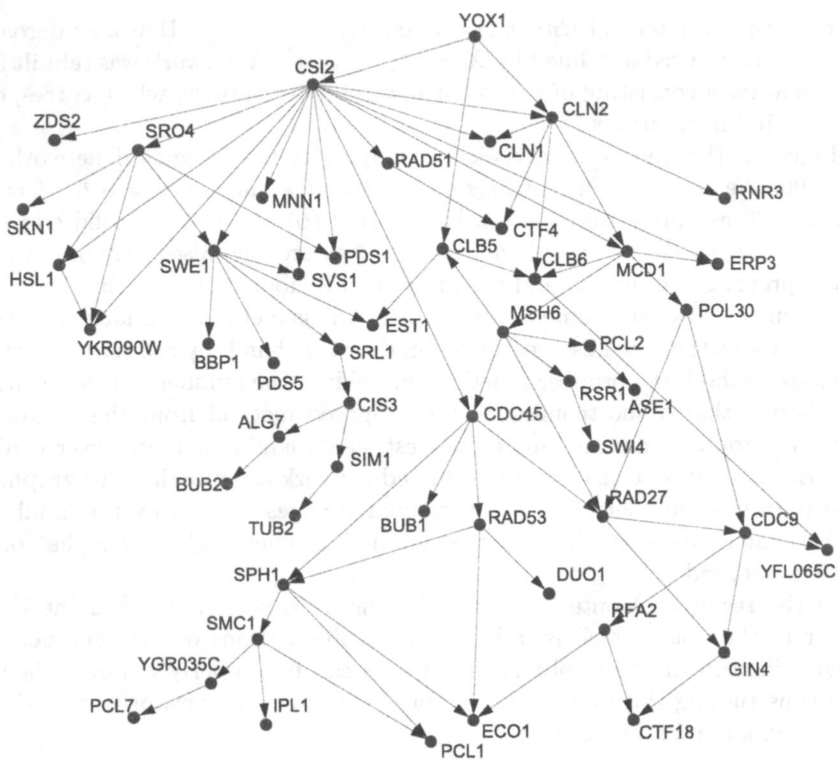

Fig. 3. Resulting cell cycle gene network obtained by the proposed method

between *YOX1* and the late G1-cyclins and B-cyclins more precisely. An interesting relationship is found that the cell wall biogenesis gene *CSI2* triggers many genes in our resulting network, while only three genes are regulated by *CSI2* in the network of Imoto *et al.* [9]. Although the function of *CSI2* as a structural component of the chitin synthase 3 complex is still unclear, Friedman *et al.* [5] showed that *CLN2*, *RNR3*, *SVS1*, *SRO4* and *RAD51* are highly correlated each other and *CSI2*-*SRO4* pair works together as cell wall regulation at the plasma membrane. In addition, the yeast protein function assignment [21] by Marcotte *et al.* [13] shows that *CSI2* is functionally linked with vanadate sensitive suppressor *SVS1*. Based on the resulting network in Figure 3, we may suggest that *CSI2* interacts many genes and may play an important role in cell cycle. Based on those observations, our proposed methods are capable of extracting causal relationships between genes effectively from gene expression data.

4 Discussion

We proposed the use of the moving boxcel median filter and the residual bootstrap for constructing Bayesian networks aimed at robust estimation of gene networks from microarray data. Main difficulties of the estimation of gene networks based on microarray data are caused by the outliers and the nonlinearity of the relationships between genes. We solved these problems by the proposed method and attained an accurate gene network rather than the previous method.

We consider the following problems as our future topics: We set the parameters α, M and λ empirically. These parameters, however, could affect the resulting networks and we need to develop a suitable criterion for choosing them from a statistical point of view. Recently, researches have been focused on using multiple types of genomic data such as binding site information, protein-protein interaction and so on, together with microarray data for extracting more reliable information [11, 14, 18, 20]. We would like to extend our method to handle such genomic data for estimating more accurate gene networks.

References

1. Akutsu, T., Miyano, S., Kuhara, S.: Identification of genetic networks from a small number of gene expression patterns under the Boolean network model. Pac. Symp. Biocomput. **4** (1999) 17–28.
2. Chen, T., He, H.L., Church, G.M.: Modeling gene expression with differential equations. Pac. Symp. Biocomput. **4** (1999) 29–40.
3. De Hoon, M.J.L., Imoto, S., Kobayashi, K., Ogasawara, N., Miyano, S.: Inferring gene regulatory networks from time-ordered gene expression data of *Bacillus subtilis* using differential equations. Pac. Symp. Biocomput. **8** (2003) 17–28.
4. Efron, B., Tibshirani, R.J.: An Introduction to the Bootstrap. Chapman & Hall/CRC. (1993).
5. Friedman, N., Linial, M., Nachman, I., Pe'er, D.: Using Bayesian network to analyze expression data. J. Comp. Biol. **7** (2000) 601–620.
6. Hartemink, A.J., Gifford, D.K., Jaakkola, T.S., Young, R.A.: Combining location and expression data for principled discovery of genetic regulatory network models. Pac. Symp. Biocomput. **7** (2002) 437–449.
7. Horak, C.E., Luscombe, N.M., Qian, J., Bertone, P., Piccirrillo, S., Gerstein, M., Snyder, M.:Complex transcriptional circuitry at the G1/S transition in *Saccharomyces cerevisiae*. Genes & Development **16** (2002) 3017–3033.
8. Huberman, J.A.: Cell cycle control of S phase: a comparison of two yeasts. Chromosoma **105** (1996) 197–203.
9. Imoto, S., Goto, T., Miyano, S.: Estimation of genetic networks and functional structures between genes by using Bayesian network and nonparametric regression. Pac. Symp. Biocomput. **7** (2002) 175–186.
10. Imoto, S., Kim, S., Goto, T., Aburatani, S., Tashiro, K., Kuhara, S., Miyano, S.: Bayesian network and nonparametric heteroscedastic regression for nonlinear modeling of genetic network. J. Bioinform. Comp. Biol. **1**(2) (2003) 231–252.
11. Imoto, S., Higuchi, T., Goto, T., Tashiro, K., Kuhara, S., Miyano, S.: Combining microarrays and biological knowledge for estimating gene networks via Bayesian networks. J. Bioinform. Comp. Biol. **2**(1) (2004) 77–98.
12. Jensen, F.V.: An Introduction to Bayesian Networks. University College London Press, (1996).

13. Marcotte, E.M., Pellegrini, M. Thompson, M.J., Yeates, T.O., Eisenberg, D.: A combined algorithm for genome-wide prediction of protein function. Nature **402** (1999) 83–86.
14. Nariai, N., Kim, S., Imoto, S., Miyano, S.: Using protein-protein interactions for refining gene networks estimated from microarray data by Bayesian networks. Pac. Symp. Biocomput. **9** (2004) 336–347.
15. Ott, S., Imoto, S., Miyano, S.: Finding optimal models for small gene networks. Pac. Symp. Biocomput. **9** (2004) 557–567.
16. Pe'er, D., Regev, A., Elidan, G., Friedman, N.: Inferring subnetworks from perturbed expression profiles. Bioinformatics **17** (ISBM2001) S215–S224.
17. Pramila, T., Shawna, M., GuhaThakurta, D., Jemiolo, D., Breeden, L.L.: Conserved homeodomain proteins interact with MADS box protein Mcm1 to restrict ECB-dependent transcription to the M/G1 phase of the cell cycle. Genes & Development **16** (2002) 3034–3045
18. Segal, E., Wang, H., Koller, D.: Discovering molecular pathways from protein interaction and gene expression data. Bioinformatics **19** (ISMB2003) i264–i272.
19. Spellman, P.T., Sherlock, G., Zhang, M.Q., Iyer, V.R., Anders, K., Eisen, M.B., Brown, P.O., Botstein, D., Futcher, B.: Comprehensive identification of cell cycle-regulated genes of the yeast *Saccharomyces cerevisiae* by microarray hybridization. Mol. Biol. Cell **9** (1998) 3273–3297.
20. Tamada, Y., Kim, S., Bannai, H., Imoto, S., Tashiro, K., Kuhara, S., Miyano, S.: Estimating gene networks from gene expression data by combining Bayesian network model with promoter element detection. Bioinformatics **19** (ECCB2003) ii227–ii236.
21. `http://www.doe-mbi.ucla.edu/Services/GPofYPF/yeastlist.html`

Spatial Modeling and Simulation of Diffusion in Nuclei of Living Cells

Dietmar Volz[1], Martin Eigel[1], Chaitanya Athale[1], Peter Bastian[2], Harald Hermann[3], Constantin Kappel[1], and Roland Eils[1]

[1] Div. Theoretical Bioinformatics, German Cancer Research Center (DKFZ), Im Neuenheimer Feld 580, D-69120 Heidelberg, Germany
r.eils@dkfz-heidelberg.de
http://www.dkfz.de/ibios/index.jsp

[2] Interdisciplinary Center of Scientific Computing (IWR), Im Neuenheimer Feld 368, D-69120 Heidelberg, Germany

[3] Div. Cell Biology, German Cancer Research Center (DKFZ), Im Neuenheimer Feld 580, D-69120 Heidelberg, Germany

Abstract. The mobility of fluorescently labelled molecules in the interphase nucleus has been increasingly employed to investigate the spatial organization of the interchromosomal space. We suggest an improved two-dimensional anisotropic diffusion model to address the inhomogeneous nature of nuclear organization, which is at odds with the generally applied 'well-mixed' compartmental assumption. To consider the transfer function of the imaging system, we derived a modified fundamental solution of the two-dimensional, time-dependent diffusion equation. The model was validated through comparison of the forward simulation results with fluorescence recovery after photobleaching experiments using nuclear localization signal (NLS) - tagged YFP recorded by confocal laser scanning microscopy. To improve the fit error in the vicinity of the nuclear boundary, we suggest an isotropic diffusion model with Neumann boundary condition accounting for the exact shape of the nuclear boundary. The suggested approach is a first step towards diffusion tomography of the cell nucleus.

1 Introduction

The mammalian cell nucleus, presumably devoid of any membrane bound compartments, is intricately compartmentalized [1]. This was also shown in studies where chromosome specific labelling techniques were used (for review see [2]). Aspects of the functional architecture have also been under investigation [3, 4, 5, 6, 7]. Since the advent of the Green Fluorescent Protein (GFP) [8], the study of the transport kinetics of tagged proteins inside the live cell nucleus have become a widely applied method used to address this question [9, 10]. It has been shown that molecules of very different functions all appeared to move rapidly [11]. These studies used models of passive transport inside the cell nucleus by fitting the dynamics to a diffusion equation. This homogeneous, isotropic diffusion

V. Danos and V. Schachter (Eds.): CMSB 2004, LNBI 3082, pp. 161–171, 2005.

model assumes that the nucleus can be considered a 'well-mixed' compartment, i.e. tracks of molecules are assumed to be solely triggered by some uncorrelated physical obstructions like e.g. the chromatin and other nuclear entities significantly influencing the dynamics. No assumptions on binding or other effects are therefore included in such a model. Also, no hypotheses about an active transport of these molecules are made [12].

In our model the generally used diffusion constant is replaced by a diffusion tensor. Thus, a homogeneous but anisotropic dynamics can be described [13, 14]. It will be shown that particularly at the boundary the fit error for the boundary-free diffusion model is increased by several orders of magnitude. This finding indicates that the estimated model parameters are affected by a boundary artefact that is not negligible.

Therefore, in a further step we applied an isotropic model including the nuclear boundary. Since there does not exist a fundamental solution for a diffusion equation including a boundary of arbitrary shape, the arising system of equation and boundary condition was solved numerically by a sophisticated solver (UG) for partial differential equations (PDEs). The combination of both approaches builds a 'simulation-tool-chain'. The reconciliation of the experimental data with both the coarse boundary-free model and the more elaborate boundary-including model will be shown to result in more reliable parameter estimates.

2 Materials and Methods

2.1 Cell Culture, DNA Construct and Transfection

Human adrenal cortex carcinoma Sw13 cells [15] were cultured on circular coverslips in Dulbeccos Minimum Essential Medium (DMEM, GIBCO) with 10% fetal calf serum (fcs) and antibiotics without phenol red. They were then transfected with the NLS tagged pEYFP-c1 construct (Clontech) using FuGene 6 (Roche, Mannheim, Germany).

2.2 Fluorescence Confocal Microscopy

2D images were recorded over time using a laser scanning confocal microscope (Leica LSM SP2) equipped with a detachable 37 ° C temperature controlled chamber and a 100x Plan-Apochromat NA oil immersion objective. The YFP channel used 514 nm Ar Laser for excitation and detection was at 526 nm. The pinhole settings of 1 Airy unit implies a Full Width Half Maximum of 614.5 nm. The images were scanned at a line frequency of 1000 lines/s and a time delay of 40 ms between each image.

Before starting to record a sequence, a rectangular image capture region was determined and the whole region also served as the bleach region (Figure1B). The dimensions of the image were 512 x 64 pixels with a resulting size of 59.5 x 7.4 μm.

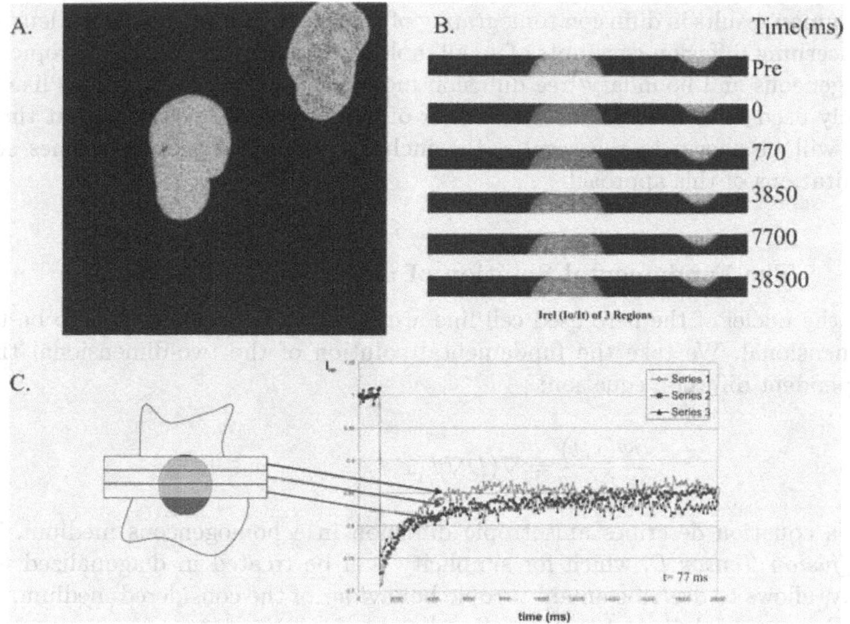

Fig. 1. The experimental data: A 2D section of a typical cell transfected with NLS-YFP in the nucleus (A). The time series of images bleached and imaged for recovery of signal (B). The recovery curves from the bleaching experiment (C)

2.3 Kinetic measurements by Fluorescence Recovery After Photobleaching (FRAP)

The images collected for the FRAP experiment (Figure 1A) involved 15 time steps of imaging with a 5% attenuated laser (pre-bleach), 10 frames imaged with 100% laser intesity (bleach) and typically 1000 frames at 5% laser intensity (post-bleach) (for a review on FRAP see [16]). A MATLAB program (The MathWorks Inc., USA) using threshold dependent segmentation was then used to read in the time series of images (Figure 1B) and derive the grey value intensities (Figure 1C) and the sizes of the segmented regions. This data was used for all subsequent fits.

3 Theory and Implementation

The FRAP dynamics are first described by a diffusion model with a respective model equation that can be solved by a fundamental solution Eq. 1. The class of diffusion models that can be solved by such a fundamental solution are necessarily made up of equations that do not consider the boundary domains of respective media. Nevertheless, they are easy to handle and computationally cheap, both in the forward as well as in the inverse direction. Here, the term 'inverse' means the search for optimal model parameters to fit to given experimental data.

Common results in diffusion tomography of the cell as well as the cell nucleus, e.g. concerning diffusion constants of small molecules rely on the use of isotropic, homogeneous and boundary-free diffusion models [10]. Early publications like [12] likely used these models due to the lack of computational power at that time. As will be shown in the results, the inclusion of boundaries overcomes some limitations of this approach.

3.1 The Fundamental Solution of the Diffusion Equation

As the nuclei of the here used cell line are flat they were considered to be two-dimensional. We take the fundamental solution of the two-dimensional time-dependent diffusion equation:

$$\frac{\partial \Phi(x,t)}{\partial t} = \nabla(D\nabla\Phi(x,t)) \quad , \qquad x \in \mathbb{R}^2 \tag{1}$$

This equation describes anisotropic diffusion in a homogeneous medium. The *Diffusion Tensor* D, which for simplicity will be treated in diagonalized form only, allows to describe an anisotropic behaviour of the considered medium. The fundamental solution of this equation in two dimensions is:

$$\Phi(x,t) = \frac{1}{4\pi\sqrt{d_1 d_2}} \exp\left(\frac{x_1 D^{-1} x_2}{4t}\right) \tag{2}$$

The parameters d_1 and d_2 are the entries of the two-dimensional diffusion tensor D. The Gaussian bell function, which corresponds to the given solution, has to be convolved by the coordinates of the total bleaching area $B(x)$. The total intensity is therefore given by:

$$\Phi_{tot}(x,t) = \Phi(x,t) * B(x) \quad . \tag{3}$$

The reconciliation with the experimental FRAP data needs a normalization procedure in order to get a scale-free signal that can be fitted to Eq. 3. Similar to a proposed normalization procedure [12], we chose a normalization that allows to accomplish the requested convolution numerically. The intensity of the laser beam is assumed to be spatially constant throughout the bleaching area. This assumption is confirmed by the experimental data. Therefore, no weighting function has been taken into account in the convolution. In addition, three by three pixels have been merged to one averaged pixel and data of 10 consecutive time points have been merged to get one averaged frame. The normalization term reads:

$$I_N(x,t) = \frac{I(x,t_{pre}) - I(x,t_{post_i})}{I(x,t_{pre}) - I(x,t_{post_{end}})} \tag{4}$$

The normalized signal I_N is a dimension-less construct of the pre-bleach signal $I(x,t_{pre})$ and the subsequent post-bleach signals $I(x,t_{post_i})$.

Taking the accordingly normalized experimental data, reconciliation is performed with an equally normalized model equation built up from the convolved fundamental solution (Eq. 3). The normalized model term now reads:

$$\Phi_{N,tot}(x,t) = \frac{\sum_i \left\{ \frac{1}{t+s} \exp\left[-\frac{1}{4(t+s)} \cdot (x_1 - B(x_{1,i}))D^{-1}(x_2 - B(x_{2,i})) \right] \right\}}{\sum_i \left\{ \frac{1}{1+s} \exp\left[-\frac{1}{4(t_{end}+s)} \cdot (x_1 - B(x_{1,i}))D^{-1}(x_2 - B(x_{2,i})) \right] \right\}}, \tag{5}$$

where $B(x_{1,i}, x_{2,i})$ is a set of coordinate pairs which make up the bleaching area. When fitting $\Phi_{N,tot}(x,t)$ to the normalized intensity values $I_N(x,t)$, the parameter s adapts the model to a possible delay between the first post-bleach image and the supposed initial post-bleach time point t_1 in the model. All computation has been done in MATLAB, using standard routines like 'fminsearch' for the optimization and a summation of results of Eq. 2 over the bleaching area to accomplish the convolution procedure in Eq. 5. The fit results will be given in the 'results' section.

The existence of fundamental Eq. 2 in this respect gives an easy way to calculate the forward problem for diffusion Eq. 1 and the inverse problem of parameter estimation by optimization.

3.2 Inclusion of the Nuclear Geometry in the Model

The software package *UG* (*U*nstructured *G*rids) served as a platform for solving the diffusion problem within the boundary of an experimentally examined cell nucleus. UG is a toolbox for the numerical solution of partial differential equations on unstructured meshes using multigrid methods. It is freely available from (http://cox.iwr.uni-heidelberg.de/~ug/). This framework provides a broad collection of ready to use problem discretizations as part of its problem class libraries. In addition, the UG library provides geometric and algebraic data structures as well as a large number of mesh manipulation, numerical algorithm and visualization functions. Being implemented in the C programming language, it can be adapted flexibly to specific differential equations, domains and discretization methods [17].

To achieve a diffusion model with boundary conditions we had to implement several additional modules within UG. As a first step, we **(a)** implemented a specific discretization by application of a problem class from the UG problem class library. The convection-diffusion finite volume implementation was used as a basis for the simulation. It was adjusted to represent a purely two-dimensional instationary diffusion equation discretization as stated above, including the diagonal diffusion tensor D. Both entries herein were set to the same value resulting in a global isotropic diffusion within the domain defined by the selected nucleus.

Then **(b)** implementation of the main application that defines the domain description, boundary conditions and coefficient functions was carried out. For polygonal domain definition approx. 100 sampling points were selected manually from the boundary of the examined nucleus in an overview image. In between these points, coordinates were interpolated linearly by the domain loader that was implemented. The multigrid was initialy triangulated by defining a mesh consisting of four triangular elements spread at the upper and lower border, two convex quadrilaterate elements in between. These elements are also called subdomains of the coarsest grid Ω_0. Uniform refinement of the entire grid was performed four times resulting in a five-level multigrid with about 10^4 elements on the finest grid Ω_4. Neumann condition with $\frac{\partial \Phi}{\partial n}(x,t) = 0$ was applied to the entire boundary $\partial\Omega$. The Neumann condition can be interpreted as conservation of mass within the domain due to impermeable object boundaries.

Next, a **(c)** UG script controlling the simulation was developed for glueing together problem class, boundary and domain. Definitions of the simulation parameters were included, consisting of the time solver, Gauss-Seidel pre- and post-smoothers and the discretization scheme. Multigrid V-cycles were performed. Initial parameters for the domain were set, i.e. homogenous concentration values $\Phi(x, t_{post_1}) = \rho_{post_1}$ for the vertical extent of the bleach area $y_{lower} < y < y_{upper}$, $\Phi(x, t_{post_1}) = \rho_{pre}$ for the remaining domain. For each time step the calculated solution matrix for Ω_4 was written out and converted into a matrix sequence readable by MATLAB for further investigation.

4 Results

We studied two different cell nuclei and a window of parameter estimation with a size of 6 by 6 averaged pixels. Tables 1 and 2 show values of tensor entries for

Table 1. Diffusion constants $[\mu m^2/s]$ estimated in subdomains of regions having a high fit-error. Regions #1 and #5 are situated on bottom and top of each nucleus, respectively, and therefore in direct vicinity of the boundary

#Cell	#Region	Subdomain (x-interval)	Diff. Const. (x-axis)	Diff. const. (y-axis)	fit-error
2	5	[137,143]	1.19	0.55	65.2
3	1	[70,76]	0.38	0.68	14.3
3	5	[92,98]	0.0158	1.051	13.3

Table 2. Diffusion constants $[\mu m^2/s]$ estimated in subdomains of regions having a low fit-error. Regions #2 and #3 are situated in the middle of each nucleus. The y-intervals of each subdomain are situated in the middle of each bleaching stripe

#Cell	#Region	Subdomain (x-interval)	Diff. Const. (x-axis)	Diff. const. (y-axis)	fit-error
2	2	[119,125]	0.446	0.0930	2.00
3	3	[88,94]	0.0713	0.0701	2.12
3	2	[62,68]	0.00635	0.0289	2.06
3	2	[68,74]	0.0486	0.0665	1.83

selected locations of parameter estimation using the anisotropic diffusion model without boundary conditions.

Apparently, regions with high and regions with low fit error, respectively, are discernable. Regions with high fit error mainly occur in the vicinity of the boundary. The respective boundary-induced error is mapped to the tensor entries in a

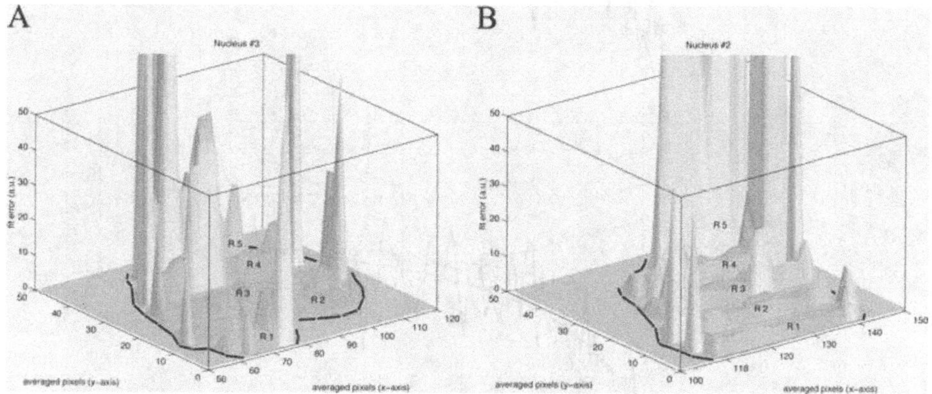

Fig. 2. (A) Fit error for cell C3 as a function of the location of parameter estimation within each of five bleaching regions (R1-R5). The black line shows the nuclear boundary. The spatial resolution of the fit error has been achieved by varying the location of the window of parameter estimation as described in tables 1 and 2. (B) Fit error for cell C2

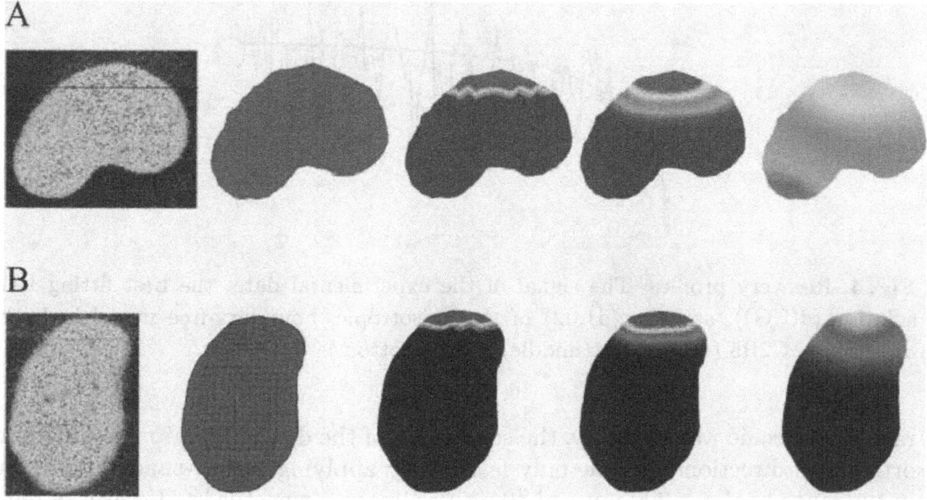

Fig. 3. (A) Simulation of FRAP experiments visualized by UG. From left to right: The fluorescing cell C3, the discretized cell, three stages of a UG simulation. (B) Simulation of FRAP experiments for cell C2

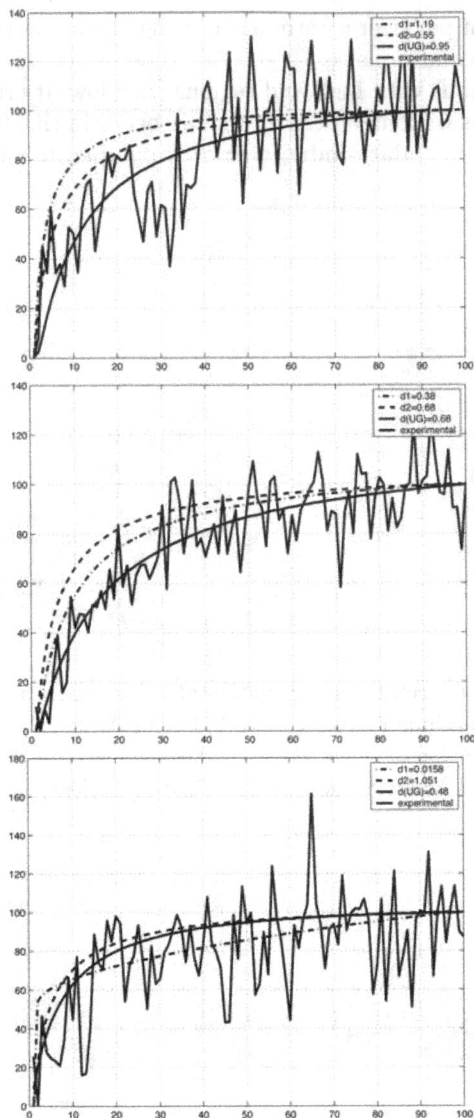

Fig. 4. Recovery profiles: The signal of the experimental data, the best fitting UG solution (d(UG)), and fits (d1,d2) of the anisotropic, boundary-free model in both directions (C2R5 (top), C3R1(middle), C3R5(bottom))

rather unspecific way. Notably, the separation of the dynamics into two different orthogonal directions became only feasible by applying a time-space resolution in the range of about 200 nm and 40 ms per image, respectively. However, a reliable separation of present boundary artefacts such that only one of both entries in the diffusion tensor would be affected could not be achieved. Regions with low fit error show some slight spatial dependence of at least one of both tensor entries. These findings can be explained by the fact that the maximum value of the Gaussian distribution consists of an average of both directions, resulting in a cross-talk between the two directions.

An overview of the spatially resolved fit error for both cells is given in figures 2 A and 2 B. The variation in the non-boundary regions might be due to the fact that the local geometry changes in those parts of the nucleus. This in turn leads to fluctuations in the balancing dynamics of the molecules. The cause for such inhomogeneity might be the differential distributions of chromatin in the nucleus in various regions inducing a pseudo-boundary or soft boundary.

The above findings indicate that the estimated model parameters are affected by a boundary artefact that cannot be resolved by an anisotropic diffusion model without considering boundary conditions. Stable model improvements therefore require the consideration of the nuclear geometry. Figures 3 A and 3 B show a forward simulation using an isotropic diffusion model with a Neumann boundary condition considering the exact shape of each nucleus. The fit results shown in Figure 4 for the boundary-including diffusion model and the boundary-free anisotropic model, respectively, illustrate a comparison between both approaches. Apparently, for regions with a high fit error the UG solution yields a better fit to the data in the first time-slice of the recovery than both of the tensor-triggered directional flows.

5 Conclusion

In summary, a smoother recovery dynamics can be achieved by including the boundary into the model rather than with a coarse boundary-free model.

Further investigations are required regarding the spatial characteristics of the cell medium. The spatially resolved mathematical model already allows for simulating an inhomogeneous diffusion process. To acquire meaningful results, techniques for aligning simulated and observed data have to be included. Eventually, this will lead to a reduction of the global simulation error in the inverse solution by iteratively optimizing spatially resolved diffusion coefficients. A temporal dependence of the spatially resolved diffusion coefficient could also be taken into account with this approach. A good initial guess for parameters will be of high relevance since the optimization is typically a complex task producing widely ramified solution approximations.

The analysis of the spatial and temporal variations in the intranuclear structure e.g. by changing chromatin densities will be an important next step. In combination with an improved spatial and temporal resolution in the FRAP experiments by applying new techniques of high resolution microscopy such as 4Pi-microscopy [18] will give rise to a spatially resolved representation of the nuclear structure with regard to molecular diffusion characteristics. Such studies will also help to improve the transport model of molecules. In future studies, additional transport terms will be taken into account thus addressing transport characteristics deviating from diffusion dynamics.

Diffusion dynamics give a first approximation of the complex dynamics presumably including active transport patterns or channel-like structures within the cell nucleus. A higher temporal resolution of the data is therefore essential, especially in the time window immediately after bleaching. The examination of the possibly inhomogeneous nature of the medium, both in experiment and simulation, will be an essential step to unravel the complex internal structure of the cell nucleus and its implication for protein dynamics.

Our work is in line with other studies in a similar direction [19] suggesting the need for forward simulations when estimating mobility. Our study is an important first step towards molecular diffusion tomography of the cell nucleus, combining sophisticated methods of Scientific Computing with the potential of advanced experimental assays.

Acknowledgements

We are grateful to Daniel Gerlich, EMBL Heidelberg, for fruitful discussions, Michaela Reichenzeller (DKFZ) for the plasmid and Karsten Richter (DKFZ) for help with the confocal microscope.

References

1. Monneron, A, Bernhard, W.: Fine structural organization of the interphase nucleus in some mammalian cells. J. Ultrastruct. Res. **27** (1969) 266–288
2. Cremer, T. and Cremer, C.: Chromosome territories, nuclear architechture and gene regulation in mammalian cells. Nat. Rev. Genet. **2** (2001) 292–301
3. Cremer, T., A. Kurz, R. Zirbel, S. Dietzel, B. Rinke, E. Schröck, M. R. Speicher, U. Mathieu, A. Jauch, P. Emmerich, H. Scherthan, T. Ried, C. Cremer and P. Lichter: Role of chromosome territories in the functional compartmentalization of the cell nucleus. Cold Spring Harbour Symposia on Quantitative Biology LVIII (1993) 777–792
4. Lamond, A. I. and Earnshaw, W. C.: Structure and function in the nucleus. Science **280** (1998) 547–553
5. Misteli, T.: Cell biology of transcription and pre-mRNA splicing: nuclear architecture meets nuclear function. J. Cell Sci. **113** (200) 1841–1849
6. Dundr, M, Misteli, T.: Functional architecture in the cell nucleus. Biochem. J. **163** (2001) 509–17
7. Barboro, P., D'Arrigo, C., Mormino, M., Coradeghini, R., Parodi, S., Patrone, E. and Balbi, C.: An intranuclear frame for chromatin compartmentalization and higher-order folding. J. Cell Biochem. **88** (2003) 113–120
8. Chalfie, M., Tu, Y., Euskirchen, G., Ward, W. W., Prasher, D. C.: Green fluorescent protein as a marker for gene expression. Science **263** (1994) 802–805
9. Misteli, T.: Protein dynamics: implications for nuclear architecture and gene expression. Science **291** (2001) 843–847
10. Phair, R.D. and Misteli, T.: Kinetic modeling approaches in *in vivo* imaging. Nat. Rev. Mol. Cell Biol. **2** (2001) 898–907

11. Phair, R.D. and Misteli, T.: High mobility of proteins in the mammalian cell nucleus. Nature **404** (2000) 604–609
12. Axelrod, D., Koppel, D. E., Schlessinger, J., Elson, E. and Webb, W. W.: Mobility Measurement by Analysis of Fluorescence Photobleaching Recovery Kinetics. Biophys. J. **16** (1976) 1055–1069
13. Crank, J.: The Mathematics of Diffusion, Oxford University Press, Oxford, 1976
14. Weickert, J.: Anisotropic Diffusion in Image Processing, Teubner, Stuttgart, 1997
15. Hedberg, K. K. and Chen, L. B.: Absence of intermediate filaments in a human adrenal cortex carcinoma-derived cell line. Exp. Cell Res. **163** (1986) 509–517
16. Reits, E. A. and Neefjes, J. J.: From fixed to FRAP: measuring protein mobility and activity in living cells. Nat. Cell Biol. **3** (2001) E145–E147
17. Bastian, P., Birken, K., Lang, S., Johannsen, K., Neu N., Rentz-Reichert, H. and Wieners, C.: UG: A flexible software toolbox for solving partial differential equations. Computing and Visualization in Science **1** (1997) 27
18. Nagorni, M. and Hell, S. W.: 4Pi-confocal microscopy provides three-dimensional images of the microtubule network with 100- to 150-nm resolution. J. Struct. Biol. **123** (1998) 236–247
19. Siggia. E.D., Lippincott-Schwartz, J., Bekiranov, S.: Diffusion in Inhomogenous Media: Theory and Simulation Applied to Whole Cell Photobleach Recovery. Biophys. J. **79** (2000) 1761–1770

The Biochemical Abstract Machine BIOCHAM

Nathalie Chabrier-Rivier, François Fages, and Sylvain Soliman

Projet Contraintes, INRIA Rocquencourt,
BP105, 78153 Le Chesnay Cedex, France
{Nathalie.Chabrier-Rivier, Francois.Fages, Sylvain.Soliman}@inria.fr
http://contraintes.inria.fr

Abstract. In this article we present the Biochemical Abstract Machine BIOCHAM and advocate its use as a formal modeling environment for networks biology. Biocham provides a precise semantics to biomolecular interaction maps. Based on this formal semantics, the Biocham system offers automated reasoning tools for querying the temporal properties of the system under all its possible behaviors. We present the main features of Biocham, provide details on a simple example of the MAPK signaling cascade and prove some results on the equivalence of models w.r.t. their temporal properties.

1 Introduction

In networks biology, the complexity of the systems at hand (metabolic networks, extracellular and intracellular networks, networks of gene regulation) clearly shows the necessity of software tools for reasoning globally about biological systems [1]. Several formalisms have been proposed in recent years for modeling biochemical processes either qualitatively [2, 3, 4] or quantitatively [5, 6, 7, 8, 9]. State-of-the-art tools integrate a graphical user interface and a simulator, yet few formal tools are available for reasoning about these processes and proving properties about them. Our focus in Biocham has been on the design of a biochemical rule language and a query language of the model in temporal logic, that are intended to be used by biologists.

Biocham has been designed in the framework of the ARC CPBIO on "Process Calculi and Biology of Molecular Networks" [10] which aims at pushing forward a declarative and compositional approach to modeling languages in Systems Biology. Biocham is a language and a programming environment for modeling biochemical systems, making simulations, and checking temporal properties. It is composed of :

1. a rule-based language for modeling biochemical systems, allowing patterns and constraints in the definition of rules;
2. a simple simulator;
3. a powerful query language based on Computation Tree Logic CTL;
4. an interface to the NuSMV [11] model checker for automatically evaluating CTL queries.

V. Danos and V. Schachter (Eds.): CMSB 2004, LNBI 3082, pp. 172–191, 2005.

The use of Computation Tree Logic (CTL) [12] for querying the temporal properties of the system provides an alternative technique to numerical models based on differential equations, in particular when numerical data are missing. The model-checking tools associated to CTL automate reasoning on all the possible behaviors of the system modeled in a purely qualitative way. The semantics of Biocham ensures that the set of possible behaviors of the model over-approximates the set of all behaviors of the system corresponding to different kinetic parameters.

Biocham shares several similarities with the Pathway Logic system [4] implemented in Maude. Both systems rely on an algebraic syntax and are rule-based languages. One difference is the use in Biocham of CTL logic which allows us to express a wide variety of biological queries, and the use of a state-of-the-art symbolic model checker for handling the complexity of highly non-deterministic models.

The first experimental results of this approach for querying models of biochemical networks in temporal logic have been reported in [13, 14], on a qualitative model of the mammalian cell cycle control [15, 16] and in [14] on a quantitative model of gene expression [9]. In this paper we describe the Biocham system which provides a modeling environment supporting this methodology.

The next section defines the syntax of Biocham objects, rules and patterns, and their semantics. The following section describes the CTL query language and the expression of biological queries. Then we detail Biocham functionalities on a simple model of the MAPK signaling cascade. In section 5 we discuss the comparison of different models of given biological systems and show two equivalence results w.r.t. CTL properties. Section 6 reports our on-going experience in applying inductive logic programming techniques to learning reaction rules from temporal properties, and learning rule patterns from a given set of reaction rules. Finally we compare our approach with related work and conclude on the perspectives of this work.

2 Syntax and Semantics

2.1 A Simple Algebra of Biochemical Compounds

Biocham manipulates formal objects which represent chemical or biochemical compounds, ranging from ions, to small molecules, macromolecules and genes. Biocham objects can be used also to represent control variables and abstract biological processes.

Syntax:

```
  object = molecule | abstract
molecule = name | molecule-molecule | molecule~{name,...,name}
         | gene | ( molecule )
    gene = #name
abstract = @name
```

In the simplest and the most flexible syntactical form, a molecule is simply given a name. Multimolecular complexes are denoted with the linking operator -. This binary operator is assumed to be associative and commutative, hence the order of the elements in a complex does not matter. Note that the same hypothesis is made in Pathway Logic [4] and other systems [17]. In the cases where one would like to distinguish between different orders of association, one can denote the different complexes with specific names. A third syntactical form serves to write modified forms of molecules, like attaching the set of phosphorylated sites with the operator $\sim$. Several sets can be attached. The order of the elements is irrelevant.

Example 1. `RAF`, `RAF-RAFK` and `RAF~{p1}-MEK` are valid Biocham notations for, respectively, a Raf protein, its complex with a Raf-kinase, and an activated form of Raf in the complex with the MAPK/ERK kinase. `(RAF-MEK)~{p1}` is another notation for the same phosphorylated form of the complex without making precise the constituent which is phosphorylated. Precising or not the phosphorylated constituent defines two formally different complexes. It is also possible to specify more precisely the phosphorylation, for instance by naming its sites, as in `RAF~{ser338,tyr341}`, but then again, this defines a formally different molecule.

The fourth syntactical form is used to denote genes or gene promotors, with a name beginning with #. These objects are assumed to be *unique*, which has a consequence on the way reactions involving such objects are interpreted by Biocham, as explained in the next section.

Example 2. `DMP1-#p19ARF` can be used to denote the binding of protein DMP1 on the promotor of the gene producing protein p19ARF noted `#p19ARF`.

The same assumption of uniqueness is made on abstract objects that are noted with a '@'. Abstract objects can be used to represent particular phases of a process, complete subsystems or abstract biological processes.

Example 3. `@UbiPro` can be used to denote the Ubiquitine/Proteasome subsystem and write this formal object as a catalyst in degradation reaction rules.

2.2 Reaction Rules

Biocham reaction rules are used primarily to represent biochemical reactions. They can be used also to represent state transitions involving control variables or abstract processes, or to represent the main effects of complete subsystems such as protein synthesis by DNA transcription without introducing RNAs in the model.

Syntax:

```
reaction = name: reaction
         | solution => solution
         | solution =[object]=> solution
         | solution =[solution => solution]=> solution
         | solution <=> solution
         | solution <=[object]=> solution
solution = _ | object | solution + solution | ( solution )
```

A solution is thus a sum of objects, the character _ denotes the empty solution. The order and multiplicity of molecules in a solution are ignored, only the presence or absence of objects are considered.

The following abbreviations can be used for reaction rules: `A<=>B` for the two symmetrical rules, `A=[C]=>B` for the rule `A+C=>B+C` with catalyst molecule `C`, and `A=[C=>D]=>B` for the rule `A+C=>D+B`.

Example 4. `RAF + RAFK => RAF-RAFK` is a complexation rule.
`MEK =[RAF~{p1}]=> MEK~{p1}` is a phosphorylation rule with catalyst `RAF~{p1}`. This rule is equivalent to `MEK + RAF~{p1} => MEK~{p1} + RAF~{p1}`.

A reaction transforms one solution matching the left-hand side of the rule, into another solution in which the objects of the right-hand side have been added. The molecules in the left-hand side of the rule which do not appear in the right-hand side may be non-deterministically present or consumed in the resulting solution. This convention defines the *boolean abstraction* of stoichiometric models used in Biocham. It reflects the capability of Biocham to reason about all possible behaviors of the system with unknown concentration values and unknown kinetics parameters [1, 3].

Following the uniqueness assumption, molecule parts marked as "genes" with the '#' notation, or any compound built on such a molecule (such as `DMP1-#p19ARF` for instance) are not multiplied. These objects remain unique and they are deterministically consumed in the form in which they appear in the left-hand side of the rule. The same goes for control variables, noted with a '@', which are deterministically consumed.

Biocham has also a rich pattern language with constraints which is used to specify molecules and sets of reaction rules in a concise manner, it is detailed in the next section.

2.3 Patterns

Patterns introduce the special character ? and variables noted with a name beginning with a $ to denote unspecified parts of a molecule. These variables can be constrained with simple set constraints.

Example 5. `list_rules(RAF~?-? + ? => ?).`

This command contains a pattern matching all rules reacting with any form (phosphorylated or complexed) of RAF.

`list_rules(? =[RAFK]=> ?).`

This pattern is matching all rules involving the catalyst RAFK, i.e. having RAFK in their left and right-hand sides, even if they were not written with the catalyst notation.

With some restrictions, patterns can be used to define reaction rules. This is not the case of the above example as the patterns can match unspecified (unconstrained) molecules. Two constructs are provided in Biocham to specify the domains of variables either globally or locally to a rule:

- `declare`: allowing to define all the phosphorylation sites of a molecule;
- `where`: imposing local constraints on variables in a rule declaration.

Example 6. `declare MEK~{{},{p1},{p2},{p1,p2}}.`

```
declare MAPK~parts_of({p1,p2}).
...
MEK~$P + RAF~{p1} <=> MEK~$P-RAF~{p1}
where p2 not in $P.
and   $cycA in {cycA, cks1-cycA}
```

In this context, MEK is declared to have two phosphorylation sites `p1, p2` and that all combinations are possible, MAPK has two sites too, with again all combinations allowed.

Then appears a reaction pattern which specifies the complexation or decomplexation of some forms of MEK not already phosphorylated on p2. This pattern is used to define reactions. There are 2 possible forms (not containing p2) for `$P` to combine with the 2 possibilities for the direction of the rule. This reaction pattern thus expands into 4 reaction rules.

Section 4 contains a Biocham model of the MAPK signaling cascade written with a set of 16 reaction rule patterns which expand into 30 rule instances.

2.4 Kripke Semantics

A Biocham model is a set of reaction rules given with an initial state. As any rule pattern can be expanded into a set of reaction rules, there is no loss of generality in considering only reaction rules. The formal semantics of a Biocham model is a Kripke structure that is a mathematical structure which provides a firm ground for :

- comparing different modeling formalisms and languages,
- comparing different models of a same biological system,
- importing models from other sources,
- and designing and implementing automated reasoning tools.

A *Kripke structure* K is a triple (S, R, L) where S is the set of states, $R \subseteq S \times S$ is a total relation (i.e. for any state $s \in S$ there exists a state $s' \in S$ such that $(s, s') \in R$) called transition, and $L : S \to 2^A$ is a labeling function over the set of atomic propositions A, which associates to each state the set of atomic propositions true in that state. A path in K starting from a state s_0 is an infinite sequence of states $\pi = s_0, s_1, ...$ such that $(s_i, s_{i+1}) \in R$ for all $i \geq 0$.

Clearly, one can associate to a Biocham model a Kripke structure, where the set of states S is the set of all tuples of boolean values denoting the presence or absence of the different biochemical compounds (molecules, genes and abstract processes), the transition relation R is the union (i.e. disjunction) of the relations associated to the reaction rules (which will be noted $\rightarrow$), and the labeling function L simply associates to a given state the set of biochemical compounds which are present in the state. Reaction rules in Biocham are asynchronous in the sense that one reaction rule is fired at a time (interleaving semantics), hence the transition relation is the union of the relations associated to the reaction rules On the other hand, in a synchronous semantics for Biocham, the transition relation would have been defined by intersection. The choice of a synchronous semantics has been rejected in Biocham as it would bias fundamental biological phenomena such as the masking of a relation by another one and the resulting inhibition or activation of biological processes. Note that as explained in Sect. 2.2, the boolean abstraction of enzymatic reactions used in Biocham associates several transitions to a single Biocham reaction rule, one for each case of possible consumption of the molecules in the left-hand side of the rule.

That Kripke structure defines the semantics of a Biocham model as a non-deterministic transition system where the temporal evolution of the system is modeled by the succession of transition steps, and the different possible behaviors of the system are obtained by the non-deterministic choice of reactions.

3 Querying Biocham Models in Temporal Logic CTL

Thanks to its simple Kripke semantics, Biocham supports the use of the Computation Tree Logic CTL [12] as a query language for querying the temporal properties of Biocham models. This methodology introduced in [13, 14] is implemented in Biocham with an interface to the state-of-the-art symbolic model checker NuSMV [11].

CTL basically extends propositional logic used for describing states, with operators for reasoning over time and non-determinism. Several temporal operators are introduced in CTL: $X\phi$ meaning ϕ is true at next transition, $G\phi$ meaning ϕ is always true, $F\phi$ meaning finally true, and $\phi U\psi$ meaning ϕ is always true until ψ becomes true. For reasoning about non-determinism, two path quantifiers are introduced: $A\phi$ meaning ϕ is true on all paths, $E\phi$ meaning ϕ is true on some path. In CTL, all temporal operators must be immediately preceded by a path quantifier (e.g. $AFG\phi$ is not in CTL, but $AF(EG\phi)$ is).

CTL is expressive enough to express a wide range of biological queries. Simplest queries are *about reachability*: is there a pathway for synthesizing a protein P, $EF(P)$? *About pathways*: can the cell reach a state s while passing by another state s_2, $EF(s_2 \wedge EF(s))$? Is state s_2 a necessary checkpoint for reaching state s, $\neg E((\neg s_2)\ U\ s)$? Can the cell reach a state s without violating certain constraints c, $E(c\ U\ s)$? Is it possible to synthesize a protein P with-

Table 1. Inductive definition of the truth relations $s \models \phi$ and $\pi \models \phi$ in a given Kripke structure K

$s \models \alpha$	iff $\alpha \in L(s)$,
$s \models E\psi$	iff there is a path π from s such that $\pi \models \psi$,
$s \models A\psi$	iff for every path π from s, $\pi \models \psi$,
$\pi \models \phi$	iff $s \models \phi$ where s is the starting state of π,
$\pi \models X\psi$	iff $\pi^1 \models \psi$,
$\pi \models F\psi$	iff there exists $k \geq 0$ such that $\pi^k \models \psi$,
$\pi \models G\psi$	iff for every $k \geq 0$, $\pi^k \models \psi$,
$\pi \models \psi U \psi'$	iff there exists $k \geq 0$ such that $\pi^k \models \psi'$ and $\pi^j \models \psi$ for all $0 \leq j < k$.

out creating nor using protein Q, $I \Rightarrow E(\neg Q\ U\ P)$? *About steady states and permanent states*: is a certain (partially described) state s of the cell a steady state, $s \Rightarrow EG(s)$? a permanent state, $s \Rightarrow AG(s)$? Can the cell reach a given permanent state s, $EF(AGs)$? Must the cell reach a given permanent state s, $AF(AGs)$? Can the system exhibit a cyclic behavior w.r.t. the presence of a product P, $EG((P \Rightarrow EF\ \neg P) \wedge (\neg P \Rightarrow EF\ P))$? The latter formula expresses that there exists a path where at all time points whenever P is present it becomes eventually absent, and whenever it is absent it becomes eventually present. This formula is not expressible in LTL [12], where formulas are of the form $A\phi$ with ϕ containing no path quantifier.

The formal semantics of CTL in a fixed Kripke structure K is given in Table 1, as the inductive definition of the truth relation stating that a CTL formula ϕ is true at state s, written $s \models \phi$, or true along path π, written $\pi \models \phi$ (the clauses for ordinary boolean connectives are omitted). π^i denotes the suffix of π starting at s_i.

4 A Simple Example

The Mitogen-Activated Protein Kinase (MAPK) cascades are a well-known example of signal transduction, since they appear in many receptor-mediated signal transduction schemes. They are actively considered in pharmaceutical research, for their applications to cancer therapies. The MAPK/ERK pathway is indeed hyperactivated in 30% of all human cancer tumours [18].

The structure of a MAPK cascade is a sequence of activations of three kinases in the cytosol. The last kinase, MAPK, when activated, has an effect on different substrates in the cytosol but also on gene transcription in the nucleus.

Since this cascade has been studied a lot, mathematical models of it appear in most model repositories, like for instance that of Cellerator [19] or the SBML repository page [20], both coming from [21]. This cascade was also the first example treated by Regev, Silverman and Shapiro [2] in the pi-calculus process algebra which was an initial source of inspiration for our own work.

Models based on ordinary differential equations (ODE) allow us to reproduce simulation results like the one pictured out in Figure 1, where the concentration

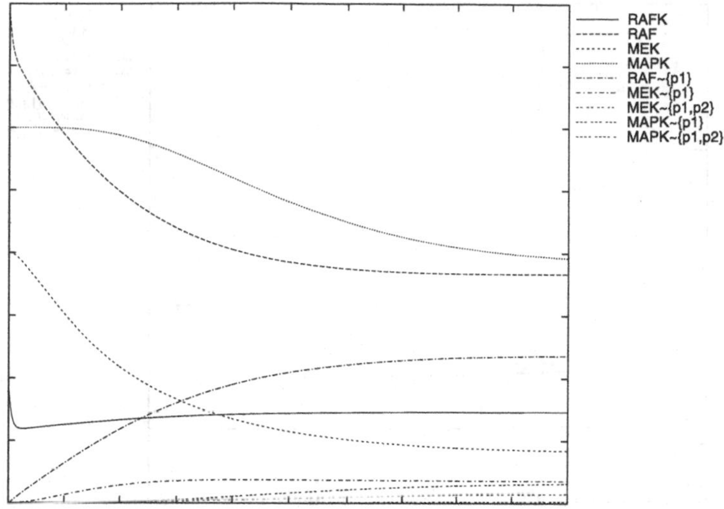

Fig. 1. Simulation result of an ODE model of the MAPK cascade

of the visualized compounds is represented on the vertical axis and time on the horizontal axis. In Figure 2, the concentrations axis has been simply split and rescaled to the maximum value for each compound.

It is possible to see from such simulations how the cascade evolves in time. It is possible to change input quantities to check for a significant change in the outcome of the simulation. Similarly, the sensitivity of the system to the values of the parameters can be checked by running different simulations with different values of the parameters.

4.1 The MAPK Cascade in Biocham Syntax

Our aim in Biocham is to introduce complementary techniques to automate reasoning on all possible behaviors of the system modeled in a purely qualitative way. Taking the above model, one sees that it is built quite directly from the enzymatic reactions and Michaelis-Menten kinetics. Abstracting the kinetics part, one gets a system of biochemical reactions that can be interpreted as a non-deterministic transition system over boolean variables denoting the presence or absence of the compounds in the signaling cascade. The semantics of Biocham (explained in Sects. 2.2 and 2.4) ensures that the set of the possible behaviors of the boolean model over-approximates the set of all behaviors of the system for all kinetic parameters' values.

Here is the full code of the MAPK example[1] in Biocham syntax. The phosphorylation sites for `MEK` and `MAPK` are declared first, and then the Biocham rules

[1] adapted from the SBML model `http://www-aig.jpl.nasa.gov/public/mls/cellerator/notebooks/MAPK-in-solution.html`

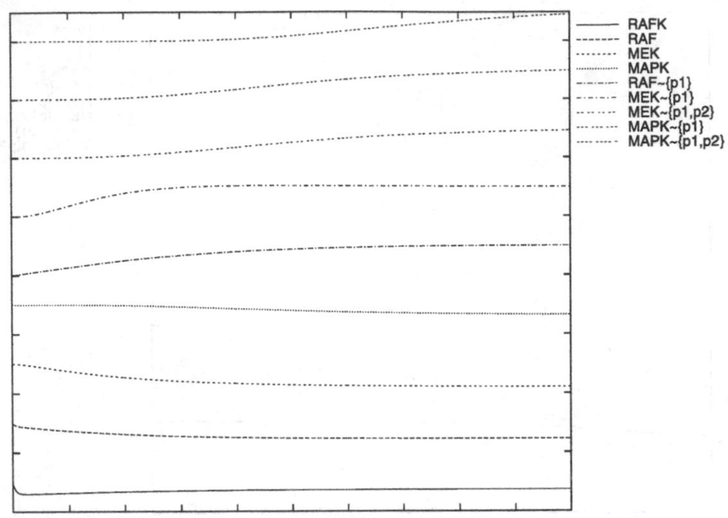

Fig. 2. Same simulation as figure 1, side by side rescaled view

are given, sometimes with pattern variables (noted **$P**) which are constrained in the **where** part of the rules. In this model, the first rules are reversible, the other ones are directional.

```
declare MEK~parts_of({p1,p2}).
declare MAPK~parts_of({p1,p2}).

RAF + RAFK <=> RAF-RAFK.

RAF~{p1} + RAFPH <=> RAF~{p1}-RAFPH.

MEK~$P + RAF~{p1} <=> MEK~$P-RAF~{p1}
   where p2 not in $P.

MEKPH + MEK~{p1}~$P <=> MEK~{p1}~$P-MEKPH.

MAPK~$P + MEK~{p1,p2} <=> MAPK~$P-MEK~{p1,p2}
   where p2 not in $P.

MAPKPH + MAPK~{p1}~$P <=> MAPK~{p1}~$P-MAPKPH.

RAF-RAFK => RAFK + RAF~{p1}.

RAF~{p1}-RAFPH => RAF + RAFPH.

MEK~{p1}-RAF~{p1} => MEK~{p1,p2} + RAF~{p1}.
MEK-RAF~{p1} => MEK~{p1} + RAF~{p1}.
```

```
MEK~{p1}-MEKPH => MEK + MEKPH.
MEK~{p1,p2}-MEKPH => MEK~{p1} + MEKPH.

MAPK-MEK~{p1,p2} => MAPK~{p1} + MEK~{p1,p2}.
MAPK~{p1}-MEK~{p1,p2} => MAPK~{p1,p2} + MEK~{p1,p2}.

MAPK~{p1}-MAPKPH => MAPK + MAPKPH.
MAPK~{p1,p2}-MAPKPH => MAPK~{p1} + MAPKPH.
```

These rule patterns define the following set of expanded reaction rules:

```
biocham: expand_rules.
1 RAF+RAFK=>RAF-RAFK.
2 RAF-RAFK=>RAF+RAFK.
3 RAF~{p1}+RAFPH=>RAFPH-RAF~{p1}.
4 RAFPH-RAF~{p1}=>RAF~{p1}+RAFPH.
5 MEK+RAF~{p1}=>MEK-RAF~{p1}.
6 MEK-RAF~{p1}=>MEK+RAF~{p1}.
7 MEK~{p1}+RAF~{p1}=>MEK~{p1}-RAF~{p1}.
8 MEK~{p1}-RAF~{p1}=>MEK~{p1}+RAF~{p1}.
9 MEKPH+MEK~{p1}=>MEKPH-MEK~{p1}.
10 MEKPH-MEK~{p1}=>MEKPH+MEK~{p1}.
11 MEKPH+MEK~{p1,p2}=>MEKPH-MEK~{p1,p2}.
12 MEKPH-MEK~{p1,p2}=>MEKPH+MEK~{p1,p2}.
13 MAPK+MEK~{p1,p2}=>MAPK-MEK~{p1,p2}.
14 MAPK-MEK~{p1,p2}=>MAPK+MEK~{p1,p2}.
15 MAPK~{p1}+MEK~{p1,p2}=>MAPK~{p1}-MEK~{p1,p2}.
16 MAPK~{p1}-MEK~{p1,p2}=>MAPK~{p1}+MEK~{p1,p2}.
17 MAPKPH+MAPK~{p1}=>MAPKPH-MAPK~{p1}.
18 MAPKPH-MAPK~{p1}=>MAPKPH+MAPK~{p1}.
19 MAPKPH+MAPK~{p1,p2}=>MAPKPH-MAPK~{p1,p2}.
20 MAPKPH-MAPK~{p1,p2}=>MAPKPH+MAPK~{p1,p2}.
21 RAF-RAFK=>RAFK+RAF~{p1}.
22 RAFPH-RAF~{p1}=>RAF+RAFPH.
23 MEK~{p1}-RAF~{p1}=>MEK~{p1,p2}+RAF~{p1}.
24 MEK-RAF~{p1}=>MEK~{p1}+RAF~{p1}.
25 MEKPH-MEK~{p1}=>MEK+MEKPH.
26 MEKPH-MEK~{p1,p2}=>MEK~{p1}+MEKPH.
27 MAPK-MEK~{p1,p2}=>MAPK~{p1}+MEK~{p1,p2}.
28 MAPK~{p1}-MEK~{p1,p2}=>MAPK~{p1,p2}+MEK~{p1,p2}.
29 MAPKPH-MAPK~{p1}=>MAPK+MAPKPH.
30 MAPKPH-MAPK~{p1,p2}=>MAPK~{p1}+MAPKPH.
```

The Biocham rules can be exported to a `.dot` file for use with the Graphviz[2] visualization suite. The generated map is depicted in Figure 3.

[2] `http://www.research.att.com/sw/tools/graphviz/`

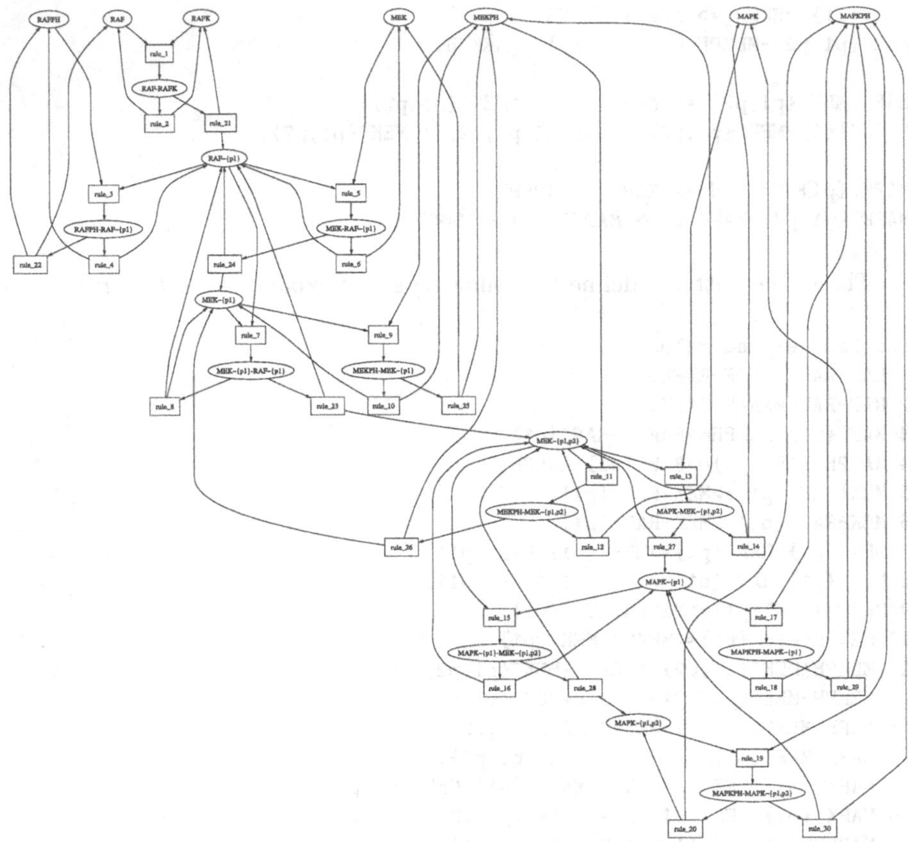

Fig. 3. Interaction map generated from BIOCHAM rules for the MAPK cascade

4.2 Simulation

Since a Biocham model is highly non-deterministic, simulations are randomized, which means that at each time step, one of the possible reactions happens. An initial state can be defined by taking present the following molecules:

```
present({
RAFK,
RAF,
MEK,
MAPK,
MAPKPH,
MEKPH,
RAFPH
}).
```

and absent the following ones:

```
absent({?-?,?~{p1}~?}).
```

The last pattern declares the complexes (?-?) and the molecules phosphorylated at p1 (?~{p1}~?) as absent from the initial state. It is equivalent to the following sequence:

```
absent(RAF-RAFK).
absent(RAFPH-RAF~{p1}).
absent(MEK-RAF~{p1}).
absent(MEK~{p1}-RAF~{p1}).
absent(MEKPH-MEK~{p1}).
absent(MEKPH-MEK~{p1,p2}).
absent(MAPK-MEK~{p1,p2}).
absent(MAPK~{p1}-MEK~{p1,p2}).
absent(MAPKPH-MAPK~{p1}).
absent(MAPKPH-MAPK~{p1,p2}).
absent(RAF~{p1}).
absent(MEK~{p1}).
absent(MEK~{p1,p2}).
absent(MAPK~{p1}).
absent(MAPK~{p1,p2}).
```

Figure 4 depicts one random simulation of the MAPK cascade, that is one but only one possible behavior of the system at the boolean abstraction level. One can notice that this trace is not a boolean abstraction (by thresholds) of the numerical simulation. On the other hand, the numerical simulation can be abstracted in a feasible boolean trace.

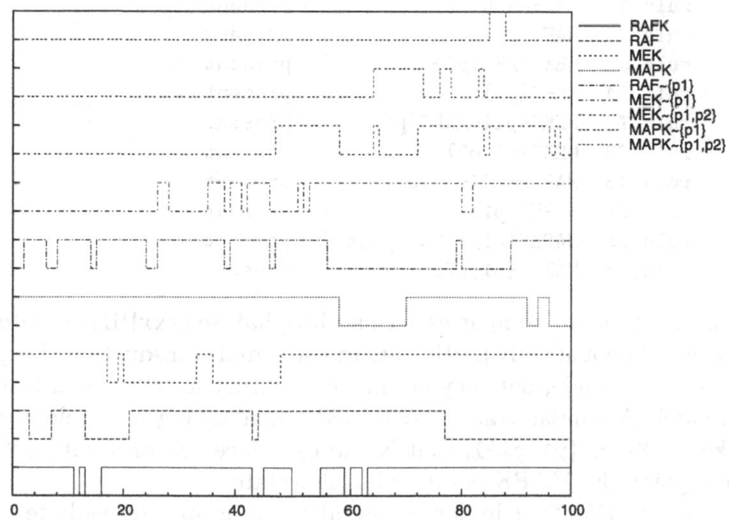

Fig. 4. Random simulation of the Biocham model of the MAPK cascade

4.3 CTL Queries

Biocham uses the Computation Tree Logic (CTL) [12] as a query language for querying the temporal properties of the system under all possible conditions. Querying a Biocham model in CTL temporal logic provides a mean to analyze exhaustively all possible behaviors of the system from first principles of enzymatic reactions, in particular when numerical data are not available.

A biological query like for example "Is the activation of the second kinase of the cascade (MEK) compulsory for the cascade ?" asks whether the phosphorylated form of MEK, noted in Biocham `MEK~{p1}`, is necessary to the production of the activated MAPK, noted `MAPK~{p1,p2}`, which is the output of the cascade, that is whether `MEK~{p1}` is a checkpoint. In Biocham, one expresses this query by the CTL formula

```
biocham: !(E(!(MEK~{p1}) U MAPK~{p1,p2}))
true
```

This formula expresses the non (!) existence (E) of a path on which `MEK~{p1}` is absent (!) until (U) `MAPK~{p1,p2}` becomes present, that is to say that `MEK~{p1}` is a checkpoint. This formula is checked automatically by the system.

The same query about a complex with a phosphatase, such as the complex `MEK~{p1}-MEKPH`, is false. These complexes are thus not checkpoints. The `why` command computes a counterexample in the form of a pathway which validates the negation of the query:

```
biocham: !(E(!(MEK~{p1}-MEKPH) U MAPK~{p1,p2}))
false
biocham: why
Step 1    Initial state
Step 2    rule 1    RAF-RAFK                   present
Step 3    rule 21   RAF~{p1}                   present
Step 4    rule 5    MEK-RAF~{p1}               present
Step 5    rule 24   MEK~{p1}                   present
Step 6    rule 7    MEK~{p1}-RAF~{p1}          present
Step 7    rule 23   MEK~{p1,p2}                present
Step 8    rule 13   MAPK-MEK~{p1,p2}           present
Step 9    rule 27   MAPK~{p1}                  present
Step 10   rule 15   MAPK~{p1}-MEK~{p1,p2}      present
Step 11   rule 28   MAPK~{p1,p2}               present
```

This means that the complexes with a phosphatase (xxxPH) are intermediate products that do not strictly participate in the signal transduction. They are here to regulate the cascade, but they are not mandatory for the signal transduction in this model. A similar trace is obtained when asking a simple accessibility query like `EF(MAPK~{p1,p2})`, that is the existence (E) of a path on which at some time point (F) MAPK is fully phosphorylated.

It is worth noting that imposing the absence of an intermediate product is generally difficult in an ODE based simulation tool, without touching the model. Complex CTL queries thus have no natural counterpart in a numerical model and complement the information that can be deduced from interaction maps.

The largest example treated so far with Biocham is a model of the Mammalian cell cycle control [13] developed after Kohn's map [16], involving 500 proteins and genes and 147 rule patterns which expand into 2733 rule instances. The computational results reported in [13] show the feasibility of this approach on such large examples as the CTL queries can be evaluated in a few seconds using state-of-the-art symbolic model-checking tools.

5 Comparing Models

5.1 Importing Biochemical Models from Other Formalisms

Since the basic building block of a Biocham model is an (enzymatic) reaction, it is quite easy to import any model based on such reactions into Biocham. This is the case of most graphical map-based models, but also of some ODE models, derived from the mass-action law or Michaelis-Menten kinetics. A well known source of such models is KEGG [22], which provides (graphical) maps of metabolic and signaling pathways. Biocham has been designed to provide such maps with a simple yet precise semantics.

In this respect, the Biocham project is part of the workpackage entitled "Towards a Bioinformatics Semantic Web" in the EU network REWERSE[3].

5.2 CTL-Based Equivalence of Models

There are usually lots of different possible models for a same system, depending on the level of detail and of the available knowledge of that system. Even at the level of a single enzymatic reaction, there are already two common ways to specify it into Biocham:

either by detailing the formation of an unstable complex:

```
A + B <=> D
D => A + C
```

or directly, as

`A + B => A + C` (which can also be written `B =[A]=> C`)

The first model is a direct translation of common ODE numerical models. However when converting to Biocham, the second model may be more natural as it is simpler. We provide below some equivalence results w.r.t. the CTL properties of both modelings.

We suppose that the complex `D` only appears in the above rule, as otherwise the simplification is obviously not correct. Let us write K_1 for the Kripke structure associated with the first modeling, and K_2 for that associated with the second. These structures define two truth relationships: $\models_1$ and $\models_2$.

Proposition 1 (Reachability). *Let ϕ be an atomic CTL formula, if $I \models_2 EF(\phi)$ then $I \models_1 EF(\phi)$.*

[3] The 6th EU Framework Programme Network of Excellence REWERSE stands for REasoning on the WEb with Rules and SEmantics, see `http://www.rewerse.net`

Moreover, if A *and* B *do not appear negatively (i.e. under an odd number of negations) in* ϕ *and* D *does not appear at all in* ϕ*, then* $I \models_1 EF(\phi)$ *implies* $I \models_2 EF(\phi)$.

Proof. Let us first detail the transitions associated with the Biocham rules. In K_1 we have:

1. A + B → A + B + D
2. A + B → A + D
3. A + B → B + D
4. A + B → D
5. D → A + B + D
6. D → A + B
7. D → A + C + D
8. D → A + C

In K_2:

1. A + B → A + B + C
2. A + B → A + C

The first implication is straightforward, since whenever there exists a path in K_2 there exists a path in K_1, which is identical except for the transitions using A + B → A + C that can be replaced by two transitions: A + B → D and D → A + C, and those using A + B → A + B + C that can be replaced by A + B → B + D and D → A + C.

For the second implication, if the path of K_1 making ϕ true ends in a state without D, then it is easy to mimick it in K_2 with transitions 1 and 2. Otherwise, one can reason by induction on the transition to prove that the previous state without D also made ϕ true. For each of the transitions 1 to 4 (of K_1) if the right part $\models \phi$ then the left part $\models \phi$. For instance for transition 2, we have A + B + E → A + D + E, where A + D + E $\models \phi$, but since D does not appear in ϕ and B does not appear negatively in ϕ we have A + B + E $\models \phi$. If the last step is the use of 5 or 7, we do not need the induction: since D does not appear in ϕ, replacing it with respectively 6 or 8 is enough to get a path ending without D and making ϕ true. □

Proposition 2 (Checkpoints). *Let* $\neg E(\neg\phi U\psi)$ *be a checkpoint formula, i.e.* ϕ *and* ψ *are atomic formulae describing states,*

if A *and* B *do not appear negatively in* ψ *and* D *do not appear positively in* ψ*, then*

$I \models_2 \neg E(\neg\phi U\psi)$ *implies* $I \models_1 \neg E(\neg\phi U\psi)$.

if A *and* B *do not appear negatively in* ϕ *and* D *do not appear positively in* ϕ*, then*

$I \models_1 \neg E(\neg\phi U\psi)$ *implies* $I \models_2 \neg E(\neg\phi U\psi)$.

Proof. Let first remind the meaning of $I \models \neg E(\neg\phi U \psi)$: it means that there exist no path π leading to ψ such that ϕ is false in each state of π.

For the first implication, let us consider a path π leading to ψ in K_1; if π ends in a state where `D` is absent, then one can consider a path π' in K_2, obtained in the same way as in the previous proposition. This path is still leading to ψ in K_2, contains fewer states, ϕ is true in at least one of those states, and thus in one of the states of π. If π ends in a state where `D` is present, the only difference between that state and the previous (or next) state where `D` was not present amounts to the appearance of `D` and possible disappearance of `A` and `B` and none of those can be checked by ψ (since it is not positive in `D` nor negative in `A` or in `B`), thus ψ was already true in the state where `D` was not present and we can apply the above reasoning to get a state where ϕ is true.

For the reverse, let π be a K_2 path leading to ψ, there exists a K_1 path π' leading to the same state, obtained as in the previous proposition. We know that some state s of π' makes ϕ true, and if s is also in π then we are done. Otherwise s is a state such that `D` is present and once again the only differences with the previous state without `D` amount to `D`, `A` and `B` and because of the hypotheses made for ϕ, ϕ was already true in the state without `D`, and thus ϕ is true in a state of π. □

5.3 Enriching Models

To go one step further, there is the need to encompass models written in different formalisms and languages. The CMBSlib [23] web site[4] has been created as an open repository of computational models of biological systems, in order to:

- compare different *models* expressed in the same formalism,
- compare different *formalisms* and *tools* for a same model,
- cross-fertilize modeling experience and language issues between designers.

This library currently includes models of biological processes obtained from the literature and by translation from KEGG maps or ODE models into different formalisms. It is open to all contributions in any (ascii) format and in most exotic formalisms.

6 Learning

6.1 Learning Reaction Rules from Temporal Properties

With such a simple syntax and semantics for describing reaction rules in Biocham, it is possible to apply learning techniques to reaction rules discovery. We have done some preliminary experiments using the inductive logic programming system Progol [24] for the automatic discovery of missing Biocham reaction rules in a simple model of the cell cycle with 10 variables, given a set of accessibility

[4] `http://contraintes.inria.fr/CMBSlib`

properties. The basic experiment consists in furnishing a set of examples of accessibility relations and a set of counterexamples, and letting the inductive logic program search for a set of reaction rules satisfying the accessibility properties of the system. In the first phase of validation of the learning technique, where we are, the models we use are known models, from which we compute a set of temporal properties, and remove one or more reaction rules to check whether the missing rules can be recovered by learning from the temporal properties.

More generally, the basic idea is to specify the intended or observed temporal properties of the system with CTL formulas, and apply learning techniques such as inductive logic programming, in order to correct the model by suggesting to add or modify Biocham rules in the model[5].

6.2 Learning Patterns as Generalizations of Existing Rules

The same kind of learning techniques, namely inductive logic programming, can also be applied to the search of generalizations of existing rules, or even of appearing compounds, by the means of Biocham's pattern language.

Since the patterns allow basically rules with variables and constraints on these variables, it is quite straightforward to try and learn such patterns from existing models. Here again the status is that of preliminary experiments, but there is much hope in using this technique to complete partial models.

7 Related Work

High-throughput technologies addressing cell functions at a whole genome scale are revolutionizing cell biology. The challenge of virtual cell projects is to map molecular interactions within the cell, and to build virtual cell models predicting the effects of a drug on a given cell.

Virtual cell environments, like for instance the Virtual Cell project [25] or Cellerator [19], maintain a library of models of different parts of the cell, among different living organisms. ODE models typically range from a tenth of variables to 50 variables like in the Budding Yeast cell Cycle model of [26]. On the other hand, qualitative models represented by interaction maps allow for the global modeling of a large number of interacting subsystems.

A formal model of the Mammalian cell cycle control has been developed in the ARC CPBIO [13, 10] after Kohn's map [16]. This model transcribed in Biocham involves 500 proteins and genes and 147 reaction rule patterns which expand into 2733 reaction rule instances. Performance results of CTL querying in this model are reported in [13]. Symbolic model checking techniques used in Biocham are efficient enough to automatically evaluate CTL queries about biochemical networks of several hundreds or thousands of rules and variables [13, 14]. It is worth noting however that this is far below the size of digital circuits

[5] We investigate this approach in the 6th PCRD EU project APRIL 2 "Applications of Probabilistic Inductive Logic Programming", `http://www.aprill.org`.

that the same model checking algorithms can treat. The reason for this discrepancy in performance comes from the high level of non-determinism which results from the competition between reaction rules and the soup aspect of biochemical solutions.

The Pathway Logic of [4] is close to Biocham for the algebraic representation of cell compounds and the representation of molecular interactions by rewriting rules. However, the boolean abstraction used in Biocham and the state-of-the-art symbolic model checker NuSMV permit the handling of potentially larger models. The choice of CTL for expressing biological queries provides also more expressiveness than LTL, which is used in Pathway Logic. Much can be gained by exchanging Biocham and Pathway Logic models, cross-fertilizing our modeling experiences and comparing language issues in particular w.r.t. the pattern language. The CMBSlib open repository [23] has been created for this purpose as well as for comparison with very different formalisms.

Combining ODE models with purely qualitative models like current Biocham models is an important issue for managing the complexity of concurrent interacting models. This combination is under investigation within the framework of non-deterministic hybrid systems.

8 Conclusion and Perspectives

Biocham is a free software[6] for modeling biochemical processes and querying these models in temporal logic. The largest example treated so far is a model of the mammalian cell cycle control [13] after Kohn's diagram [16]. Other models have been imported from interaction maps available on the Web and ODE models. This shows the simplicity of the scheme and the flexibility of this approach.

Our first experiments for learning reaction rules from a partial model and reachability properties of the system are encouraging. We are still in the phase of validating the learning method based on Inductive Logic Programming. The next phase will be, in collaboration with biologists, to try to apply learning techniques to the discovery of new reaction rules.

Currently, Biocham is primarily oriented towards the qualitative modeling of biochemical processes and the querying of the temporal properties of boolean models. This approach can be generalized however to numerical models by relying on constraint-based model checking techniques [14]. In this extension, called Biocham2, variables can denote real values expressing the concentrations of molecules, and rules are extended with constraints to denote the relationship between the old and the new values of the variables. In particular, biochemical systems described by differential equations can be handled in this framework using time discretization methods, and can be combined with boolean models. The modeling power of such non-deterministic hybrid systems is under investigation.

[6] Biocham system can be downloaded from `http://contraintes.inria.fr/BIOCHAM`

Acknowledgments

This work benefited from various discussions with our colleagues of the ARC CPBIO, in particular with Alexander Bockmayr, Vincent Danos and Vincent Schächter, and of the European project APRIL 2, in particular with Stephen Muggleton.

References

1. Ideker, T., Galitski, T., Hood, L.: A new approach to decoding life: Systems biology. Annual Review of Genomics and Human Genetics **2** (2001) 343–372
2. Regev, A., Silverman, W., Shapiro, E.Y.: Representation and simulation of biochemical processes using the pi-calculus process algebra. In: Proceedings of the sixth Pacific Symposium of Biocomputing. (2001) 459–470
3. Nagasaki, M., Onami, S., Miyano, S., Kitano, H.: Bio-calculus: Its concept, and an application for molecular interaction. In: Currents in Computational Molecular Biology. Volume 30 of Frontiers Science Series. Universal Academy Press, Inc. (2000) This book is a collection of poster papers presented at the RECOMB 2000 Poster Session.
4. Eker, S., Knapp, M., Laderoute, K., Lincoln, P., Meseguer, J., Sönmez, M.K.: Pathway logic: Symbolic analysis of biological signaling. In: Proceedings of the seventh Pacific Symposium on Biocomputing. (2002) 400–412
5. Matsuno, H., Doi, A., Nagasaki, M., Miyano, S.: Hybrid petri net representation of gene regulatory network. In: Proceedings of the 5th Pacific Symposium on Biocomputing. (2000) 338–349
6. Hofestädt, R., Thelen, S.: Quantitative modeling of biochemical networks. In: In Silico Biology. Volume 1. IOS Press (1998) 39–53
7. Alur, R., Belta, C., Ivanicic, F., Kumar, V., Mintz, M., Pappas, G.J., Rubin, H., Schug, J.: Hybrid modeling and simulation of biomolecular networks. In Springer-Verlag, ed.: Proceedings of the 4th International Workshop on Hybrid Systems: Computation and Control, HSCC'01. Volume 2034 of Lecture Notes in Computer Science., Rome, Italy (2001) 19–32
8. Ghosh, R., Tomlin, C.: Lateral inhibition through delta-notch signaling: A piecewise affine hybrid model. In Springer-Verlag, ed.: Proceedings of the 4th International Workshop on Hybrid Systems: Computation and Control, HSCC'01. Volume 2034 of Lecture Notes in Computer Science., Rome, Italy (2001) 232–246
9. Bockmayr, A., Courtois, A.: Using hybrid concurrent constraint programming to model dynamic biological systems. In Springer-Verlag, ed.: Proceedings of ICLP'02, International Conference on Logic Programming, Copenhagen (2002) 85–99
10. ARC CPBIO: Process calculi and biology of molecular networks (2002–2003) http://contraintes.inria.fr/cpbio/.
11. Cimatti, A., Clarke, E., Enrico Giunchiglia, F.G., Pistore, M., Roveri, M., Sebastiani, R., Tacchella, A.: Nusmv 2: An opensource tool for symbolic model checking. In: Proceedings of the International Conference on Computer-Aided Verification, CAV'02, Copenhagen, Danmark (2002)
12. Clarke, E.M., Grumberg, O., Peled, D.A.: Model Checking. MIT Press (1999)
13. Chabrier, N., Chiaverini, M., Danos, V., Fages, F., Schächter, V.: Modeling and querying biochemical networks. Theoretical Computer Science **To appear** (2004)

14. Chabrier, N., Fages, F.: Symbolic model cheking of biochemical networks. In Priami, C., ed.: CMSB'03: Proceedings of the first Workshop on Computational Methods in Systems Biology. Volume 2602 of Lecture Notes in Computer Science., Rovereto, Italy, Springer-Verlag (2003) 149–162
15. Chiaverini, M., Danos, V.: A core modeling language for the working molecular biologist. In Priami, C., ed.: CMSB'03: Proceedings of the first Workshop on Computational Methods in Systems Biology. Volume 2602 of Lecture Notes in Computer Science., Rovereto, Italy, Springer-Verlag (2003) 166
16. Kohn, K.W.: Molecular interaction map of the mammalian cell cycle control and DNA repair systems. Molecular Biology of Cell **10** (1999) 703–2734
17. Maimon, R., Browning, S.: Diagrammatic notation and computational structure of gene networks. In Yi, T.M., Hucka, M., Morohashi, M., Kitano, H., eds.: Proceedings of the 2nd International Conference on Systems Biology, Online Proceedings (2001) http://www.icsb2001.org/toc.html.
18. Kolch, W., Kotwaliwale, A., Vass, K., Janosch, P.: The role of raf kinases in malignant transformation. In: Expert Reviews in Molecular Medicine. Volume 25. Cambridge University Press (2002) http://www.expertreviews.org/02004386h.htm.
19. Shapiro, B.E., Levchenko, A., Meyerowitz, E.M., Wold, B.J., Mjolsness, E.D.: Cellerator: extending a computer algebra system to include biochemical arrows for signal transduction simulations. Bioinformatics **19** (2003) 677–678 http://www-aig.jpl.nasa.gov/public/mls/cellerator/.
20. et al., M.H.: The systems biology markup language (SBML): A medium for representation and exchange of biochemical network models. Bioinformatics **19** (2003) 524–531 http://sbml.org.
21. Levchenko, A., Bruck, J., Sternberg, P.W.: Scaffold proteins may biphasically affect the levels of mitogen-activated protein kinase signaling and reduce its threshold properties. PNAS **97** (2000) 5818–5823
22. Kanehisa, M., Goto, S.: KEGG: Kyoto encyclopedia of genes and genomes. Nucleic Acids Research **28** (2000) 27–30
23. Soliman, S., Fages, F.: CMBSlib: a library for comparing formalisms and models of biological systems. In Danos, V., Schächter, V., eds.: CMSB'04: Proceedings of the second Workshop on Computational Methods in Systems Biology. Lecture Notes in Computer Science, Springer-Verlag (2004)
24. Muggleton, S.H.: Inverse entailment and progol. New Generation Computing **13** (1995) 245–286
25. Schaff, J., Loew, L.M.: "the virtual cell". In: Proceedings of the fourth Pacific Symposium on Biocomputing. (1999) 228–239
26. Chen, K.C., Csikász-Nagy, A., Györffy, B., Val, J., Novàk, B., Tyson, J.J.: Kinetic analysis of a molecular model of the budding yeast cell cycle. Molecular Biology of the Cell **11** (2000) 396–391

Towards Reusing Model Components in Systems Biology

Adelinde M. Uhrmacher, Daniela Degenring, Jens Lemcke, and Mario Krahmer

Department of Computer Science,
University of Rostock, D-18051 Rostock, Germany
{lin, daniela.degenring, jens.lemcke, mario.krahmer}
@informatik.uni-rostock.de

Abstract. For reusing model components, it is crucial to understand what information is needed and how it should be presented. The centrality of abstraction being inherent in the modelling process distinguishes model components from software components and makes their reuse even more difficult. Objectives and assumptions which are often difficult to explicitate become an important aspect in describing model components. Following the argumentation line of the Web Service ontology OWL-S, we propose a set of metadata which is structured into profile, process model, and grounding to describe model components. On the basis of the specific model component Tryptophan Synthase, its metadata is refined in XML. The reuse of the described model component is illustrated by integrating it into a model of the Tryptophan operon.

1 Introduction

Modelling in general requires a lot of effort, thus the question how models can be reused is a major research effort in modelling and simulation, e.g. [5,10,24,26,30]. Particularly, the availability of software methods that support a modular design of models and their widespread application has revitalised research on reusability of models. Most of the work has concentrated on the technical interoperation of simulation systems, e.g. [13], or how to build simulation systems that support a hierarchical, modular composition of models, e.g. [11,26]. However, progress on developing model components that can be reused by third parties for different objectives has been slow. One of the central reasons might be that capturing the semantics of a model component in an unambiguous way has so far been elusive.

Much of the subsequent is based on the following assertion: A model is an abstraction of a system to support some concrete objective. Thus, we follow the definition of Minsky [27] that "A Model (M) for a system (S) and an experiment (E) is anything to which E can be applied in order to answer questions about S.". As Cellier [9] points out, this defintion does not describe "models for systems" per se. A model is always related to the tuple system and experiment. A model of a system might therefore be valid for one experiment and invalid for another. In consequence of this definition, it is very unlikely to derive a model being valid for

V. Danos and V. Schachter (Eds.): CMSB 2004, LNBI 3082, pp. 192–206, 2005.

all possible experiments, unless it is an identical copy of the system and thus no longer a model. Modelling is a process of abstraction. It involves simplification and omission of details. Which simplifications are permissible and what details can be omitted, depends on simulation objectives. Therefore, all valid reuse must take the objectives of developing this specific model component into account.

In the following, we will explore possibilities to capture the meaning of a model component to support its reuse in Systems Biology. The paper will be structured as follows. First we will explain the concept of model components and take a look at related efforts in Computer Science. Afterwards, we will suggest a set of metadata to describe the syntax and semantics of model components. Within a modular, hierarchical model of the Tryptophan Synthase at cell level, we will identify possible model components for reuse. An enzyme responsible for the last reaction step of the Tryptophan synthesis will be singled out as a possible candidate. We will fill in the metadata for the identified model component and show the reuse of the model component in a different context. Afterwards, we will conclude by discussing related work in modelling and simulation and in Systems Biology.

2 Model Components

Model components and software components have much in common. Both are units of independent deployment and third-party composition. They should come with clear specifications of what they require and provide. A component should be able to plug and play with other components [20]. Software components offer a service to an unknown environment. Thus, approaches developed to facilitate the identification and discovery of suitable Web Services could possibly be exploited for our purpose. For example, the OWL-based Web Service ontology OWL-S (Ontology Web Language for Services) supplies Web Service providers with a core set of markup language constructs to describe the properties and capabilities of their Web Services in an unambiguous, computer-intepretable way [1]. Based on the Resource Description Framework (RDF) expressing simple semantic relations, DAML+OIL (DARPA[1] Agent Markup Language, Ontology Interface Layer) provides mechanisms for describing complex class and property hierarchies. The Web Services classification OWL-S (formerly DAML-S) utilises this framework to unambiguously classify software components in an ontology while including semantic information. The broad aim is to build up a so-called Semantic Web linking specifications of multiple distributed services together to be used in many applications. Central issue of this approach is to provide a precise description expressed in a formal language to support

- Automatic discovery,
- Automatic invocation,
- Automatic composition and
- Automatic monitoring of Web Services.

[1] Defense Advanced Research Projects Agency

To accomplish these goals, OWL-S is structured into three parts:

1. The service profile for advertising and discovering services. It answers the question: What does the service require of the user(s) or other agents and what does it provide for them?
2. The process model, which gives a detailed description of a service's operation and answers the question: How does it work?
3. The grounding, which provides details on how to interoperate with a service via messages and answers the question: How is it used?

Although having much in common, model components also differ from software components, because they always represent an abstraction of reality. The problem of reusing model components thus seems to be even more difficult than the reuse of software components. Due to different objectives in the modelling process, alternative models of the same physical system may equally use different abstractions. Explorative modelling, multi-resolution and multifaceted modelling emphasise the importance of developing families of models and recognising the existence of multiplicities of objectives and models as a fact of life [14,35]. Selecting a model for reuse requires an understanding of its meaning, part of which refers to its objectives, constraints as well as underlying assumptions. Those are difficult to capture, because they often are only implicitly included, and modellers might not even be aware of many of them.

Components as "a self-contained, interoperable, reusable and replaceable unit that encapsulates its internal structure and provides useful services to its environment through precisely defined interfaces" facilitate the development of models by composition as could be shown in empirical studies [33]. Thus, the number of component-based approaches and component libraries offered by commercial simulation systems is steadily increasing [6]. However, component libraries have so far been restricted to well established, highly specialised and well understood application areas of modelling and simulation offering mostly low-level, prefabricated components whose underlying assumptions and constraints are commonly known. Improving model reuse for areas like Systems Biology requires a significant development in several areas [29]. This includes

- understanding what information is needed to support reuse and how it should be presented,
- developing mechanisms to collect and record this information,
- understanding how to design for reuse,
- developing advanced search tools to locate model components and
- developing criteria to decide when model reuse is desirable.

In the following, we will focus on the questions: What information about model components is needed to support reuse not only syntactically but also semantically, and how can this information be represented? Finding answers seems more tractable if the questions are restricted to a specific domain of interest, as in our case the area of Systems Biology.

3 Defining Metadata for Model Components

The information that we distinguished as crucial in describing a model component will be stored in a set of metadata whose structure resembles OWL-S [1] and the description structure proposed in [19].

The **model component profile** contains the information for advertising and searching a model component. It should contain the overall application domain, the name of the model component, the name of the entity or process that it models, the objective of the model, a short textual description and the simulation environment or simulation formalisms it has been designed for. In Systems Biology, a link to general taxonomies like the gene ontology and the enzyme ontology can be established by the corresponding indices. In [19], the validation, assumptions and the non-domain-dependent classification is included in the profile. In contrast, we moved the former two into the "grounding" and assigned the latter to the "process model".

The **component's process model** describes how the model component works. It thus contains the non-domain-dependent classification of the model's type and a kind of abstract description of the internal processing. If a service is described, the process model is used to specify the coordination strategies based on which the service interacts with other services. Generally, there is typically no interest to reveal the "internals" of software components or services. However, as discussions among modellers showed [2], knowledge about its internals increases the trust into a model component.

The **model component's grounding** contains the interface of a model component. In some areas like embedded systems, models contain interfaces to externally running software. This is hardly the case in Systems Biology. Anyway, answering the question how a model component is used implies to answer the question how it interacts with other models, and whether and how it interacts with the user. Furthermore, before reusing a component, it needs to be parametrised. How to use a model component is also restricted by the underlying assumptions and constraints. They therefore have to be described as well.

4 Tryptophan Synthase — Selection of a Model Component

The Tryptophan Synthase model proposed by [16] has been developed in JAMES [32], which supports modular, hierarchical modelling. Thousands of enzyme models are responsible for converting the incoming metabolites IGP, G3P, Indole and Serine to their products including Tryptophan. Their inputs and outputs are served by a single bulk solution model managing the current amount of freely floating metabolites.

Figure 1 shows part of the GUI of JAMES, which is currently under development [7]. In the upper left corner, the overall hierarchical structure of the model is shown. In the lower left corner, the behavior of an α-subunit is specified as a

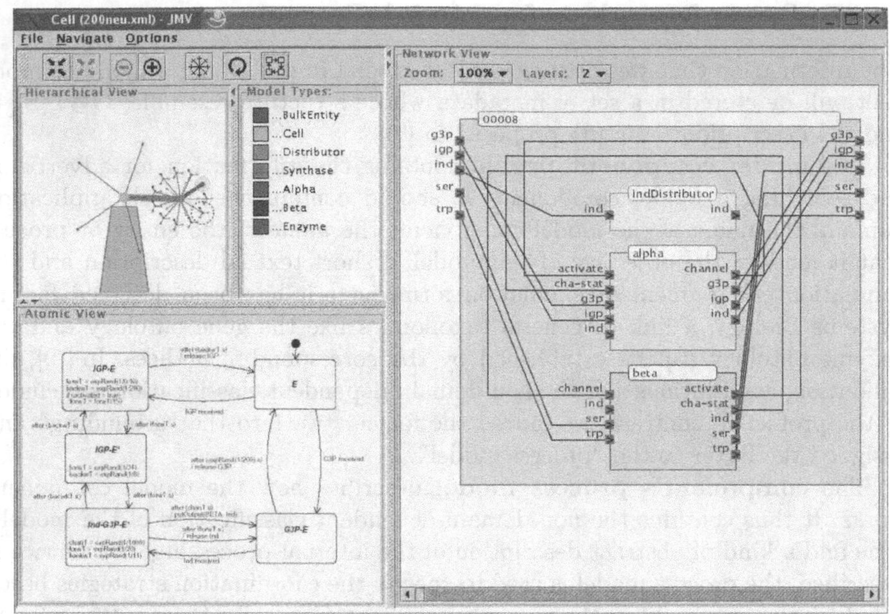

Fig. 1. Screen shot showing the three different perspectives [7]

statechart. At the right hand side, we see the structure of an enzyme with its subunits and their interaction via a static channel.

When identifying model components being eligible for reuse, we face three potential candidates: the population of enzymes, the bulk solution and single enzymes. The bulk solution serves as an experimental frame to the thousands of enzymes. Thus, candidates for reuse are either the population of homogeneous enzymes or one single enzyme. This leads us to the problem of selecting a suitable granularity of components to be stored in a reuse library. If components are at too low a level, the modelling process resembles coding from scratch. If model components are high-level aggregates, then their reuse is limited. Independent of the application domain, the simulation system already offers specialised, coupled model components that can be parametrised to contain an arbitrary number of model components of certain classes interacting in a homogeneous manner. In addition, the properties of the population are entirely specified by its members, the single enzymes, which comprise two subunits, the α- and the β-subunit interacting with each other via the static channel. Thus, from their complexity, they seem to form suitable building blocks for models.

5 Metadata for the Model Component Tryptophan Synthase

Based on the example of the Tryptophan Synthase, the metadata of this model component, whose overall structure has been discussed in Sect. 3, will be refined and filled in an XML format. XML is widely used for storing and exchanging models. For a general discussion see [17] and for Systems Biology see Cell Markup Language (CellML) [12] and Systems Biology Markup Language (SBML) [3]. In addition, an XML specification will allow to exploit multimedia databases for an efficient storage and retrieval of model components [23].

The current name in the profile (Fig. 2) reflects the fact that several models of Tryptophan Synthase might exist, which might vary referring to the model formalims used or with respect to the model's abstraction level. For example, `AAA` refers to the pathway, i.e. aromatic amino acid biosynthesis, `Trp` is a shortform for the Tryptophan Synthase, `Ecoli` is the organism — as the properties of enzymes differ between different organisms. To keep the figure simple, the `name` of the

```
<profile>
  <name>Enzyme.AAA.Trp.Ecoli.DEVS.Model_1</name>
  <application_domain>Enzymology;Systems Biology</application_domain>
  <domain_dependent_classification>
    GO:0004834 ; EC:4.2.1.20
  </domain_dependent_classification>
  <text_description>
    Tryptophan Synthase is an enzyme classified by the E.C. number:
    EC 4.2.1.20. The described enzyme stems from the organism E.coli.
    It produces Tryptophan and Glycerole-3-phosphate and
    consumes Serine, Indole-glycerole-3-phosphate.
    The model component is developed using the DEVS formalism.
  </text_description>
  <objective>
   Analysing the static channeling effect; Analysing the behavior of
   single enzymes.
  </objective>
  <responsible_persons>
    <person function="developer">
      <name>Daniela Degenring</name>
      <e-mail>daniela.degenring@informatik.uni-rostock.de</e-mail>
    </person>
  </responsible_persons>
  <simulation_environment>
    <simulation_system>
      <name>James</name>
      <version type="direct_match">CoSA 1.0</version>
      <subclass_of>
         <name>DEVS-Simulators</name>
         <information type="URI">
           http://www.sce.carleton.ca/faculty/wainer/standard/
         </information>
      </subclass_of>
    </simulation_system>
    <platform>
      <name>Java</name>
      <version type="same_or_above">1.4.1_03</version>
    </platform>
  </simulation_environment>
  <executable type="uri">
    http://www.informatik.uni-rostock.de/~dd012/models/trpsynth.jar
  </executable>
  <references type="uri">
    http://www.informatik.uni-rostock.de/~dd012/models/trpsynth.bib
  </references>
</profile>
```

Fig. 2. The profile of the model component

```
<process>
  <modelling_and_simulation_classification>
    <separated_model_and_simulation>yes</separated_model_and_simulation>
    <model>
      <class>Discrete-Event</class>
      <formalism>DEVS</formalism>
      <topology>Coupled model</topology>
      <scale_of_variables>Qualitative</scale_of_variables>
      <scale_of_variables>Semi-quantitative</scale_of_variables>
      <type_of_events>Situation-triggered</type_of_events>
      <type_of_events>Time-triggered</type_of_events>
      <stochastics>
        <applied>yes</applied>
        <distribution>Exponential</distribution>
      </stochastics>
      <world_view>Process-based</world_view>
      <world_view>Statecharts</world_view>
    </model>
  </modelling_and_simulation_classification>
  <metamodel>
    <type>Statechart</type>
    <figure type="uri">
      http://www.informatik.uni-rostock.de/~dd012/models/trpsynth/doc/chart
    </figure>
    <animation type="uri">
      http://www.informatik.uni-rostock.de/~dd012/models/trpsynth/doc/chart.avi
    </animation>
  </metamodel>
  <expected_effects_of_parameter_change>
    <text>
      Effects of changing tunnel-capacity see Degenring (2003)
      For other effects see Anderson (1995)
    </text>
  </expected_effects_of_parameter_change>
</process>
```

Fig. 3. The process model of the model component

```
...
<xsd:element name="model">
  <xsd:complexType>
    <xsd:sequence>
      <xsd:element name="class" minOccurs="1" maxOccurs="unbounded">
        <xsd:simpleType>
          <xsd:restriction base="xsd:string">
            <xsd:enumeration value="Continuous"/>
            <xsd:enumeration value="Discrete-event"/>
            <xsd:enumeration value="Discrete-stepwise"/>
          </xsd:restriction>
        </xsd:simpleType>
      </xsd:element>
      ...
    </xsd:sequence>
  </xsd:complexType>
</xsd:element>
...
```

Fig. 4. Exemplary XML Schema fragment for defining the type of model

model component is represented by a plain string of characters. However, to enable an automatic model retrieval, a more structured way of defining the `name`'s constituents is needed, e.g. by a hierarchy of XML tags. In the same way, the `domain_dependent_classification` has to be refined, e.g. by using XML namespaces.

The model's classification is part of the process model (Fig. 3). It refers to the type of model, e.g. whether it is a continuous, discrete-event or discrete-

```
<grounding>
  <model_component_to_model_component>
    <input type="materialistic" unit="1">IGP</input>
    ...
    <input type="materialistic" unit="1">Serine</input>
    <output type="materialistic" unit="1">Tryptophan</output>
    ...
    <output type="materialistic" unit="1">Indole</output>
    <output type="informational">Enzyme.dead</output>
  </model_component_to_model_component>
  <model_component_to_model_user>
    <output type="model_state">Internal phase</output>
    <output type="statistical">Tunnel occupation</output>
  </model_component_to_model_user>
  <invariant type="m_c_to_m_c">
    <sbml xmlns="http://www.sbml.org/sbml/level1" level="1" version="1">
      <model name="Tryptophan Synthesis">
        <reaction name="IGP + Ser -> G3P + Trp">
          <listOfReactants>
              <specieReference specie="IGP" stoichiometry="1"/>
              <specieReference specie="Ser" stoichiometry="1"/>
          </listOfReactants>
          <listOfProducts>
              <specieReference specie="Trp" stoichiometry="1"/>
              <specieReference specie="G3P" stoichiometry="1"/>
          </listOfProducts>
        </reaction>
      </model>
    </sbml>
  </invariant>
  <integration_with_other_models>
    <text>
      as part of a population within a multi-level, or micro
      model for analysing purposes,
      single enzyme combinations for education purposes
    </text>
    <documented_use_of_model_component>
      <text>
        Degenring (2003); Degenring (2004)
      </text>
  </integration_with_other_models>
    .
    .
    .
```

Fig. 5. The interface specification as part of the grounding the model component

stepwise model. Again, the `model`-tag contains just a plain string of characters to define the classification. In order to give an unambiguous description, it would be desirable for simulationists to agree upon some common and precise terms to identify and specify these properties (Fig. 4).

In the process model (Fig. 3), the tag `metamodel` refers to a metamodel, which forms a model of the executable model component. The metamodel shall provide a more abstract, often visual representation to the user, thus facilitating the understanding of a components' potential dynamics. In JAMES, if the model component is sufficiently simple, it can directly be specified as a statechart and no further abstraction via a separate metamodel is needed. However, often the statechart will form an abstraction and thus can be interpreted as a metamodel of the underlying atomic model in JAMES. In this context, statecharts are only one possibility to

```
    .
    .
    .
  <parametrisation>
    <parameter minValue="gt0" maxValue="not_yet_explored"
                              default="0.02222" unit="sec">
      E-Ser_to_E~AA
    </parameter>
    ...
    <parameter minValue="gt0" maxValue="not_yet_explored"
                              default="0.001" unit="sec">
      Ind-G3P-E*_to_Channel
    </parameter>
    <parameter minValue="gt0" maxValue="not_yet_explored"
                              default="150" unit="1">
      IGP-E_to_IGP-E*_Speed-up
    </parameter>
  </parametrisation>
 <underlying_assumptions>
    <text>
      The model deals with molecule numbers instead of concentrations.
      Due to the transformation from a differential equation to a
      stochastic
      discrete-event model according to Gillespie (1976), more than 100
      model components of this type should possibly be included for
      analysing purposes.
    </text>
  </underlying_assumptions>
  <validation>
    <text>
      Reproduction of original experimental and simulation results
      Anderson (1995).
      Additionally, the effect of the tunnel capacity was investigated
      with plausible results Degenring (2003).
      Ongoing work: implementation of the model into a Trp-operon model
      according to Santillan (2001).
    </text>
  </validation>
</grounding>
```

Fig. 6. The parametrisation, assumptions, and validation as part of grounding the model component

define a metamodel: e.g. live sequence charts, e.g. [21], Petri Nets, e.g. [28], graphs, e.g. [22], and even pseudocode are alternatives. Regarding the metamodel, it is important that it contains a declarative and easy to understand description of the underlying model. However, what is easier to understand depends on the context of the user. Alternative descriptions should therefore be included. Developing different metamodels of a single model component requires a lot of effort and has to be balanced to the effort of directly inspecting its source code.

The main part of the grounding is concerned with the specification of the interface (Fig. 5). Most inputs and outputs are of type `materialistic` which means that they are not multiplied if they are sent to multiple addressees. In contrast, information is a non-consumable ressource and thus can be multiplied. Single molecules form the inputs and outputs of our model component. Between those inputs and outputs of a model component, invariants are defined. In the case of our enzyme, the mass conservation law holds between consumed and produced species, which has been specified according to the SBML [3] suggestion for chemical reactions.

Model components include a notion of time. In simulation, we typically distinguish between three types of time: the physical time, which refers to the time of the modelled physical system; the simulation time, which refers to the representation of the physical time within the simulation; and the wall-clock time, which refers to the time that progresses during executing the simulation. In the above example, one unit of simulation time corresponds to one second in physical time. The model component gives a stochastic, discrete-event description of the behaviour of an enzyme. Thus, methods to estimate parameters and to analyse the output of stochastic, discrete-event simulation have to be considered. For a discussion of implications of using stochastic simulations in Systems Biology see e.g. [25].

6 Reuse of the Model Component

The major role of the Tryptophan Synthase lies in the enzyme cascade for the production of Tryptophan. The Tryptophan Synthase is produced by transcription and translation of the Tryptophan operon. The process of transcription and translation is regulated by the amount of Tryptophan. So we could imagine to combine our model component with other enzymes to capture the entire aromatic amino-acid biosynthesis, or we could integrate it with a model of the Tryptophan operon.

As Trp-enzyme catalyses the last reaction step in the Tryptophan synthesis, the model component can be combined with other enzyme models, which proliferate the substrates of the reaction, i.e. mainly the IGP Synthase or rather the Serine Synthase, and corresponding enzymes. A coupling with enzymes, which consume the built products, G3P and Tryptophan, like the G3P Dehydrogenase catalysed reaction (EC 1.1.99.5) or the Trp-tRNA-synthesising enzyme, is also possible and would direct the focus to their products. In addition, the products of the Tryptophan model component can interact with other model components as effectors. The inhibiting influence of Tryptophan onto the transcription of the Trp-operon via the process of attenuation or gene repression is e.g. a known fact. Since all metabolites participate at the same time in other reactions, no direct coupling between various enzymes is proposed but an indirect interaction via a model like our bulk solution might prove beneficial. The model of the bulk solution records the amount of available metabolites, enzymes, genes, and mRNA, and calculates the frequency of collissions.

Whereas the above combinations act on more or less the same time scale abstraction, analysing the regeneration of the Trp Synthase requires to combine the model component Trp Synthase with other models that describe the operon and thus act at a different time scale. Here the model component Trp-operon produces the model component Trp Synthase during the process of transciption and subsequent translation. This type of model requires that the simulation system supports variable structure models, i.e. models that change their composition and interaction during simulation as e.g. JAMES does.

We integrated several hundreds of the model component Tryptophan Synthase into a model that describes the transcription dynamics of the Tryptophan operon focussing on gene regulation and monitoring switch activities. First experiments executed in JAMES were able to reproduce the results documented in [31], which was no big surprise considering the description of the model component Tryptophan Synthase. The model component Tryptophan Synthase had been validated under similar assumptions, i.e. assuming comparatively high numbers of the Tryptophan model component, and the integration with the operon has even been foreseen by the model component's description. Dealing with different time scales is the virtue of discrete-event simulation, so no errors should have been induced by the simulation engine. The only question was, whether a simpler model of the enzyme would not have sufficed, as the role of the channel was not the objective of the current simulation study. However, this question has still to be explored. The integration of the current model component would allow to test more sophisticated hypotheses about structural interdependencies within the overall gene expression process. For example, if the objective had been to test the hypothesis how a damaged gene affects the capacity of the channel, the model component Tryptophan Synthase with its explicit subunits and the channel would have appeared most suitable.

7 Related Work in Systems Biology

Major efforts in modelling and simulation are aimed at supporting the reuse of models and entire simulations. The High Level Architechture (HLA) [13] is a general purpose architecture for simulation reuse and interoperability. Developed under the leadership of the Defense Modelling and Simulation Office (DMSO), it provides a standard means (IEEE 1516) for individual simulations or federates to interoperate in a federation since 1996. HLA is mostly concerned with synchronising different simulations via the Run-Time Infrastructure (RTI) and thus with the question of interoperation [8]. However, for HLA to work, the interoperable federates share a common Federation Object Model (FOM) document that describes the types of information they are exchanging. This metadata not only includes information needed for synchronisation and coordination by the RTI software such as classes, attributes, and interactions, but also includes descriptive information such as the objective and sponsor of the federation or federate. HLA focuses on the concrete synchronisation of different simulation systems rather than on the retrieval of models.

Other major research efforts are directed towards improving the exchange of models between simulation systems in Systems Biology. Representatives are the Systems Biology Markup Language (SBML) and the Cell Markup Language (CellML) [12]. Whereas SBML is meant to support basic biochemical network models, CellML covers a more general field of application including electrophysiological and mechanical models as well as biochemical pathway models. Both provide in addition to parameter definitions information about the equations of the underlying biological processes such as reaction mechanisms. Those can be executed by simulation systems supporting an SBML or CellML interface. Thus, a complete, unambiguous description of the model is required to enable these different simulation systems to execute it.

Additionally, CellML also supports the assignment of metadata to facilitate the reuse of model components by providing background information. Similarly to our approach, e.g. information on the species is given by referring to established ontologies and making use of the biological databases on the web. The underlying assumptions are covered in the limitations and validation slots. The type of model is characterised by referring to an ontology of mathematical problems as CellML and SBML focus on continuous system models.

In contrast, our approach is not restricted to the continuous realm. It therefore becomes important to explicitly represent what model formalism is used. Whereas continuous models can typically be described by differential or mathematical equations, no such general exchange format does exist for models belonging to different modelling formalisms. To execute and thus eventually reuse a model, a link to its specifation and simulation engine or its entire implementation is often unavoidable. Thus at this point, the reuse of general models reveals a Web Service characteristic. Current efforts like multi-formalism modelling are directed towards facilitating the reuse of models of different formalisms [34,15]. The idea is to provide a meta description of these formalisms. So in the future, it might be possible to include the specifiation of the model and a meta description of the formalism used.

8 Conclusion

The reuse of components promises to facilitate the design of models. We suggested a set of metadata to describe model components and illustrated its use with the model component Tryptophan Synthase. The overall structure of the metadata was influenced by the ontology for Web Services OWL-S, whose features profile, process model and grounding have been redefined in this context. The metadata are structured to distinguish between information used for retrieving a component, i.e. profile, information about how the component works, i.e. the process model, and the specification of its interface, i.e. grounding. As do [29], we believe that a key part of a model description must involve capturing the objectives, assumptions and constraints under which the original models were developed in a form that can be searched and analysed. Currently, our proposed metadata contains still many text slots. For an advanced search that would be able to identify model components of interest, for consistency checks to recognise incompatibilities among model components, and to help determining the fidelity of the entire model, the proposed metadata present only a first step.

Compared to approaches like SBML and CellML, our approach focuses on the retrieval of components rather than on supporting the exchange of models between simulation systems. Our long term goal is similar to that of OWL-S: providing meta information about a model in a formal, rigorous manner to support the automatic discovery and composition of model components. For the final execution of models, we will depend on efforts like SBML and CellML or, as our approach is not restricted to a particular modelling and simulation paradigm, on efforts like multi-formalism modelling and simulation.

As in the case of modelling in general, we are confronted with the problem to provide as little information as needed but not less. Many modellers might find the inspection of the source code more informative. Of course, this would not allow to manage model component repositories in an effective and efficient manner. Still the question is, whether researchers are likely to use and maintain these repositories, and whether we now understand what information is needed to support reuse of model components. The answer can not be given by us but can only found in discussions with third-party users. Therefore, an integration into efforts like CellML or SBML is mandatory.

Acknowledgements

We would like to thank the anonymous and non-anonymous referees Jan Himmelspach and Mathias Röhl for their comments on an earlier version of this paper. Part of the research presented in this paper has been funded by the DFG.

References

1. http://www.daml.org/services/owl-s/1.0/.
2. http://www.dagstuhl.de/04041/.
3. http://www.sbml.org/docs/.
4. K.S. Anderson, A.Y. Kim, J.M. Quillen, E. Sayers, X.J. Yand, and E.W. Miles. Kinetic characterization of channel impaired mutants of tryptophan synthase. *The Journal of Biological Chemistry*, 270(50):29936–29944, 1995.
5. J. Aronson and P. Bose. A model-based aproach to simulation composition. In *Proceeding of the Fifth Symposium on Software Reusability*, pages 73–82, 1999.
6. F. Barros and H. Sarjoughian, editors. *Special Issue on Component-Based Modelling and Simulation.* Simulation - Transactions of the SCS Simulation. Sage, 2004.
7. S. Biermann, A.M. Uhrmacher, and H. Schumann. Supporting multi-level models in systems biology by visual methods. In *Proceedings of European Multi-Simulation Conference*, page submitted, 2004.
8. A. Buss and L. Jackson. A comparison of hla, corba, and rmi. In *Proceedings of the 1998 Winter Simulation Conference*, pages 819–825, 1998.
9. F. E. Cellier. *Continuous System Modeling.* Springer, New York, 1992.
10. G. Chen and B. K. Szymanski. Object-oriented paradigm: component-oriented simulation architecture: toward interoperability and interchangeability. In *Proceedings of the 2001 Winter Simulation Conference*, pages 495–501, 2001.
11. G. Chen and B. K. Szymanski. Cost: A component-oriented discrete event simulator. In *Proceedings of the 2002 Winter Simulation Conference*, pages 776–782, 2002.
12. A.A. Cuellar, C.M. Lloyd, P.F. Nielsen, D.P. Bullivant, D.P. Nickerson, and P.J. Hunter. An overview of cellml 1.1, a biological model description language. *Simulation - Transactions of the SCS*, 79(12):740–747, 2003.

13. J. Dahmann, R. Fujimoto, and R. Weatherly. The dod high level architecture: An update. In *Proceedings of the 1998 Winter Simulation Conference*, pages 797–804, 1998.
14. P.K. Davis, J.H. Bigelow, and J. McEver. Exploratory analysis and a case history of multi-resolution, multiperspective modeling. Technical Report RP-925, RAND, 2000.
15. J. de Lara and H. Vangheluwe. AToM3: a tool for multi-formalism and meta-modelling. In *European Joint Conference on Theory And Practice of Software (ETAPS), Fundamental Approaches to Software Engineering (FASE)*, pages 174 – 188, 2002.
16. D. Degenring, M. Röhl, and A. M. Uhrmacher. Discrete event simulation for a better understanding of metabolite channeling- a system-theoretic approach. In Priami C., editor, *Lecture Notes in Computer Sciences*, volume 2602, pages 114–126. Springer Verlag Heidelberg, 2003.
17. P.A. Fishwick. Using xml for simulation modeling. In *Proceedings of the 2002 Winter Simulation Conference*, 2002.
18. D. T. Gillespie. A general method for numerically simulating the stochastic time evolution of coupled chemical reactions. *Journal of Computational Physics*, 22:403–434, 1976.
19. M. Heisel, J. Luethi, A.M. Uhrmacher, and E. Valentin. A description structure for simulation model components. In *Proceedings of the Summer Simulation Conference*, page submitted, 2004.
20. B. Hnich, T. Jonsson, and Z. Kiziltan. On the definition of concepts in component based software development. Technical report, University Department of Information Science, Uppsala, Sweden, 2000.
21. N. Kam, D. Harel, H. Kugler, R. Marelly, A. Pnueli, E.J. Hubbard, and M.J. Stern. Formal modelling of c. elegans development: A scenario based approach. In Priami C., editor, *Lecture Notes in Computer Sciences*, volume 2602, pages 4–20. Springer Verlag Heidelberg, 2003.
22. H. Kitano. A graphical notation for biochemical networks. *Biosilico*, 1(5):169–176, 2003.
23. M. Klettke and H. Meyer. *XML & Datenbanken - Konzepte, Sprachen und Systeme.* DPunkt Verlag, 2003.
24. G. Mackulak and F. Lawrence. Effective simulation model reuse; a case study for amhs modeling. In *Proceedings of the 1998 Winter Simulation Conference*, pages 979–984, 1998.
25. M. Marhl. Transition from stochastic to deterministic behaviour in dependence on the divergence of systems. In *3rd Workshop on Computaiton of Beiochemical Pathways and Genetic Networks*, pages 49–58. Logos Verlag, 2003.
26. Y. Miller, J. A. Ge and J. Tao. Component-based simulation environments: Jsim as a case study using java beans. In *Proceedings of the 1998 Winter Simulation Conference*, pages 373–381, 1998.
27. M. Minsky. Models, Minds, Machines. In *Proc. IFIP Congress*, pages 45–49, 1965.
28. M. Nagasaki, A. Doi, H. Matsuno, and S. Miyano. Genomic object net: a platform for modeling and simulating biopathways. *Applied Bioinformatics*, 2003.
29. C. M. Overstreet, R.E. Nance, and O. Balci. Issues in enhancing model reuse. In *First International Conference on Grand Challenges for Modeling and Simulation*, 2002.

30. E. Page and J. Opper. Observations on the complexity of composable simulation. In *Proceedings of the 1999 Winter Simulation Conference*, pages 553–560, 1999.
31. M. Santill'an and M. C. Mackey. Dynamic regulation of the tryptophan operon: A modeling study and comparison with experimental data. *Proceeding of the National Academy of Sciences of the USA*, 98(4):1364–1369, 2001.
32. A. M. Uhrmacher, P. Tyschler, and D. Tyschler. Modeling Mobile Agents. *Future Generation Computer System*, 17:107–118, 2000.
33. E. C. Valentin, A. Verbraeck, and H. G. Sol. Effect of simulation building blocks on simulation model development. In *Proceedings of the International Conference of Technology, Policy and Innovation*, pages CD–ROM proceedings, 2003.
34. H. Vangheluwe and J. de Lara. Meta-models are models too. In *Proc. of the Winter Simulation Conference*, pages 597 – 605, 2002.
35. B.P. Zeigler. *Multifacetted Modelling and Discrete Event Simulation.* Academic Press, London, 1984.

VICE: A VIrtual CEll

D. Chiarugi[1], M. Curti[1], P. Degano[1], and R. Marangoni[1]

Dipartimento di Informatica, Università di Pisa
Via Filippo Buonarroti,2, I-56127 Pisa, Italy
dvchiaru@tin.it, {curtim, degano, marangon}@di.unipi.it

Abstract. We report on the specification and analysis of VICE, a hypothetical cell with a genome as basic as possible. We used an enhanced version of the π-calculus and a prototype running it to study the behaviour of VICE. The results of our experimentation *in silico* confirm that our virtual cell "survives" in an optimal environment and shows a behaviour similar to that of real prokaryotes.

1 Introduction

A main problem of contemporary biology is understanding the dynamics of genes and proteins inside the cellular molecular machinery, when they give rise to a living organism [15]. Unfortunately, nowadays there are no experimental techniques able to track the dynamics of the complete metabolome of a cell. A promising approach is to represent all the known relationships between the elements in a metabolome *in silico*, so building up a sort of a *virtual cell* [11, 17].

The choice of the organism to model is crucial, because even simple biological entities, like bacteria, have a complexity extremely high; their simulation thus requires huge computational resources. Thus, we designed a hypothetical organism, possessing a very basic genome. To do this, we refined a previous work [19] that proposes a basic prokaryote genome by eliminating duplicated genes and other redundancies from the smallest known bacterial genomes. We further modified this proposal and we obtained a very basic prokaryote-like genome, which only contains 180 different genes. We wrote the metabolic pathways for this hypotetic organism, and we obtained a theoretical description its molecular machinery. Our goal is to use this hypotetical genome to represent a *whole* living organism, called VICE, which can be seen as a simplified prokaryote.

The contribution from the computer science side comes from the observation that cell's mechanisms and global computing applications are closely related [23]. Biological components and organisms can be seen as processes and as networks of processes, respectively, while cell interactions are represented as communications between processes [21, 24].

To describe VICE we chose a formalism typically used to specify the behaviour of networks of processes, following [3, 4, 5, 21, 23, 24, 22]. More precisely, we adopted an enhanced operational semantics for the π-calculus [18, 4, 5] as a simulation tool to handle the complexity of biological systems within a uniform

V. Danos and V. Schachter (Eds.): CMSB 2004, LNBI 3082, pp. 207–220, 2005.

framework. A computation of a network of processes representing a biological cell can then be interpreted as the description of the combination of several metabolic pathways, and graphically displayed. In this way, a biological system is described through a formal language. Also, its dynamic behaviour is mechanically extracted from the textual, mathematical, compositional description above. The graphical representation of system behaviour shows then each step of the reactions involved, at the level of abstraction chosen by the designer. Furthermore, a solid theory exists for the π-calculus, upon which techniques and tools for mechanically analysing system behaviour have been built; all these can then be profitably reused in the biological field.

An important feature of our non-conventional semantics is its capability of describing various aspects of systems orthogonally, among which qualitative ones, like *causal* or *locational* relations between actions [5], or *stochastic* information [2, 20]. Also, these aspects are computed mechanically. Indeed, we built an iterpreter of the π-calculus based on the above semantics, and a small set of prototypical analysis tools.

We exploited the tools mentioned above to study the behaviour of VICE in interphase. Since we modelled a *whole* cell, we are able to study *all* its metabolic pathways. In particular we tested that VICE is able to "survive" in a "normal" environment with enough water and nutrients. Also, we found that its behavior is similar to that of real prokaryotes acting *in vivo* under similar circumstancies, at least for the environmental conditions investigated.

Further investigations to asses our proposal will certainly include the simulation of the effects of perturbations induced on the metabolic network, like the knocking down of a gene or the presence of enzyme inhibitors. These investigations are particularly important from a biological point of view; we stress that they require the description of a whole cell, as we did, rather than that of some of its metabolic pathways.

The next section intuitively presents the main features of our enhanced π-calculus. Section 3 discusses the biological objects of VICE and the choices we made to design it. In section 4 we report on the experiments made *in silico* and on their interpretation in biological terms. Appendix A intuitively describes the pathways of VICE modelled so far; Appendix B displayes the formal specification of two pathways: Glycolysis and Glycolipid.

2 The Calculus

With the help of a biochemical example, we briefly show how to use the enhanced π-calculus for modelling a biological cell. We shall use the enhanced reduction semantics proposed in [4], that refines previous work on the SOS enhanced operational semantics by Degano and Priami [6]. It permits to express both qualitative and quantitative aspects of process behaviour. In particular, here we extend the proposal in [4] with stochastic information along the line of [2, 20].

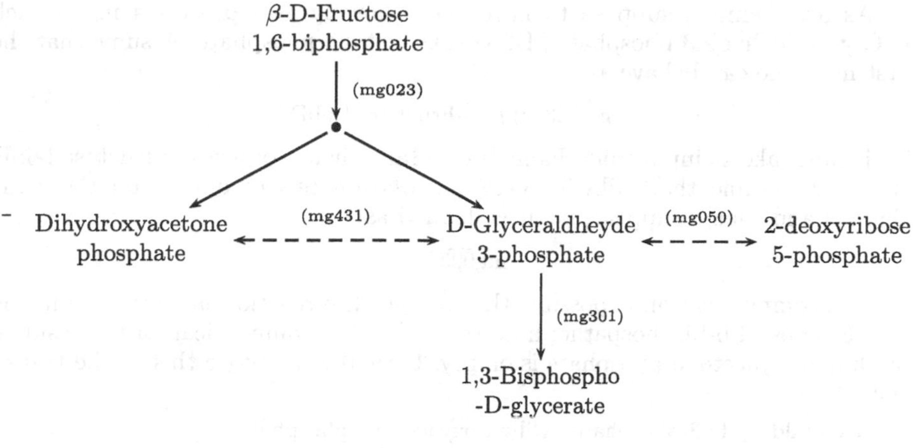

Fig. 1. A fragment of the Glycolysis Pathway

For the sake of readability, below we shall introduce the enhanced reduction semantics in a manner as intuitive as possible. Therefore, we shall omit most of the formal definitions (e.g. those of $\equiv$ and $\triangleright$ occurring in Table 1), and refer the interested reader to [5] for a precise presentation, more details and explanations.

Consider the (portion of) Glycolysis shown in Fig. 1. For denoting metabolites we use their IUPAC names, whereas for enzymes we use the code in [9], to which we refer the reader. Here, we only recall that in Fig. 1 we used: (i) mg023 for fructose-bisphosphate aldolase, (ii) mg431 for phosphoglycerate mutase, (iii) mg301 for glyceraldehyde 3-phosphate dehydrogenase, and (iv) mg050 for 2-deoxyribose-5-phosphate aldolase. The Appendix B contains our specification of the whole Glycolisys (and of other connected oxidation reactions) in the π-calculus.

The metabolite β-D-fructose-1-6BP splits giving Dihydroxyacetone phosphate and D-Glyceraldehyde 3-phosphate. In turn, Dihydroxyacetone phosphate, catalyzed by the enzyme mg431, becomes D-Glyceraldehyde 3-phosphate. This can behave in three different manners, becoming either the metabolite 1,3-Bisphospho-D-glycerate via mg301, or dihydroxyacetone phosphate via mg431, or 2-deoxyribose 5-phosphate via mg050.

We specify molecules as *concurrent processes* $A_1, \ldots, A_n$ put in parallel, and written as $A_1 \mid \ldots \mid A_n$. Each process can perform input/output actions on given *channels*. The basic mechanism of *communication* allows processes to exchange information, flowing from the *sender* to the *receiver*. The input/output actions can be sequentially composed with the "." operator. In turn, we use channels to model catalyzing enzymes.

A reaction between two molecules catalyzed by an enzyme is then represented by a communication. Of course, the molecules involved are modelled by two processes that perform complementary input/output actions on the channel that models the catalyzing enzyme.

As an example, suppose to have the following two processes in parallel: D-Glyceraldehyde_3-phosphate | Dihydroxyacetone_phosphate. Assume that the first molecule can behave as

$$\text{mg023}(x).\beta\text{-D-Fructose\_1,6bP}$$

i.e. it can make an input from channel$mg023$and then produceβ-D-fructose-1-6bP; similarly, assume that Dihydroxyacetone phosphate can output on the same channel and then disappear, i.e. it is defined as

$$\overline{\text{mg023}}\langle a \rangle.$$

Then a communication is possible that models the reaction *aldolase*, producing β-D-Fructose 1,6-bisphospathe; note that after the communication the residual of Dihydroxyacetone phosphate is empty. Formally, we write this as the transition:

$$\text{D-Glyceraldehyde\_3-phosphate} \mid \text{Dihydroxyacetone\_phosphate} \xrightarrow[R]{\langle ||_0 mg023(x), ||_1 \overline{\text{mg023}}\langle a \rangle \rangle} \beta\text{-D-fructose-1-6bP} \mid \mathbf{0}$$

understanding that a communication occurred, made by an input from the left hand-side process (w.r.t. the | operator) and an output made by the right hand-side process. The transition transforms the system at the left of the arrow to that at its right (note that in this simple example there is no real exchange of information from the sender and the receiver, but they only synchronize as the dummy value a is never used afterwards).

Besides input/output actions, we also have (a family of) *silent* actions, denoted by τ_i. We use them to represent (composite) cell activities, the detail of which we are not interested in. For readability, we shall always index silent actions with the involved enzyme. For example,

$$\beta\text{-D-fructose-1-6bP} = \tau_{mg023}.\ (\text{Dihydroxyacetone\_phosphate} \mid \text{D-Glyceraldehyde\_3-phosphate})$$

represents that β-D-fructose-1-6bP spontaneausly splits in two molecules called Dihydroxyacetone phosphate and D-Glyceraldehyde 3-phosphate, in presence of the enzyme mg023.

Mutually exclusive behaviour of the same reactant is modelled by a *non-deterministic choice*, denoted by +. For example, D-Glyceraldehyde 3-phosphate (defined below) has either of the following four behaviour, according to which enzymes is present: (i) execute a τ_{mg301} becoming 1,3-Bisphospho-D-glycerate, or (ii) execute a τ_{mg431} becoming Dihydroxyacetone phosphate, or (iii) execute a τ_{mg050} becoming 2-Deoxyribose 5-phosphate, or (iv) interact with Dihydroxyacetone phosphate through the *mg023* channel (see above for the definition of D-Glyceraldehyde 3-phosphate).

$$\begin{aligned}
\text{D-Glyceraldehyde_3-phosphate} &= \tau_{mg301}.\text{1,3-Bisphospho-D-glycerate}+\\
&\quad \tau_{mg431}.\text{Dihydroxyacetone_phosphate}+\\
&\quad \tau_{mg050}.\text{2-Deoxyribose-5Phosphate}+\\
&\quad \text{mg023}(x).\beta\text{-D-fructose-1-6bP}\\
\text{Dihydroxyacetone_phosphate} &= \tau_{mg431}.\text{D-Glyceraldehyde_3-phosphate} + \overline{\text{mg023}}\langle a \rangle
\end{aligned}$$

Iteration of the same behaviour is specified through the definition of a process, whose name occurs within the definition itself (*constant definition*). For example, we can (abstractly) specify the aquisition/release of energy by the ATP/ADP molecules as the following two mutually recursive constants:

$$\mathrm{ATP} = \tau_{aq}.\mathrm{ADP}$$
$$\mathrm{ADP} = \tau_{rl}.\mathrm{ATP}$$

where τ_{aq} and τ_{rl} model the exchange of energy (we omit here the involved inorganic ortho-phosphate).

The basic difference between the standard π-calculus and the enhanced one is the notion of *address*. Addresses, ranged over by ϑ, are sequences of $||_0$ and $||_1$, and represent a *unique identifier* for the *sequential* processes of a whole process (roughly speaking, those sub-process with an action or a summation as top-level operator). As an example, we write below β-D-Fructose 1,6-bisphosphate with the relevant addresses:

$$\beta\text{-D-Fructose\_1-6bP} = \tau_{mg023}.\,(||_0\text{Dihydroxyacetone\_phosphate} \mid ||_1\text{D-Glyceraldehyde\_3-phosphate})$$

Intuitively, the subprocess (and all its subprocesses, if any) that lays at the right of the | have an address that begins with $||_0$; similarly those on the left get addresses beginning with $||_1$; in this way one knows that a subprocess with address $||_0$ comes from Dihydroxyacetone phosphate. The user is not required to write addresses, as these are mechanically attached to the relevant processes.

We assume on processes the standard structural congruence $\equiv$ and omit here its definition. Intuitively, it permits to see the processes as floating in a solution; among other properties, it considers thus $A \mid B \equiv B \mid A$ (i.e. | is commutative) and $A \mid (B \mid C) \equiv (A \mid B) \mid C$ (i.e. | is associative).

The dynamics is specified by a labelled transition system: processes perform transitions in sequence, giving rise to *computations*. As said, each transition models a biological reaction (or an abstraction of it, modelled as an internal move). For example, consider the following system made of two copies of β-Fructose 1,6-bisphosphate (the *initial state*) and the following four transitions (leading to the *final state*) of one of its computations (**0** is the terminated process):

β-D-fructose_1-6bP | β-D-fructose_1-6bP

$$\xrightarrow{||_0\tau_{mg023}}_R$$

(D-Glyceraldehyd_3-phosphate | Dihydroxyacetone_phosphate) | β-D-fructose_1-6bP

$$\xrightarrow{||_1||_1\tau_{mg301}}_R$$

(1,3-Bisphospho-D-glycerate | Dihydroxyacetone_phosphate) | β-D-fructose_1-6bP

$$\xrightarrow{||_1\tau_{mg023}}_R$$

(1,3-Bisphospho-D-glycerate | Dihydroxyacetone_phosphate) |
(D-Glyceraldehyde_3-phosphate | Dihydroxyacetone_phosphate)

$$\xrightarrow{\langle ||_0||_1\overline{\mathrm{mg023}}\langle a\rangle,\, ||_1||_0 mg023(x)\rangle}_R$$

(1,3-Bisphospho-D-glycerate | **0**) | (β-D-fructose_1-6bP | Dihydroxyacetone_phosphate)

Table 1. Non-interleaving reduction semantic

$$\textbf{Com}: \ (R+\vartheta||_i\vartheta_0 x(w).P)|(S+\vartheta||_{1-i}\vartheta_1\overline{x}\langle y\rangle.Q) \xrightarrow{\vartheta\langle||_i\vartheta_0 x(w),||_{1-i}\vartheta_1\overline{x}\langle y\rangle\rangle}_R P\{y/w\}|Q$$

$$\textbf{Par}: \frac{P \xrightarrow{\theta}_R P'}{P|Q \xrightarrow{\theta}_R P'|Q} \qquad \textbf{Res}: \frac{P \xrightarrow{\theta}_R P'}{(\nu a)\ P \xrightarrow{\theta}_R (\nu a)\ P'}$$

$$\textbf{Tau}: \vartheta\tau.P \xrightarrow{\vartheta\tau}_R P \qquad \textbf{Struct}: \frac{Q \equiv P \quad P \xrightarrow{\theta}_R P' \quad P' \equiv Q'}{Q \xrightarrow{\theta}_R Q'}$$

$$\textbf{Const}: \frac{P\{y/x\} \xrightarrow{\theta}_R P'}{\vartheta X(y) \xrightarrow{\vartheta\theta}_R \vartheta \rhd P'} \qquad X(x) \stackrel{def}{=} P$$

A transition is obtained by applying the inference rules given in Table 1. In this way we defined a so-called *reduction semantics*, which however differs from the standard one because of the labels transitions carry. The labels record the syntactic context in which actions took place, besides the actions themselves. ¿From these labels we can infer many aspects of a computation. Here we are mainly interested in the stochastic one.

We briefly survey below how we derive stochastic information from process computations, and we refer the reader to [20, 2] for a formal presentation. The idea is to assign a "cost" to a transition by only looking at its label, and then to derive stochastic measures via standard techniques. Here, for "cost" of a transition we mean the probability of occurrence of the corresponding reaction. This value is to be defined in agreement with some biological considerations that we summarise below.

We used the constants K_M of the "Michaelis-Menten" kinetics [13, 14] to constructs reaction rates, so to link each transition to a measurable biological parameter. Another aspect taken into account is related to the two classes of reaction typical of metabolic pathways. When considering the mass-action ratio r at the steady-state we either have:

- *Near-equilibrium* reactions, where r is close to the equilibrium constant, and the rates of a reaction and of its reverse are close; or
- *Non-equilibrium* reactions, where rates of direct and reverse reactions greatly differ.

Our choice reflects the *control strength* of the enzymes involved. This quantifies the impact of the activity of an enzyme on the overall flux of a pathway: the greater the control strength, the more perturbated is the flux when the enzyme in inhibited [10, 12].

Intuitively we assign a cost to a π-calculus transition as follows. We consider first the execution of an action π on a dedicated machine that has only to perform π, and we estimate the corresponding rate r. Then, we take into account the syntactic context in which the action π will occur. Indeed, the context represents the environment in which the corresponding reaction occurs. The actual cost of a transition is finally given by a suitable combination of the estimate of the action performed, of the quantities of the reactants involved, and of the effects due to the operators of the context.

Now, from the transition system labelled with costs, we can derive a Continuous Time Markov Chain. In this way, the behaviour of a biological system, specified as a(n enhanced) π-calculus process, is represented by a stochastic process, which can be analysed with standard tools.

3 VICE: The VIrtual CEll

We chose the organism to model so to keep its genome as reduced as possible, while maintaining it close to that of a real prokaryote. Also, we would like to design a *living* organism. Our starting point has been the hypothetical Minimal-Gene-Set (MGS) proposed in [19, 16], which has been obtained by eliminating redundancies in known small bacterial genomes.

We performed a functional screening on MGS, manually searching for its basic pathways, assuming it in an optimal environment for metabolites and physiological conditions. In particular, the environment considered has enough essential nutrients, and it is shaped to dilute or remove all the potentially toxic catabolites; also competition and other stressing factors are completely banned. According to our analysis, we further eliminated some genes. Also, we found that certain critical steps were missing in pathways, and so we introduced two other genes not originally contained in MGS.

In order to keep small the specification of VICE in the π-calculus, we made a few further sligth simplifications. Typically, we grouped in a single entity all the multi-enzymatic complexes, when acting as a single cluster.

Our virtual cell possesses the following main features:

- VICE relies on a complete glycolytic pathway for the oxydation of Glucose to Pyruvate and reduced-NAD. Pyruvate is then converted to acetate which, being a catabolyte, can diffuse out of the cell. A transmembrane reduced-NAD dehydrogenase complex catalyzes the oxydation of reduced-NAD; this reaction is coupled with the synthesis of ATP through the ATP synthase/ATPase transmembrane system. This set of reactions enables VICE to manage its energetic metabolism.
- VICE has a Pentose Phosphate Pathway, composed by enzymes leading to the synthesis of Ribose Phosphate and 2-Deoxyribose Phosphate.
- For lipid metabolism, VICE has enzymes for glycerol-fatty acids condensation, but no pathways for fatty acid synthesis. So, these metabolites must be taken from the outside.
- VICE has no pathways for amino acid synthesis and, therefore, we assume all amino acids be present in the environment.
- Thymine is the only nucleotide VICE is able to synthesize *de novo*; the other nucleotides are provided by "salvage pathways".

- VICE possesses a proper set of carriers for metabolites uptake:
  - a Glycerol Uptake Facilitator Protein;
  - a PtsG System for sugars uptake;
  - an ACP carrier protein for fatty acids uptake;
  - a broad specificity amino acids uptake ATPase; and
  - broad specificity permeases for other essential metabolites uptake.
- VICE possesses the necessary enzymes for protein synthesis, including DNA-transcription and translation. VICE possesses also the whole machinery necessary for DNA synthesis.
- All the nucleotide biosyntethic pathways are present in our model, so VICEis equipped with the means for cell reproduction; however, at the present stage we have not designed these activity.

Some metabolites are considered to be ubiquitary, typically water, inorganic phosphate, some metal ions, Nicotinammide. Their concentration in the external or internal environment is supposed as constant and not significantly affected by cellular metabolism.

Summing up, VICE can take metabolites from the external environment using the set of permeases and ATPases specified above. Among the pathways of our virtual cell, there is Glycolysis: glucose and fructose taken from the outside are oxidized helding energy in the form of ATP and reduced-NAD. Pyruvate, the last metabolite of conventional Glycolysis, becomes then acetate which, in turn, diffuses out of the cell. VICE "imports" fatty acids, glycerole and some other metabolites, e.g. Choline, and uses them for the synthesys of tryglicerides and phospholipids; these are essential components of the plasma membrane. Our virtual cell is also able to synthetize DNA, RNA and proteins; the needed metabolites are mostly taken from the external environment or synthesized along its own pathways (e.g., Thymine and Ribose).

4 Results

We briefly discuss below the adequacy and the results of our proposal. First, we shall describe the experiments made *in silico*, and we shall then interpret their outcome. The results suggest that VICE can "live" in an optimal environment and that it mimicks some biological behaviour.

4.1 Biological Consistency

For the specification of VICE, we selected some basic metabolic pathways, following the functional approach mentioned in Section 3. To confirm that our proposal is safe, we checked that our virtual cell can use all the pathways choosed. This means that VICE endowes sufficient components in its genome. We also made sure that all these components are involved in some simulation runs, and so all of them are necessary. We carried on these tests assuming that VICE acts in the ideal environment discussed in Section 3. These experiments *in silico* have been performed varying the following conditions:

- the amount $\mathcal{C}$ of metabolites, in particular of Glucose 6-phosphate; this estimate of metabolite concentration is modelled by letting the computation start from an initial process that has the specifications of VICE in parallel with that of a fixed number of metabolites. We actually used 100, 200, 300 and 500 copies of Glucose 6-phosphate.
- the time interval of the observation $\mathcal{T}$; this is modelled by fixing the number of transitions. Actually, we considered computations made of 20K, 30K and 40K transitions.

For each pair $(\mathcal{C}, \mathcal{T})$, we run our interpreter and obtained a computation $\xi_{(\mathcal{C},\mathcal{T})}$. The time spent in each run ranges from a few seconds for $\mathcal{C} = 100$ and $\mathcal{T} = 20K$ to about three hours for $\mathcal{C} = 500$ and $\mathcal{T} = 40K$, on a 1.2 GHz processor.

We then made an exhaustive search on the computations, and we found that all the patways have been followed with almost the same probability. Also, all the components of VICE result necessary, as they have been used during some computations. Consequently, we confirm the safety of our hypothetical prokaryote, or at least of its formal specification.

4.2 Interpretation of the Results

In order to check that VICE behaves more or less like a living organism, we examined the initial and final state of each computation $\xi_{(\mathcal{C},\mathcal{T})}$. It turns out that these states differ because a certain amount of Glucose 6-phosphate has been oxydized producing energy and other metabolites. This confirms that VICE behaves like a "living" cell.

We also compared some aspects of the behaviour of VICE with that of real prokaryotes acting *in vivo* in similar circumstancies [14]. In particular, we investigated the glycolytic pathway, on which the literature has a huge quantity of biological data. A first analysis shows that the distribution of the metabolites along the (various occurrences in the computations of the) glycolytic pathway of VICE significantly matches with those of real organisms reported in the literature. To strengthen evidence of that, we suitably fixed the distribution of metabolites in the initial state of each computation. This distribution is chosen in accordance with the one exhibited by real prokaryotes when they stay in their steady state. The experimental results do confirm the trend of VICE glycolysis towards the real one. Recall that we assumed that the occurrence of a transition corresponds to the production of a fixed quantity of a specific metabolite. In our simulation, we assumed for example that the transition t:

$$\beta\text{-D-fructose\_1-6bP} \xrightarrow[R]{||_0\tau_{mg023}}$$
$$\text{D-Glyceraldehyd\_3-phosphate} \mid \text{Dihydroxyacetone\_phosphate}$$

produces 2×10^{-3} nMoles of D-Glyceraldehyd_3-phosphate and Dihydroxyacetone_phosphate (of course, this transition can occour when the β-D-fructose_1-6bP is plugged in any context; see the computation in Sect. 2). Therefore, we can count the number n of occurrences of t within a selected computation to estimate the amount of metabolites produced. Moreover, recall also that a transition describes the catalysis of a certain enzyme. In the transition t above, the enzyme is the aldolase mg023. The number of times mg023 involved in a transition t along the chosen computation is also n and it turns out to be proportional to the flow rates through the reaction catalyzed.

Our experimental results and the real ones are in Figure 2. Part (a) displayes the number of transitions through the various steps of VICE Glycolysis (in arbitrary units). Part (b) shows the specific flow rates estimated *in vivo* in prokaryotes, and measured in $meq\frac{C}{g\_of\_cells \times h}$.

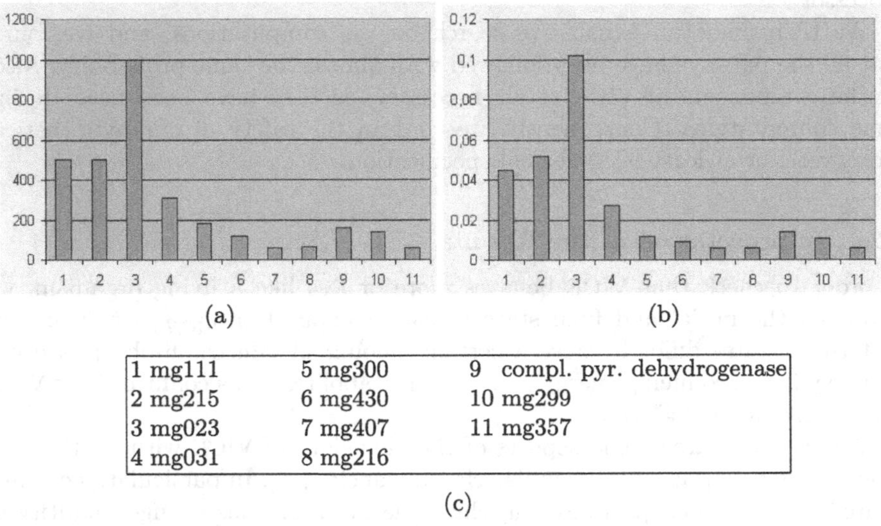

(a) (b)

1 mg111	5 mg300	9 compl. pyr. dehydrogenase
2 mg215	6 mg430	10 mg299
3 mg023	7 mg407	11 mg357
4 mg031	8 mg216	

(c)

Fig. 2. The trend of VICE glycolysis (a) towards the real one (b). Table (c) shows the correspondence between indexes on x-axis and enzymes. Their codes are in [9]

Other virtual cells have been proposed in different projects aimed at representing, understanding and testing metabolic networks of real organisms. Perhaps the most known are:

- The metabolic network of a generic eukariote, implemented through E-CELL, an open-source Web-available simulator and differential solver [11], and
- virtual *E. coli*, which has been studied in order to describe the genetic control network and the metabolic fluxes of *Escherichia Coli.* This virtual cell is mainly based on quantitative differential equation description of biochemical reactions [1, 7, 8].

The above mentioned virtual cells do satisfy the minimal requirements (homeostasis and energetic balance) for a biological cell, just as it is the case for VICE.

Also, the biochemical dynamics (roughly, the time course of metabolites concentrations along a pathway) of all the three virtual cells are consistent with biochemical experimental data. Even though these virtual cells have been formalized in quite different approaches, their overall behaviour generally agree with each other (and with the real behaviour).

We are confident that our approach can be further pushed to specify in the (enhanced) π-calculus other aspects of real cells, in particular activities more complex than the basic biochemical ones described here. In particular, we are extending our tool in order to describe *in silico* complex interactions between pathways. Hopefully, this could lead to detect new pathways, not experimentally known, to be then confirmed *in vitro*.

References

1. E. Almaas and B. Kowács *et al.* Global organization of metabolic fluxes in the bacterium *Escherichia Coli. Nature*, 427:839–843, 2004.
2. C. Bodei, M. Buchholtz, P. Degano, M. Curti, C. Priami, F. Nielson, and H. Riis Nielson. Performance evaluation of security protocols specified in LySa. In *Procs. of the 2nd W/S on Quantitative Aspects of Programming Languages, ENTCS*, 2004.
3. M. Chiaverini and V. Danos. A core modeling language for the working molecular biologist. In Corrado Priami, editor, *Procs. of the* 1st *Int. W/S on Computational Methods in Systems Biology*, volume 2602 of *LNCS*. Springer, 2003.
4. M. Curti, P. Degano, and C.T. Baldari. Causal π-calculus for biochemical modelling. In *Procs. of the* 1st *Int. W/S Computational Methods in Systems Biology*, volume 2602 of *LNCS*. Springer, 2003.
5. M. Curti, P. Degano, C. Priami, and C.T. Baldari. Modelling biochemical pathways through enhanced π-calculus. *Theoretical Computer Science*, To appear, 2004.
6. P. Degano and C. Priami. Enhanced operational semantics. *ACM Computing Surveys*, 28(2):352–354, 1996.
7. J.S. Edwards, R.U. Ibarra, and B.O. Palsson. In silico predictions of *Escherichia coli* metabolic capabilities are consistent with experimental data. *Nature Biotechnology*, 19:125–130, 2001.
8. J.S. Edwards and B.O. Palsson. The *Escherichia coli* mg1665 in silico metabolic genotype: its definition, characteristics and capabilities. *Proceedings of the National Academy of Sciences*, 97:5528–5533, 2000.
9. C.M. Fraser *et al.* The minimal gene complement of mycoplasma genitalium. *Science*, 270(1):397–403, 1995.
10. Thomas M. Devlin *et al. Textbook of Biochemistry, fifth edition.* Wiley and Sons, Inc., 2002.
11. Tomita Masaru *et al.* E–CELL: software environment for whole–cell simulation. *Bioinformatics*, 15:72–84, 1998.
12. D.A. Fell. *Understanding the control of metabolism.* Portland Press, London, United Kingdom, 1997.
13. A. Fersht. *Structure and Mechanism in Protein Science: A Guide to Enzyme Catalysis and Protein Folding.* Freeman, 1999.
14. G.G. Hammes and P.R. Shimmel. *The Enzymes, vol. 2.* P.D. Boyer (New York Academic Press), 1970.
15. H. Kitano. *Foundations of System Biology.* MIT Press, 2002.

16. E.V. Koonin. How many genes can make a cell: The minimal-gene-set concept. *Annuual Review Genomics and Human Genetics*, 01:99–116, 2000.
17. L.M. Loew and J.C. Schaff. The virtual cell: a software environment for computational cell biology. *Trends Biotechnology*, 19(10):401–406, Oct. 2001.
18. R. Milner. *Communicating and Mobile Systems: the π-calculus.* Cambridge Univ. Press, 1999.
19. A.R. Mushegian and E.V. Koonin. A minimal gene set for cellular life derived by comparison of complete bacterial genomes. *Proceedings of National Academy of Science USA*, 93:10268–10273, 1996.
20. C. Nottegar, C. Priami, and P. Degano. Performance evaluation of mobile processes via abstract machines. *IEEE Transactions on Software Engineering*, 27, 10:867–889, 2001.
21. C. Priami, A. Regev, W. Silverman, and E. Shapiro. Application of a stochastic passing-name calculus to representation and simulation of molecular processes. *Information Processing Letters*, 80:25–31, 2001.
22. A. Regev, E.M. Panina, W. Silverman, L. Cardelli, and E. Shapiro. Bioambients: An abstraction for biological compartments. *Theoretical Computer Science*, To Appear, 2004.
23. A. Regev and E. Shapiro. Cells as computations. *Nature*, 419:343, 2002.
24. A. Regev, W. Silverman, and E. Shapiro. Representation and simulation of biochemical processes using the π-calculus process algebra. In *Pacific Symposium of Biocomputing (PSB2001)*, pages 459–470, 2001.

Appendix A

The whole metabolic machinery of VICE is composed both by complete biochemical pathways, such as glycolysis, and by single-step reactions (e.g. reduced-NAD oxidation). Belonging to a pathway or not, all these biochemical reactions are interconnected to form a network, and should be seen as a single functional unity. However, for the sake of readability, in the presentation below we arbitrary choose to describe the metabolic network of VICE as subdivided into functional groups.

Carbohydrate Oxidative Metabolism. The three pathways that form this group constitute the energy sources for VICE. They mainly consist in oxidation reactions to provide our virtual cell with the needed chemical energy stored in some metabolites. These pathways have been designed on ancient and basic bacterial metabolism, similar to that of *Mycoplasma genitalium.* Therefore, they result less complex and efficient than their homologous in higher organisms, e.g., than the Krebs' Cycle. Consequently, at the end of these pathways of VICE, molecules are only partly oxidized; however, the energy production suffices for all the other energy-consuming pathways. In more detail, the three pathways are:

- **Glycolysis**, consisting of 8 steps. This pathway starts with glucose or fructose and ends with pyruvate; part of the chemical energy of sugars is converted into ATP and reduced-NAD.
- **Pyruvate Metabolism.** Pyruvate generated by Glycolysis is further oxidized, yielding reduced-NAD and acetate, which can diffuse out of the cell.

- **Reduced-NAD Oxidation and ATP Synthesis.** Reduced-NAD is oxidized by a transmembrane protein (mg275) yielding NAD and protons. This reaction is coupled with the *pumping* of protons toward the cell exterior; an ATPsynthase/ATPase transmembrane complex can perform ATP synthesis using the potential energy of the proton pumping.

We shall present in Appendix B the formal specification of this group of pathways.

Lipid Metabolism. This set of pathways leads to the synthesis of some glycerolipids and glycerophospholipids, starting from exogenous complex monomers (fatty acids, choline, glycerol). VICE only uses lipids for structural purposes; indeed it has no pathways for fatty acid oxidation. Appendix B contains the formal specification of this group of pathways, roughly described below.

Glycerol 3-phosphate represents the starting point of two different pathways: one leading to the synthesis of triacilglycerols, and the other leading to the synthesys of glycerophospholipids. Glycerol 3-phosphate, obtained by phosphorilating exogenous glycerol (imported via the Glycerol Uptake Facilitator Protein mg033), becomes phosphatidic acyd in a two-steps reaction patway, involving the Acyl Carrier Protein as acyl donor. Phosphatidic acyd can be further converted in phosphatidylglycerol or phosphatidylethanolamine which, are glycerophospholipids. Triacylglycerols derives from phosphatidic acyd after a two steps reaction chain.

Nucleotide Metabolism. Like any basic bacteria, VICE has no pathways for *de novo* synthesis of nucleotides. Therefore, we allow our virtual cell to syntethize ribonucleotides (and, consequently deoxyribonucleotides), using the so-called "salvage pathway", which starts from ribose-5-phosphate, ATP and nitrous bases; ribose-5-phosphate is produced in the Penthose Phosphate Pathway (see below), while nitrous bases are taken from outside, using a proper carrier. Instead, VICE has no salvage pathway for thymidine triphosphate synthesis; this nucleotide is then obtained starting from uridine monophosphate.

Protein Synthesis. VICE is not equipped for amino acids synthesis, but it can synthesize proteins. The RNA-translation mechanism is designed as a process in which the multimeric ribosomal complex is acting as a single unity.

DNA/RNA Synthesis. The synthesis of nucleic acids in a real biological cell involves a large number of enzymes which act as a single physical unity, and this is our choice for VICE. It uses RNA synthesis to drive subsequent protein synthesis. Also, VICE can synthesize DNA, even though the cell replication has not been designed yet.

Metabolite Uptake. VICE possesses all the necessary carriers to uptake the different metabolites it needs, among which:

- PTS system, to uptake and simultaneously phosphorilate several sugars (at this stage of design, VICE is only able to metabolize glucose and fructose);
- Glycerol Uptake Facilitator: a protein which uptakes glycerol, starting the lipid metabolism;

- ATP dependent transport system, for amino acids uptake;
- Nitrous bases transport system;
- Acyl Carrier Protein, for fatty acids uptake;
- Phosphate permease, Na+ATPase, cationic/metal ATPase, for the uptake of phosphate, Na+ and cationic and metal ions, respectively.

Other Pathways. The following pathways are also needed:

- **Penthose Phosphate Pathway.** In VICE, this pathway is reduced to five steps, and leads to the production of ribose-5-phosphate, used in the nucleotide synthesis.
- **Coenzyme synthesis.** NAD and FAD are synthesized starting from monomers taken from the external medium, using proper carriers.

Appendix B

Glycolysys pathway and pyruvate-to-acetate pathway

α-D-Glucose_6P = τ_{mg111}.β-D-Fructose_6P
β-D-Fructose_6P = τ_{mg111}.α-D-Glucose_6P + τ_{mg215}.β-D-Fructose_1-6bP
β-D-Fructose-1-6_bP = τ_{mg215}.β-D-Fructose_6P +
  τ_{mg023}.(Dihydroxyacetone_phosphate | D-Glyceraldehyde_3-phosphate)
Dihydroxyacetone_phosphate = τ_{mg431}.D-Glyceraldehyde_3-phosphate +
  $\overline{mg023}\langle a\rangle$
D-Glyceraldehyde_3-phosphate = τ_{mg301}.1,3-Bisphospho-D-glycerate +
  τ_{mg431}.Dihydroxyacetone_phosphate + τ_{mg050}.2-Deoxyribose_5P +
  $mg023(x)$.β-D-Fructose_1-6bP
1,3-Bisphospho-D-glycerate = τ_{mg301}.D-Glyceraldehyde_3-phosphate +
  τ_{mg300}.3-Phospho-D-glycerate
3-Phospho-D-glycerate = τ_{mg300}.1,3-Bisphospho-D-glycerate +
  τ_{mg430}.2-Phospho-D-glycerate
2-Phospho-D-glycerate = τ_{mg430}.3-Phospho-D-glycerate +
  τ_{mg407}.Phosphoenolpyruvate
Phosphoenolpyruvate = τ_{mg407}.2-Phospho-D-glycerate + τ_{mg216}.Pyruvate
Pyruvate = τ_{mg216}.Phosphoenolpyruvate + $\tau_{dehydrogenase}$.AcetylCoA
AcetylCoA = $\tau_{dehydrogenase}$.Pyruvate + τ_{mg299}.Acetyl_phosphate
Acetyl_phosphate = τ_{mg299}.AcetylCoA + τ_{mg357}. Acetate

Lipid Metabolism

Glycerol_extracellular = τ_{mg033}.Glycerol
Glycerol = τ_{mg033}.Glycerol_extracellular + τ_{mg038}.Glycerol_3-phosphate
Glycerol_3-phosphate = τ_{mg038}.Glycerol + τ_{acp}.1-Acylglycerophosphate +
  $\overline{mg114}\langle a\rangle$
1-Acylglycerophosphate = τ_{acp1}.Phosphatidic_acyd
Phosphatidic_acyd = $\overline{mg437}\langle b\rangle$ + $\tau_{phosphatase}$.Diacylglycerol
CTP = $mg437(x)$.CDP_Diacylglycerol
CDP_Diacylglycerol = $mg114(x)$.(Phosphatidylglycerophosphate | CMP)
Phosphatidylglycerophosphate = $\tau_{phosphatase}$.Phosphatidylglycerol
Diacylglycerol = τ_{acp2}.Triacylglycerol

Biological Domain Identification Based in Codon Usage by Means of Rule and Tree Induction

Antonio Neme[1] and Pedro Miramontes[2]

[1] IIMAS, UNAM, México
neme@uxmcc2.iimas.unam.mx
[2] Facultad de Ciencias, UNAM, México

Abstract. There are three domains in living nature: archaea, bacteria and eukarya. It has been shown, trough a number of multivariate tools, that codon usage, a 64 dimensional vector that stablishes how often a given organism makes use of each codon, is related to domain. Another method is proposed here based in rule and tree induction from codon usage of several organisms. It is shown that domain can be identified trough codon usage and a simple set of rules. Two methods were applied, $CN2$ and $C4.5$. Obtained rules describe data better than other methods, in the sense that are topological interpretable and have phenomenological meaning.

1 Introduction

Codon usage is the preference shown by organisms to use a certain synonymous codon to code amino acids. 18 out of 20 amino acids are coded by more than one synonymous codon and the fact some organisms (or genes) prefer a given codon to code for a certain amino acid is known as *codon bias* [5].

Organisms may be tought of as points in 64−dimensional space, accordingly to codon usage data. The distribution shown by them, thus, codon bias, has been a subject of intense research in molecular biology. Codon bias has not been explained. Several theories have been proposed but there is not a general explanation for it [11, 8]. Each organism may be represented by its codon usage vector, that contains the frequency per ten thousand of each codon.

Grantham used principal componet analysis in [3] to show that codon usage is related to biological domain. Using a self-organizing map, more evidence has been given to show that, in general, codon usage is related to biological domain, with a few counterexamples of special organisms (like *Th. Maritima* and *U. Urealiticum*) that does not seem to follow the expected pattern [7].

A set of understandable rules that classifies properly a group of organisms may be a better tool for molecular biologists to explain codon bias on domain basis, because important variables (frequency of each codon) and its relationships are explicitly settled. In this work, we obtain a set of rules that properly identifies the domain an organism belongs to.

V. Danos and V. Schachter (Eds.): CMSB 2004, LNBI 3082, pp. 221–224, 2005.

2 Methods

Rule and tree inference are part of an artificial intelligence field named *machine learning.* They have been extensively applied in data analysis because they are more transparent and easier to interpret than other methods, as for example, trainned neural networks or a regression models [2]. The goal is to find a function (here in form of rules) f that properly classify a set of examples X. For each individual sample vector X_i, it is associated a label or class (domain in this work), c_i. Thus, $f(X_i) = c_i$ means that applying the set of rules f to X_i the proper class will be identified.

Rule induction may be seen as a search problem: it finds a set of rules that are coherent (no classification errors) and complete (all organisms are classified). There are several algorithms for rule induction [1, 4], but the one applied here is the so called $CN2$. For a detailed explanation of this algorithm, see [2].

The tree induction method applied in this work is $C4.5$, propossed by Quinlan [9]. On it, a set of decisions is found so that each partition maximizes a *gain* criteria, based on information content. At every step, the variable that maximizes information (the number of objects correctly identified) is chosen.

3 Results

Rules obtained by applying CN2 to codon usage vector, obtained from the *Kazusa data bank* [6] of 159 organisms (most of them completely sequenced) are shown in table 1. There are 28 *archaea*, 68 bacteria and 63 eukarya. Codons

Table 1. Rules obtaided by CN2. Numbers bewteen squared brackets identify the number of organisms that satisfies conditions in the rule and belongs to domain *archaea* (first), *bacteria* (second) and *eukarya* (third)

IF UUU<0.42 and CUU<0.25 and CGU<0.02 THEN domain=archaea [13 0 0]
IF UAC>0.18 and CUC>0.35 THEN domain = archaea [7 0 0]
IF UUU<0.27 and CUG<0.29 and AGG>0.16 THEN domain=archaea [12 0 0]

IF UUU>0.18 and CAA>0.28 THEN domain = bacteria [0 22 0]
IF UAC<0.21 and GCC>0.50 THEN domain = bacteria [0 18 0]
IF CCA>0.12 and AAC<0.14 THEN domain = bacteria [0 10 0]
IF UUU>0.11 and CCA<0.09 and GGG<0.15 THEN domain = bacteria [0 13 0]
IF UGU<0.07 and CGU>0.13 THEN domains = bacteria [0 22 0]
IF 0.11<UUC<0.13 THEN domain = bacteria [0 6 0]

IF UCC>0.04 and UGC>0.1 THEN domain = eukarya [0 0 36]
IF UCA>0.15 and CCA>0.19 THEN domain = eukarya [0 0 13]
IF UUC>0.10 and UCU>0.19 and CCA> 0.10 THEN domain = eukarya [0 0 12]
IF AAC>0.77 THEN domain = eukarya [0 0 2]
IF UCC>0.2 THEN domain = eukarya [0 0 6]
IF UUG>0.19 and CGA>0.09 THEN domain = eukarya [0 0 3]

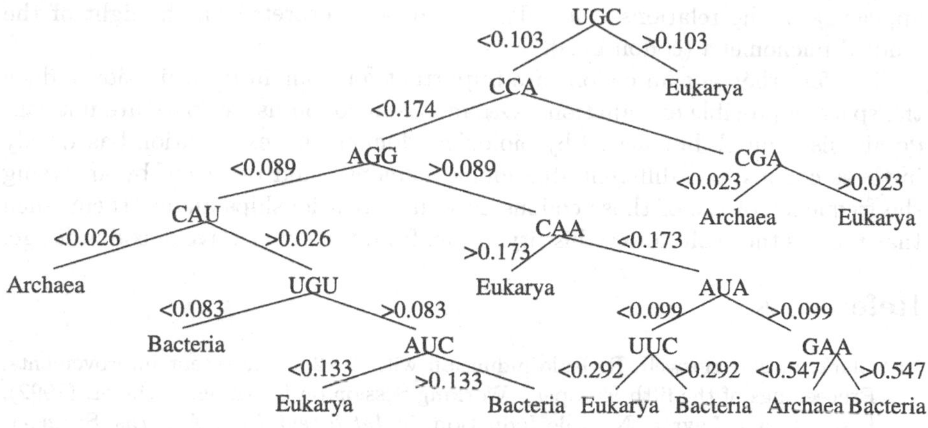

Fig. 1. Tree obtained by C4.5

are represented by its nucleotide sequence, such as AGG, meaning Adenine followed by Guanine followed by Guanine.

There are 18 rules and no classification errors were found. 22 codons, coding for 12 amino acids, were required to classify organisms by CN2 and, as is common in that algorithm [2], some examples are included in more than one rule.

The tree obtained by $C4.5$ for the same organisms is shown in Figure 1. In this tree, only 11 codons, coding for 9 amino acids, were required. There are, however, six misclassified organisms (3.8% error): three archaeas: *Archaeoglobus profundus*, identified as an eukarya, and *Methanococcus voltae* and *Methanosarcina mazei*, both identified as bacterias. Two eukaryas were misclassified: *Ostrinia nubilalis* and *Fusarium sporotrichioides*, both identified as bacteria. The only misclassified bacteria was *Buchnera aphidicola*, identified as eukarya. The rules obtained by CN2 and the tree obtained by $C4.5$ are more evidence to show that codon bias is related to biological domain. This was already known [3, 10], but what we do here is to give basis of explanation of that fact, because the rules and tree are interpretable information.

4 Conclusions

Analysis of biological data in a structured way is easier than doing so for data in form of tables or even in bidimensional maps, as those obtained by multivariate analysis. Here, we applied two formal methods to obtain structure in data for the problem of codon bias. More evidence that codon bias is affected by domains is given, but the difference is that we show a set of rules and an identification tree that may be intrepreted by specialists to explain it with more basis and with readable information. The applied methods were $CN2$ and $C4.5$.

For biologists, it may be of interest to find a pattern in codon bias dictated by domain. It is easier to look for that pattern if information is expressed in form of rules instead of looking at a graph, mainly because the variables (codons) appearing in the relations (the rules) could be interpreted on the light of the studied phenomena (codon bias).

The fact that not all codons are important for domain identification reduce the space of possible explanations. Organisms in codon usage space are not randomly distributed, but biased by biological domain. An explanation based only in those codons that differentiate among domains could be given, by analysing the frequency of use of those codons as well as relationships among them, such that reflects the evolutionary history of life, from the perspective of codon usage.

References

1. Clark, P. and Boswell, R. Rule induction with CN2: some recent improvements. Proceedings of the fifth European Working Session on Learning. Springer. (1992).
2. Flach, P. and Lavrac, N. Rule Induction, in *Intelligent Data Analysis*. Springer. 2003.
3. Grantham R, Gautier C, Gouy M, Mercier R, Pave A.. Codon catalog usage and the genome hypothesis. Nucl. Ac. Res. **1** (1980) 43-74.
4. Lavrâc, P. Flach, B. Adapting classification rule learning to subgroup discovery. Proceedings of the IEEE International Conference on Data Mining. (2002).
5. Lewin, B. Genes VII. Oxford University Press. 2000.
6. Nakamura, Y., Gojobori, T. and Ikemura, T. Codon usage tabulated from the international DNA sequence databases: status for the year 2000. Nucl. Ac. Res. **28** (2000) 292.
7. Neme, A. Codon usage and self-organizing maps. Proceedings of the Mexican Mathematical Society Meeting. 2003.
8. Powell J., Sezzi E., Moriyama E., Gleason J., and Caccone A. Analysis of a shift in codon usage in Drosophila. J. Mol. Ev. **57** (2003) Suppl. 1. 214-225.
9. Quinlan, R. C4.5 : programs for machine learning. Morgan Kaufmann. 1993.
10. Rowe G., Szabo V. and Trainor, L. Cluster analysis of genes in codon space. J. Mol. Ev. **2** (1984) 167-174.
11. Vinogradov, A. DNA helix: the importance of being *GC* rich. Nucl. Ac. Res. **7** (2003) 1838-1845.

Black Box Checking for Biochemical Networks

Dragan Bošnački

Dept. of Biomedical Engineering,
Eindhoven University of Technology,
PO Box 513, 5600 MB Eindhoven, The Netherlands
D.Bosnacki@tue.nl

Abstract. We propose *black box checking* as a framework for analyzing biochemical networks. Black box checking was originally introduced by Peled, Yannakakis and Vardi in the context of formal verification of concurrent systems as a strategy that combines model checking and testing, as two main techniques in that area. Based on the natural analogy between biochemical networks and concurrent systems we argue that black box checking can be used to design and perform experiments in a systematic manner, and also to learn about the network underlying mechanisms. We also discuss potential applications with emphasis on forward engineering of biochemical networks.

1 Introduction

Analysis of different kind of biochemical networks is one of the main challenges of Bioinformatics today. One of the main problems in this task is the combinatorial explosion which occurs as a consequence of the networks complexity. In this paper we suggest a framework to tackle this problem. The framework can also be useful in designing experiments in a more efficient way, as well as understanding the network underlying mechanisms.

There exists a natural analogy between metabolic networks and (concurrent) systems with unknown structure. Both can be considered as black boxes with an unknown internal working about which we want to check some property, i.e., verify some hypothesis about their behavior. Thus, in this work we propose to use a paradigm called Black box checking as a framework for analysis of genetic regulatory networks.

Black box checking was originally introduced by Peled, Yannakakis and Vardi [11, 10] as a strategy that combines two mainstream approaches for increasing the quality of software and hardware systems: model checking and testing. In model checking one checks the property against a system model by exhaustive exploration of the state space, i.e., all possible model behaviors. On the other hand, testing is used to check if a given model of the system conforms with the system implementation. This is done by covering only part of the possible system execution sequences. Black box combines these two approaches plus Angluin's algorithm for machine learning [1] in order to test whether the system implementation with unknown structure satisfies some given property.

V. Danos and V. Schachter (Eds.): CMSB 2004, LNBI 3082, pp. 225–230, 2005.

In the literature there are several attempts to apply either model checking (e.g. [2, 6]) or machine learning (e.g. [5, 14]) separately to problems related to biological processes. To the best of our knowledge there are no previous publications that combine these two approaches, and in particular that apply Black box checking, in a biological context.

2 Preliminaries

Here we give only a brief informal overview of model checking and testing – the two cornerstones of the Black box checking framework. We also discuss modal logics that are used to formally specify the property of the system which we want to check. See for instance [8] for the formal definitions.

2.1 Transition Systems

We use Black box checking to check properties of a system that can be represented as a finite transition system. A *finite transition system* is a tuple $S = (Q, \hat{q}, A, \delta)$ where Q is a finite set of *states*, $\hat{q} \in Q$ is the *initial* state, A is a finite set of *actions* and $\delta : Q \times A \to Q$ is a partial transition function. The set A models the *inputs*, i.e., the "interface buttons" of the black box. An action $a \in A$ is *enabled* from a given state $q \in Q$, if $\delta(q, a)$ is defined, i.e., there exists $q' \in Q$ such that $q' = \delta(q, a)$. The intuition is that with the transition defined with $\delta(q, a)$ the system S passes from state q to q' by executing the action a. An execution sequence $a_1 a_2 \dots a_n$ is enabled from $q \in Q$ if there exists a corresponding sequence of states $q_0 q_1 \dots q_n$ such that for $1 \leq i \leq n$, $q_i = \delta(q_{i-1}, a_i)$, i.e, the actions are enabled in each of the corresponding states.

2.2 LTL, Model Checking and Conformance Testing

We assume that the properties we want to check about biochemical networks, i.e., the black boxes, are formalized using the Linear Temporal Logic (LTL). Linear Temporal Logic is interpreted on infinite execution sequences of the transition systems and allows us to express various properties on concurrent systems. For instance, the fact that some action of the transition system is always eventually executed, or that the system always stabilizes (stays forever) in some given state.

Roughly speaking, model checking is an automated technique that, given a model of the system and some property, checks whether the model satisfies the property. Compared to the other (semi)automated formal techniques (for instance, deductive methods, like theorem provers) model checking is relatively easy to use. The specification of the model is very similar to programming and as such it does not require much additional expertise from the user. The verification procedure is completely automated and often takes only several minutes. Another important advantage of the method is that, if the verification fails, the possible erroneous behavior of the system can be reproduced. This significantly facilitates the location and correction of the errors.

Unlike model checking which is applied to the model of the system, (conformance) testing is performed on the system implementation. Also, in conformance testing we do not try to cover all execution sequences of the implementation. Instead, most of the time we are satisfied with checking some subset of sequences which are good representatives, i.e., they ensure a "satisfactory coverage" of the system executions that increases our confidence that the implementation satisfies given properties (conforms to the model).

2.3 Black Box Checking

The Black box checking strategy is a modification of Angluin's algorithm for machine learning [1]. Again here we give only a brief overview of the main concepts and ideas. The reader is referred to [11, 10] for more details.

We use Angluin's algorithm to learn the structure of the system, i.e., build a model in an incremental way. In its basic form the algorithm starts with a trivial model which is an empty transition system. Based on the answers of an oracle who provides executions (counterexamples) that distinguish the model from the actual implementation, better approximations, i.e., models, of the actual system are built iteratively. The role of the oracle is played by a conformance testing of the model against the implementation. (This black box testing is done under the assumption that we know the maximal number of states of the system implementation (black box) as well as the actions that can be performed by the implementation - i.e., the different box interface keys that can be pressed.)

A straightforward solution to the conformance testing would be to use the Vasilevskii-Chow algorithm [15, 7] which is standard for such tasks. Unfortunately this algorithm is computationally expensive (exponential in the number of states). So, the crucial idea of the Black box checking is to avoid using the Vasilevskii-Chow algorithm as much as possible and to replace it with model checking applied on the current model of the system. This is possible in the cases when the model checking algorithm finds a counterexample in the model which is not a counterexample in the actual implementation. This counterexample distinguishes the implementation from the model and can be used as a feedback (oracle's answer) to the Angluin algorithm in order to get a better approximation.

Thus, assuming that we have a model M, we proceed roughly in the following way. We model check M. In case we find a violation of the property, this means that a counter example σ is generated which is an execution sequence in M. We check if σ is also an execution sequence of the actual system S. If this is the case we are done, because this means that S does not satisfy the property. However, if σ is not an execution sequence of S, then its smallest prefix σ' which is not accepted by S is used to distinguish between M and S - in a similar way as the counterexample from the conformance testing (i.e., the Vasilevskii-Chow algorithm) was used above. If the model checking does not provide a counterexample (meaning that the property is satisfied in the model), then we have to apply the Vasilevskii-Chow algorithm.

There are several possible variations of Black box checking that could be useful in practice, like starting from a non-trivial initial model (non-empty transition system), or removing the assumption of a known maximal number of states in the implementation (see [10]).

3 Black Box Checking in the Context of Biochemical Networks

It was already emphasized that there is a natural analogy between biochemical networks and (concurrent) systems with an unknown structure. Both can be considered as a black box with unknown internal working about which we want to check some property, i.e., verify some hypothesis about their behavior.

We assume that we have a model of the analyzed network to which the Black box checking technique can be applied. This usually means that the model is executable, i.e., it is amenable to simulation, testing and model checking. Recently this kind of models are becoming increasingly popular (e.g. [13, 2, 6]).

The crucial question is what are the actions of the biochemical network - the "interface buttons" of the black box - that we are going to observe and how is the actual testing of the black box performed. There is a wide variety of options depending on the concrete type of biochemical networks that are treated, as well as the models that are used.

For example, let us consider a signal transducing network - like the RTK MAPK pathway from [13]. We can choose the proteins which are involved in the reaction and whose concentration (presence/absence) can be measured during the experiments as observable signals. Suppose that the model checking step of the Black box strategy has generated a counterexample which is the action sequence $a_0 a_1 \ldots a_n$. We have to check if this sequence is indeed produced by the cell. We can do this by simple observation if indeed the sequence of signals/proteins occurs. The "pressing of the interface button" here amounts to measuring the concentration of the particular pathway products/signals a_i. In the same way one can check the test sequences produced by Vasilevskii-Chow algorithm, for the cases when the model checking phase does not produce counterexamples.

This approach of pure observation is adequate if the biochemical network is left to itself, i.e., it does not involve external perturbations (for instance by human intervention.) However, one can predict that in potential applications one would like to interact with the network. For instance in disease treatment one could try to recover the normal behavior of the network by applying corrective signals. Another situation when one will have to deal with an interactive network is the forward engineering of biochemical networks that we discuss below in the application section in more detail.

In those cases apparently a more interactive testing is needed. One possibility is to consider as an action a_i the pair of signals (R_i, P_i) of the network. We choose R_i, P_i such that P_i (product) is a result of the application of the reactant R_i (which possibly interacts with some other compounds in order to produce P_i). We say that a_i is executed by the "black box", if after applying R_i we get P_i as a product.

By analogous reasoning one can identify the actions and develop a testing methodology for other types of biochemical networks like genetic networks – which could be modeled as Boolean networks (e.g. [4]) – or metabolic pathways.

3.1 Potential Applications of the Black Box Checking Framework

Currently, probably the most realistic application of the Black box checking could be to generate sets of action sequences (experiments) that can be applied in a systematic way in order to deduce the network mechanisms. This can potentially shorten the time needed for the experiments and lower the costs.

Perhaps it would not be too futuristic to imagine fully automated laboratories for analyzing the properties of networks. The black box testing phase - working with the biochemical network - can be done purely by machines without any human involvement during the application of the Black box checking strategy.

One can expect that the main problem would still be the combinatorial explosion. However, it is hoped that with the constant improvement of the hardware performance and the model checking and testing algorithms one can make the approach feasible also for networks with significant complexity.

We predict an indispensable role of fully- or semi-automated techniques in so called *forward engineering* of biochemical networks. Rather than understanding of the existing natural networks by reverse engineering, the goal of forward engineering is to design in vivo networks that would perform certain tasks. This approach has been successfully applied in several cases (see [12] for an overview). Obviously forward engineering is much closer to the original domain of application of Black box checking – there is a full analogy between designing hardware/software artifacts and biochemical networks. Provided that we have an executable design (model) the forward engineered networks can be simulated, tested, or model checked in a classical way.

Forward engineering advocates that one can learn the functioning of the networks not only by observation and deduction, but also by trial and error. Once one understands the basic design principles one can postulate the underlying mechanism of the network [12]. This should happen after a limited number of experiments and Black box checking could help in reducing this number.

4 Conclusions and Future Work

The main aim of this paper is to bring to the attention of the system biology community the Black box checking strategy as a possible framework for the analysis biochemical networks. Because of the complexity of the biological networks Black box checking seems to be a nice compromise between the formality and the feasibility of the analysis. We discussed how different biological networks can be fitted into the theory and prospective applications.

As this is to a great extent still a work in progress, the first task would be to automate as much as possible the Black Box checking strategy in the context of biochemical networks. This would be definitely needed if one wants to test further the feasibility of the approach in practice. (We experimented with some small models, like the RTK MAPK signal transduction pathway from [13], using the model checker Spin [9]. In these experiments only the model checking part was automated, while the learning algorithms were applied manually, which becomes too time consuming even for very small examples.) Naturally, the fully automated approach should be tested on real world case studies.

One can try to extend Black box checking to more quantitative models like timed or hybrid automata. A stepping stone in that direction could serve the work of [3]. Finally, the Black box checking approach can be used beyond biochemical networks - in fact for any biological system that can be modeled as a finite transition system.

References

1. D. Angluin, *Learning Regular Sets from Queries and Counterexamples*, Information and Computation, 75, 87–106, 1987.
2. M. Antoniotti, A. Policriti, N. Ugel, B. Mishra, *Model Building and Model Checking Biochemical Processes* Cell Biochemistry and Biophysics, 38:271–286, 2003.
3. G. Batt, D. Bergamini, H. de Jong, H. Garavel, R. Mateescu, *Model Checking Genetic Regulatory Networks using GNA and CADP*, Model Checking of Software: 11th Spin Workshop SPIN '04, LNCS 2989, pp. 158–163, Springer, 2004.
4. J.M. Bower, H. Belouri, (eds.) *Computational Modeling of Genetic and Biochemical Networks*, MIT Press, 2001.
5. C.H. Bryant, S.H. Muggleton, S.G. Oliver, D.B. Kell, P. Reiser, R.D. King, *Combining Inductive Logic Programming, Active Learning and Robotics to Discover the Function of Genes*, Electronic Transactions in Artificial Intelligence, 6-B1(012):1–36, 2001.
6. N. Chabrier, F. Fages, *Symbolic Model Checking of Biochemical Networks*, Computer Models in System Biology, CMSB '03, LNCS 2602, pp. 149 – 162, Springer, 2003.
7. T.S. Chow, *Testing Software Design Modeled by Finite State Machines*, IEEE Transactions on Software Engineering, SE-4, 2, pp. 178–187, 1978.
8. E.M. Clarke, O. Grumberg, D.A. Peled, *Model Checking*, MIT Press, 2001.
9. G.J. Holzmann, *The Spin Model Checker: Primer and Reference Manual*, Addison-Wesley, 2003.
10. D. Peled, *Model Checking and Testing Combined*, ICALP 2003, pp. 47–63, 2003.
11. D. Peled, M.Y. Vardi, M. Yannakakis, *Black Box Checking*, Journal of Automata, Languages and Combinatorics 7(2): 225–246, 2002.
12. C. Rao, A. Arkin, *Control Motifs for Intracellular Regulatory Networks*, Annual Review of Biomedical Engineering, 3:391–419, 2001.
13. A. Regev, W. Silverman, E.Y. Shapiro, *Representation and Simulation of Biochemical Processes Using the pi-Calculus Process Algebra*, Pacific Symposium on Biocomputing 2001, pp. 459–470, 2001.
14. R. Somogyi, S. Fuhrman, X. Wen, *Genetic Network Inference in Computational Models and Applications to Large Scale Gene Expression Data*, in [4] above, 2001.
15. M.P. Vasilevskii, *Failure Diagnosis of Automata*, Kibernetika, 4: 98–108, 1973.

CMBSlib: A Library for Comparing Formalisms and Models of Biological Systems

Sylvain Soliman and François Fages

Projet Contraintes, INRIA Rocquencourt,
BP105, 78153 Le Chesnay Cedex, France
{Sylvain.Soliman, Francois.Fages}@inria.fr
http://contraintes.inria.fr

Abstract. We present CMBSlib, a library of Computational Models of Biological Systems. It is aimed at providing a list of test problems for formalisms, modeling issues and implementation issues in systems biology.

The main motivation for CMBSlib is to stimulate research on the formal modeling of biological systems, by facilitating the exchange of formal models between researchers, and by providing a forum of comparison and validation of not only models, but also modeling formalisms and implementations.

Unlike a standardization effort, CMBSlib welcomes the most exotic formalisms and models provided they attack the modeling of well documented biological systems. Models of biological systems written in any referenced formalism can be submitted to CMBSlib. No special format or standard is required.

We discuss the advantages of and problems encountered in building such a library, give an example of typical entry in the library, and most of all we invite the community to become active contributors to CMBSlib.

1 Introduction

As the first CMSB workshop proved, modeling for Systems Biology is an important new task for computer scientists, mathematicians and biologists; and to quote the Call For Papers of this second edition:

> As the field matures, it is becoming increasingly obvious that there is probably no 'one-size fits all' formal language for molecular biology, but rather several meddling paradigms, each with its strengths and weaknesses relatively to specific analytical goals.

It therefore seems that the need to compare all these formal languages, and their corresponding analytical capabilities, is also increasing.

On the other hand, most of the modeling work done around systems biology is quite difficult to transfer since lots of articles mentioning new models only describe the resulting analyses (usually a simulation plot) and the biological lessons learned from it. The models themselves often lack proper publication.

V. Danos and V. Schachter (Eds.): CMSB 2004, LNBI 3082, pp. 231–235, 2005.

We thus advocate the need for a general framework allowing to compare:

- on the one hand, different formalisms, associated tools, resulting analyses;
- on the other hand, different models of the same or related biological systems.

CMBSlib aims at becoming such a framework, and provides at the same time a global facility to store and publish models. This is in contrast to the already existing model repositories, like that of the SBML-capable tool, Cellerator [1] for instance. All model repositories are indeed oriented towards one single formalism, whether ODEs, petri-nets or process calculi [2], while CMBSlib aims at confronting various formalisms useful for Systems Biology. Furthermore, existing repositories maintain one single model of a given biological system in a given organism, there is thus no facility for comparing different models of the same biological system.

There are also attempts at unifying the current mass of languages, such as that of the BioPAX group[1] for sharing pathway information, and if such unification succeeds we would be very happy; but until then, it seems necessary to allow the use of different formalisms, description languages and tools. CMBSlib was designed for that purpose in the framework of the ARC CPBIO [3] whose more general aim is to study new languages suited to Systems Biology.

One should also remark that the creation of analogous libraries in other domains of computer science has usually resulted in a big progress in comprehension of the issues involved. These repositories are mostly benchmarking libraries, but our constraint programming origin made us aware of CSPLIB [4], a library of constraint satisfaction problems, where "representation" is also a determining factor. That existing library was an important source of inspiration for building CMBSlib.

2 Comparing Formalisms

As made clear above, there are currently many formalisms used for modeling biological systems, and it is much too early to throw away all but one.

There is however currently no existing repository trying to encompass many formalisms in order to allow the user to choose the one the most appropriate to what he wants to do with the corresponding model.

Moreover, many models were often built about one given biological system, or even similar systems, but they usually are kept separate because they are expressed using different formalisms or description languages. This separation makes it impossible to benefit in one of those models from the enrichments brought to another one; it also impeaches any meaningful comparison.

All these barriers not only impact the users of the models, since they do not know about the real reasons to choose one or another formalism, and do not benefit from cross-improvement of the existing models, they also have a negative effect on the designers of formalism who need to build test-cases and showcases

[1] `http://www.biopax.org`

from scratch and have no facility for confronting their design choices to features found useful in other formalisms from different sources.

3 Submission Guidelines

To help users submit models to the library, we provide some simple guidelines, implemented as an HTML form on CMBSlib's home page[2]:

The aim of the library is to become a useful resource for all researchers on formal methods in systems biology. We thus welcome the submission of *any formal model of biological system*: to submit a new model, one has only to fill-in the submission form on the CMBSlib web site.

The models are classified in CMBSlib by the biological system they refer to. We therefore specify all biological systems in CMBSlib using natural language and reference to survey papers in Biology.

CMBSlib may contain different models of the same biological system, either in different formalisms or even in the same formalism. *No standard format or language is required.*

As we want to help people compare their work in CMBSlib with minimum effort, we encourage users of the library to send us the URL of any tools that might be useful to others (e.g. parsers for data files, simulators, analyzers, translators, ...). All such code is placed in the library through the generosity of the authors, and comes with all the usual disclaimers.

To make comparison with previous work easier, links to articles that use these models are provided. References to articles using models of CMBSlib are thus solicited, in order to be added to the references section of CMBSlib.

Finally, to make it easy to compare new models with others, a record of results (simulation plots, query results, etc.) will be provided. To help us keep these records up-to-date, users are encouraged to send in their latest results.

4 An Example

To get an idea of what information gets stored in the library, let us take an example, that of the "Mammalian cell cycle control".

4.1 Specification

To define what this is about we first need a *specification* of the system: this usually takes the form of a few natural language sentences describing the system. For instance:

"A model of the known interactions of the mammalian cell cycle regulatory network at the molecular level."

[2] `http://contraintes.inria.fr/CMBSlib/`

4.2 References

The informal specification should always for clarity be accompanied by some references to the relevant literature, in the present case the survey paper:

"Kurt K. Kohn, Molecular Interaction Map of the Mammalian Cell Cycle Control and DNA Repair Systems. Molecular Biology of the Cell, Vol. 10, pp.2703-2734, August 1999."

Whenever possible links to databases such as PubMed[3] or PNAS Online[4] should be included, or if copyrights allow it, links to an electronic version of the article hosted (or mirrored) in the library.

4.3 Models and Analyses

Next, one can try and provide some models of that system in different *formalisms*, like $\kappa 0$ [5] and BIOCHAM [6, 7], two transition based formalisms. Then one can show what type of analyses are possible with each formalism and tool, what part of the model is correctly captured and what should still be improved. BIOCHAM for instance allows for some model-checking of CTL properties of the defined model.

It is interesting however to note that there are other models coming from the same survey of K. Kohn [8], like one using Pathway Logic [9] which uses the formal tools of Maude[5] for analysis purposes.

Those different models each improved on the original Kohn map by elucidating unclear zones while formalizing the model, some also benefited from later improvements using the literature, it would thus be very fruitful for all of them to compare the results and correct/combine what needs so.

The same can be said of most KEGG [10] maps, which have been completed and corrected by different people a number of times, but since each of those corrections resulted in a model in a different language, they were usually never compared, nor reused.

The comparisons might be in some cases limited to some informal reasoning about the analyses obtained on different models, and in the first phase of CMBSlib's life this type of exchange will probably be the most natural, and we believe already very fruitful. However in most cases some concrete guidelines can be elaborated for comparison, and the CMBSlib team will strive for such a result, with the help of the contributing modelers.

5 Conclusion

The existence of the CMSB conference and of its community already shows the need for a platform for sharing and comparing models and formalisms.

[3] `http://www.ncbi.nlm.nih.gov/entrez/`
[4] `http://www.pnas.org/`
[5] `http://maude.csl.sri.com`

CMBSlib aims at becoming such a forum of exchange about models and formalisms, by providing tools like a model repository, a mailing-list, translators, etc. It should also stimulate research on the comparison of different formalisms for expressing and analyzing a single model, and of different models for representing a single system. We believe that the resulting cross-fertilization of models and formalisms is an important step towards the difficult task of modeling issues in Systems Biology.

It is hoped that the community will contribute to the library since that is the only way it will develop, and since such a contribution will be beneficial to all contributors, whether mathematicians, computer-scientists or biologists, to improve their formalisms, tools and models.

References

1. Shapiro, B.E., Levchenko, A., Meyerowitz, E.M., Wold, B.J., Mjolsness, E.D.: Cellerator: extending a computer algebra system to include biochemical arrows for signal transduction simulations. Bioinformatics **19** (2003) 677–678 http://www-aig.jpl.nasa.gov/public/mls/cellerator/.
2. Regev, A., Silverman, W., Shapiro, E.Y.: Representation and simulation of biochemical processes using the pi-calculus process algebra. In: Proceedings of the sixth Pacific Symposium of Biocomputing. (2001) 459–470
3. ARC CPBIO: Process calculi and biology of molecular networks (2002–2003) http://contraintes.inria.fr/cpbio/.
4. Gent, I.P., Walsh, T.: CSPLIB: A benchmark library for constraints. In Jaffar, J., ed.: Proceedings of the Fifth International Conference on Principles and Practice of Constraint Programming (CP'99). Volume 1713 of Lecture Notes in Computer Science., Springer-Verlag (1999) 480–481 A longer version appeared as technical Report TR-APES-09-1999.
5. Chabrier, N., Chiaverini, M., Danos, V., Fages, F., Schächter, V.: Modeling and querying biochemical networks. Theoretical Computer Science **To appear** (2004)
6. Chabrier-Rivier, N., Fages, F., Soliman, S.: The biochemical abstract machine BIOCHAM. In Danos, V., Schächter, V., eds.: CMSB'04: Proceedings of the second Workshop on Computational Methods in Systems Biology. Lecture Notes in Computer Science, Springer-Verlag (2004)
7. Chabrier, N., Fages, F., Soliman, S.: BIOCHAM's user manual. INRIA. (2003–2004) http://contraintes.inria.fr/BIOCHAM/.
8. Kohn, K.W.: Molecular interaction map of the mammalian cell cycle control and DNA repair systems. Molecular Biology of Cell **10** (1999) 703–2734
9. Eker, S., Knapp, M., Laderoute, K., Lincoln, P., Meseguer, J., Sönmez, M.K.: Pathway logic: Symbolic analysis of biological signaling. In: Proceedings of the seventh Pacific Symposium on Biocomputing. (2002) 400–412
10. Kanehisa, M., Goto, S.: KEGG: Kyoto encyclopedia of genes and genomes. Nucleic Acids Research **28** (2000) 27–30

Combining State-Based and Scenario-Based Approaches in Modeling Biological Systems

Jasmin Fisher[1,*], David Harel[1,**], E. Jane Albert Hubbard[2,***], Nir Piterman[1,†], Michael J. Stern[3,‡], and Naamah Swerdlin[1]

[1] Dept. of Computer Science and App. Math., Weizmann Institute, Rehovot 76100, Israel
firstname.lastname@weizmann.ac.il

[2] Dept. of Biology, New York University, New York, NY
jane.hubbard@nyu.edu

[3] Dept. of Genetics, Yale School of Medicine, New Haven, CT
michael.stern@yale.edu

Abstract. Biological systems have recently been shown to share many of the properties of reactive systems. This observation has led to the idea of using methods devised for the construction (engineering) of complex reactive systems to the modeling (reverse-engineering) of biological systems, in order to enhance biological comprehension. Here we suggest to combine the two formal approaches used in our group — the state-based formalism of statecharts and the scenario-based formalism of live sequence charts (LSCs). We propose that biological observations are better formalized in the form of LSCs, while biological mechanistic models would be more natural to specify using statecharts. Combining the two approaches would enable one to verify the proposed mechanistic models against the real data. The biological observations can be compared to the requirements in an engineered system, and the mechanistic model would be analogous to the implementation. While requirements are used to design an implementation, here the observations are used to motivate the invention of the mechanistic model. In both cases consistency of one with the other must be established, by testing or by formal verification.

1 Introduction

Experimental biology is an interplay between collecting data in experiments (observations), followed by the analysis of the data and suggestion for possible mechanistic models that would explain how the system under study works, and how it gives rise to the observed phenomena. Further experiments are often preformed to test the hypothetical

* Supported by the Dov Biegun postdoctoral fellowship.
** Supported in part by NIH grant F5490-01 and ISF grant 287/02.
*** Supported in part by NIH grant R24-GM066969 and R01-GM61706.
† Supported by the John von-Neumann center for Verification of Reactive Systems.
‡ Supported in part by NIH grant R24-GM066969.

V. Danos and V. Schachter (Eds.): CMSB 2004, LNBI 3082, pp. 236–241, 2005.

mechanism. Here we propose that the dichotomy between theses two aspects of biology calls for different formal methods.

Recently, the resemblance between reactive systems (systems that continuously interact with their environment) and biological systems has been noted [7, 11]. This observation has led to the idea of using methods devised for the construction of complex reactive systems to model biological systems. The first attempt to follow this path was a modest model of T-cell activation [11], which was followed by an extensive animated model of T-cell behavior in the thymus [4, 5]. At present there is an ongoing effort to model the vulval development in the nematode *Caenorhabditis elegans* [12].

Two approaches are used in our group to model biological behavior: an **intra-object** one based on the language of **statecharts** [6] and the **Rhapsody** tool [8, 10], and an **inter-object** one that uses the more recent language of **live sequence charts** (LSCs) [3] and the **Play-Engine** tool [9]. Both languages are visual, having a clear (and formal) syntax and semantics, and both approaches enable the construction of a formal model and its execution. In Rhapsody, a state-based transition diagram (statechart) of the system under study is constructed. The tool automatically generates executable C/C++/Java code from the statecharts. In contrast, LSCs specify scenarios of behavior between objects, with varying modalities (e.g., required, possible, and forbidden scenarios), and the Play-Engine executes these directly in a way that satisfies the modalities for each.

As typical biological data is available in the form of 'condition-result' scenarios, we believe that biological observations are best formalized in the form of LSCs, which take the following general form: if X (the prechart) occurs then Y (the main chart) should too, where X and Y consist of scenarios of behavior and can be simple or complex. Indeed, in an LSC we can formalize the terms of the experiment as the conditional prechart that enables an LSC, and we can formalize the result of the experiment as a sequence of happenings resulting from that condition. On the other hand, since many biological mechanistic models are 'state-based', we feel that in such cases it would be more natural to specify mechanistic models using statecharts, since they specify the behavior of the system based on its internal (stipulated) mechanism.

Here, we propose that these two approaches complete each other and should be used together in order to model the two aspects of a common biological system simultaneously. By using a scenario-based approach to formalize the behaviors of the biological system and a state-based approach to formalize the mechanism underlying these behaviors, one can formally verify that the mechanistic model reproduces the system's real behavior. In this functional sense, the biological observations can be compared to the requirements in an engineered system, while the proposed mechanistic model would be analogous to the system's design and implementation.[1] To carry this analogy further, in an engineered system the requirements are used to help in coming up with the design and implementation, and are then used a second time to build test suites for testing the implementation against the requirements. In biological systems the observations are used to motivate the construction of mechanistic models (that serve as working hypotheses),

[1] In another sense it is the other way around: the biological observations are directly related to the actual system as is the implementation of an engineered system, whereas requirements and biological mechanistic models are both invented by humans.

and our approach enables them to be used also in testing and verification, by simulation or, e.g., model-checking. These techniques may also yield interesting predictions that should be then corroborated experimentally in the biological system.

2 Engineering Computerized Systems Versus Reverse-Engineering Biological Systems

In **engineering**, we try to produce a **design** that satisfies a set of requirements. This set of requirements is determined by our ideas about how the system should eventually work. In biological modeling we do **reverse-engineering**, trying to construct a **mechanistic model** that explains how the biological system works. This model has to fit the experimental observations.

In engineering, we formalize the requirements (emanating from the system's concept) in a formal specification language (e.g., LSCs). This formal specification not only guides the construction of the design and implementation but is later used also to check the design's correctness. In formal modeling of biology, on the other hand, the situation is different. The biological system is already 'built'. In fact, there is a running exemplar of the system. Unfortunately, we are restricted in the way we can analyze it (in particular, we are unable to access its 'blueprint'). By experimenting with the biological system (i.e., testing it under different conditions) and recording the results of these experiments we can gain knowledge about its behavior. Once we formalize these observations in the form of LSCs we get a sort of 'requirements' specification that can be used in constructing the mechanistic model. See Figure 1.

	Computerized Systems (Engineering)	Biological Systems (Reverse-engineering)
Scenario-based	Requirements	Observations
State-based	Design & Implementation	Mechanistic Model

Fig. 1. Analogy between computerized systems and biological systems

The testing phases in both cases (engineering versus reverse-engineering) possesses an interesting duality. In engineering we test the design in order to improve assurance of its quality. In biology we probe the system, a process we can also call testing, get the results, and rerun the results on the mechanistic model. Thus, in the reverse-engineered process of modeling biology a test would be run twice: once on the biological system itself and once on the mechanistic model. As a result, once we check all the tests known at a given time-point, we get what we might call (borrowing from software engineering terminology) a complete **coverage** of the desired behavior of the mechanistic model. At that time-point, there are no more tests to preform on the system until more experiments are carried out. In contrast, when engineering man-made systems the problem of determining whether the requirements are sufficient is an interesting question in its own right.

3 Regression Testing and Model Checking

The fact that when we come up with a mechanistic model there is at hand a fixed given set of tests (the biological observations) suggests that we can use **regression testing** to get a higher type of assurance of the model's correctness. In engineering, regression testing is used to compare different versions of the same design, by running the old test suites on the new version to make sure that we haven't inadvertently changed previously decided-upon desired behavior. Thus, during the development of the design, we form a collection of tests and save the results we got when running them on the present version. Once a change is made, we run the same set of tests again and make sure that the new design produces the same results. In the case of reverse-engineering a biological system the comparison made by re-running the tests is not between an old and a new version of the system, but between the real biological system and the proposed mechanistic model thereof. Here, the collection of tests is already given, as the set of observations resulting from the performed experiments.

Both in engineering and reverse-engineering performing additional ad-hoc tests can produce interesting results. In the engineering world, arbitrary tests may produce behaviors that indicate the existence of bugs, causing the need to redo the design and implementation, whereas in reverse-engineering, ad hoc tests (such as running the formalized models on additional inputs or in different ways) may produce interesting predictions regarding the behavior of the biological system, or questions that need to be resolved by further experimentation.

In engineering, we would like to use verification techniques, such as **model-checking** (see, e.g., [2]), in order to acquire greater confidence in the correctness of the design. Model-checking is a method to formally verify that all the possible behaviors of the system satisfy a given requirement, and can be used with statecharts or LSCs. In reverse-engineering of biological systems we could use model-checking to get around the main disadvantage of model execution, which is its inability to cover all possible execution scenarios, which is particularly problematic for mechanistic models that are non-deterministic. This would be done by model-checking the specification (LSCs) against the mechanistic model (statecharts), and could provide a major additional boost to the validity of the latter. One of the results of the model-checking process could be interesting predictions regarding the behavior of the actual system.

4 Implementation

We have applied the suggestion made in this paper to the mechanistic model of [14] that explains parts of the formation of the vulva in *Caenorhabditis elegans*. We have formalized the mechanistic model in the form of statecharts, and the experimental observations (that led to the suggestion of this model) as existential LSCs. We then used the Rhapsody tool [10] and its testing component — the **TestConductor** [13]. In the TestConductor, tests are given in the form of combinations of existential LSCs. Each test can then be performed individually, or the tool can be asked to produce a report on the entire behavior of the model when checked versus all the tests. Running regression testing in different stages of the development of the statecharts model enabled us to fine-tune the model to reproduce all the behaviors on which this model is based.

An interesting aspect of this particular work is that our mechanistic model is completely deterministic. Thus, testing a scenario using simulation is sufficient to make sure that the mechanistic model reproduces the behavior depicted in the scenario. We thus have full assurance that our formalization of the mechanistic model completely reproduces the data. This fact improves our confidence in the correctness of the proposed mechanistic model. A detailed description of this modeling effort will be reported separately.

5 Concluding Remarks

Based on the above, we suggest that the state-based and scenario-based approaches complete each other. We propose that biological systems should be modelled using both approaches. Observations should be formalized by inter-object scenario-based methods (in our case, LSCs using the Play-Engine tool), while the mechanisms should be formalized by intra-object state-based methods (in our case using statecharts and the Rhapsody tool). Once this is done we can simulate all the experiments carried out in practice and use the state-based model to drive the simulation that the scenarios follow. Using regression testing, we can ensure that the mechanistic model reproduces all the behaviors observed in the living system.

A connection between Rhapsody and Play-Engine is currently under development, which will enable them to work on cooperation; see [1]. Such connection will enable to verify automatically that a biological mechanistic model is consistent with the experimental observations obtained by the system. We believe this would further facilitate our understanding of biological systems and help simulate and analyze their reactive nature.

References

1. D. Barak, D. Harel, and R. Marelly. Interplay: Horizontal scale-up and transition to design in scenario-based programming. To appear, 2004.
2. E. Clarke, O. Grumberg, and D. Peled. *Model Checking*. MIT Press, 1999.
3. W. Dam and D. Harel. LSCs: Breathing life into message sequence charts. *FMSD*, 2001.
4. S. Efroni, D. Harel, and I. R. Cohen. Modeling and simulation of the thymus. *Multidisciplinary Approaches to Theory in Medicine*, 2002.
5. S. Efroni, D. Harel, and I. R. Cohen. Toward rigorous comprehension of biological complexity: modeling, execution, and visualization of thymic T-cell maturation. *Genome Res*, 2003.
6. D. Harel. Statecharts: A visual formalism for complex systems. *SCP*, 8:231–274, 1987.
7. D. Harel. A grand challenge for computing: Towards full reactive modeling of a multi-cellular animal. *Bulletin of the EATCS*, 81:226–235, 2002.
8. D. Harel and E. Gery. Executable object modeling with statecharts. *Computer*, 30(7), 1997.
9. D. Harel and R. Marelly. *Come, Let's Play: Scenario-Based Programming Using LSCs and the Play-Engine*. Springer-Verlag, 2003.
10. I-logix,inc. http://www.ilogix.com.
11. N. Kam, D. Harel, and I. R. Cohen. The immune system as a reactive system: Modeling T-cell activation with statecharts. *Bull. Math. Bio.*, 2003.

12. N. Kam, D. Harel, H. Kugler, R. Marelly, A. Pnueli, E. J. A. Hubbard, and M. J. Stern. Formal modeling of C. elegans development: A scenario-based approach. In *1st Int. Conf. on Computational Methods in Systems Biology*, February 2003.
13. M. Lettrari and J. Klose. Scenario-based monitoring and testing of real-time UML models. In *4th Int. Conf. on the Unified Modeling Language*, October 2001.
14. P. W. Sternberg and H. R. Horvitz. The combined action of two intercellular signaling pathways specifies three cell fates during vulval induction in c. elegans. *Cell*, 58(4):679–93, 1989.

Developing SBML Beyond Level 2: Proposals for Development

Andrew Finney

University of Hertfordshire, Hatfield, AL10 9AB, UK
a.finney@herts.ac.uk

Abstract. The Systems Biology Markup Language (SBML) is an XML-based exchange format for computational models of biochemical networks. SBML Level 2, whose definition was established in June 2003, includes several enhancements to the original Level 1. This paper includes a brief overview of Level 2. Several proposals are under development to extend SBML to create Level 3. These include diagrams, 2-D and 3-D spatial characteristics, arrays, model composition and multi-component chemical species. This paper describes the current proposals for the last two features.

1 Introduction

Researchers in systems biology have a wide variety of software tools available to them. The Systems Biology Markup Language (SBML) [1] was initially developed to enable the exchange of models between these tools, as part of the ERATO Kitano project in collaboration with the community of Systems Biology modelers and software developers. A multi-institutional team based at the California Institute of Technology (USA), University of Hertfordshire (UK) and Systems Biology Institute (Japan) provides editorial, organizational and technical support to the SBML development process. Today, over 40 software tools support SBML [2], and in addition, it is the standard model definition language for the DARPA BioSPICE project and the International E. coli Alliance (IECA).

SBML encodes models consisting of biochemical entities (species) linked by reactions to form biochemical networks. SBML is being developed by the community in levels, where each level extends the set of features of the language. The structures that comprise an model in SBML Levels 1 and 2 are described elsewhere [1, 2, 3]. By freezing SBML development at incremental levels, software authors can work with stable standards and gain experience with the standard before further development. The separate levels of SBML are intended to coexist.

2 Proposals for Level 3

The Systems Biology community is now developing SBML Level 3. Although SBML has been successfully adopted by many groups developing systems biology

V. Danos and V. Schachter (Eds.): CMSB 2004, LNBI 3082, pp. 242–247, 2005.

software, there exist packages that support classes of models which presently cannot be encoded in Level 2. Level 3 is intended to provide support for these tools. The community plans to introduce features in Level 3 that will add support for: (a) composing models from component submodels; (b) describing states and interaction of components of species in terms of rules rather than explicit enumerations of all possible combinations; (c) describing 2-D and 3-D spatial geometries; (d) diagrammatic representations of models; (e) enabling parameter and initial condition values to be defined separately from models; (f) allowing for alternative mathematical representations of reactions; and (g) enabling the association of terms from controlled vocabularies to be associated with SBML elements. Proposals for these features have been described at various levels of detail [2]. The following sections describe proposals for Model Composition and Multi-Component species. These are not yet part of any SBML standard.

2.1 Model Composition

The aim of the Model Composition proposal is to enable a model to be composed hierarchically from a number of submodels. Given the existence of more than one software package supporting hierarchial model composition, (for example Promot/DIVA [4] and VLX Biological Modeler from Teranode Corporation [5]) SBML ought to support the exchange of such models. One of our key aims is enable the multiple instantiation of the same submodel within an enclosing model thus enabling, for example, a model of a cell to be composed from a number of instances of the same mitochondria submodel. This specific capability is not present in CellML despite its support for hierarchical composition. The specific proposal described here [6] builds on previous proposals [7, 8].

The model composition proposal allows for a model to contain a list of submodels and a list of instances of those submodels. Each submodel is a complete SBML model thus allowing for the hierarchical assembly of models. Each instance structure instantiates a submodel (i.e. implies the existence of a complete copy of the submodel) within the immediate containing model. An instance structure refers to a submodel using XLink attributes enabling the structure to refer to a submodel that is either internal or external to the containing XML document. (XLink [9] is a standard for referring to components of XML documents.)

A system that enables just the assembly of models from submodels as described has little utility. A method for linking model components at various levels of the model instance hierarchy is required. The model composition proposal introduces `ObjectRef` structures which can refer to objects within the instance hierarchy. The proposal has defined `ObjectRef` structures to be recursive thus allowing objects at any level of the instance hierarchy to be referenced.

The `ObjectRef` structure does not use XLink attributes because the instance hierarchy does not exist in XML form but is implied by the instance structures in models. For example consider two models A and B. A contains a species S and model B contains two instances of A, a_1 and a_2. Suppose we wished to refer to the specific species S within instance a_1. We cannot use an XLink attribute because that specific species does not exist in any XML document, it only exists in the a_1 copy of A. The single XML structure for S in model A does not represent any specific species within instances of A since there can be any number of instances of A.

SBML documents have many attributes which link the various structures forming the reaction network and placing species into compartments. In the model composition proposal these links can optionally be replaced by `ObjectRef` structures thus enabling links to form between models thus enabling, for example, a reaction to be created between species in two submodel instances or for a species to be placed inside a compartment within a submodel instance. An `ObjectRef` can only refer to an object within the instance hierarchy created by the immediate containing model: since an `ObjectRef` does contain XLink attributes it cannot create arbitrary linkages. The `ObjectRef` structures and the Level 2 attributes they replace must refer to the same object types. These types are restricted as described in the SBML Level 2 specification. For example the `speciesLink` element, an instance of the `ObjectLink` type, which replaces the `species` attribute on `SpeciesReference` structures, can only refer to `Species` objects.

Potentially when submodels are combined to form a model one biochemical entity will be represented separately in more than one of the submodels. In many bioinformatics systems this is resolved using a synonym dictionary. To enable the construction and reuse of abstract or generic models a structure is required that enable the parametrization of models. In the general case the parameter and synonym requirements can be viewed as the requirement to support one object overloading another object. Under the proposal a model contains a list of `Link` structures which could be viewed either as a directed synonym dictionary or a set of parameter assignments. Each `Link` structure consists of two `ObjectRef` structures which refer to the overloading and overloaded objects. The overloading object replaces the attribute values of the overloaded object. If the objects are species then they form a single node in the reaction network. If the objects are compartments then they form a single compartment. The types of the two objects referred to by a `Link` structure must have the same type. For example a `Species` object cannot overload a `Compartment`. There is no consensus in the community of modelers concerning the ideal form of composed models and thus the proposal does not describe any further restrictions on the linkages that can be formed between objects in the instance hierarchy. However SBML has features, such as a units system, which are appropriate for further validation of model composition.

2.2 Multi-component Species

Although SBML Level 2 can encode biochemical reaction networks the following concepts are not easily represented: (a) the hierarchical description of biochemical entities through the composition of other biochemical entities or (b) the description of generalized biochemical reactions that avoid the enumeration of many species states and reactions. Biochemical entities, depending on the description, can be composed either as simple aggregation or through graphs of other biochemical entities where arcs represent kinds of bonding. SBML requires new features to enable the representation of, for example, proteins which can contain many phosphorylation states, complexes of these proteins and models of signalling pathways which contain these proteins. Several laboratories have data sets and/or software which make explicit some or all of these features of biochemical networks often in generalized form. For example the Genome KnowledgeBase [10] captures the hierarchical assembly of complexes in the context of pathways; the LANL T10 group have developed models of generalized reaction systems from which software automatically generates an ODE based representation containing very large sets of species and reactions [11]; and the StochSim environment [12] stochastically simulates the interaction of individual chemical entities each of which has a state vector where reactions form a set of complex state transitions.

This section contains proposals for a set of new structures for addition to SBML to implement the above requirements. This proposal builds on previous proposals for representing multi-component species [13, 14]. These structures, whilst primarily aimed at supporting the representation of phosphorylated protein complexes, will be capable of capturing both the detailed structure of, and the processes acting on, all types of macromolecules.

Under this proposal a model would optionally contain a set of species type structures. A given species type simply represents all biochemical entities with the same biochemical structure (that is having identical structure for the purposes of the model). In SBML Level 2 the species structure represents a pool of entities of the same type located in a specific compartment. In this proposal the type of a species structure is made explicit via a new `speciesType` attribute which refers to a species type structure. A reaction can be generalized to occur in any compartment by referring to reactants, products and modifiers by species type rather than by compartment specific species.

Under this proposal a species type can optionally contain a set of instances of other species type structures which define the composition of the containing species type. A model can be described using such a system of hierarchically contained components however, under this proposal, the species type instances can be explicitly connected i.e. a species type can describe a graph where the nodes are species type instances and the arcs are bonds. In this scheme a species type has a set of binding site structures each of which is a potential end point for a bond. A bond is simply a pair of references to binding sites on species type instances. Just as the bond structures are optional in species type structures it is not proposed here that SBML specify the level of decomposition at which a given model will operate, for example, a protein could be described as a single indivisible object or as a sequence of amino acids.

Whilst the structures described above capture a significant amount of information that cannot be made explicit in SBML Level 2 they do not provide any facilities for representing reactions generalized to apply to classes of species types. With just these structures an accurate model would still have to contain an enumeration of all the species type structures that could occur in the modelled system. In many modelled systems the number of structures required is so large that both this representation scheme and SBML Level 2 become impractical. To solve this problem, under this proposal, reactions can be generalized to apply to classes of species types. The complete set of species and species type structures are then implied from the reactions rather than fully enumerated. In this context the species structures contained in an SBML document define the initial state and boundary conditions of the system. The species type structures define a set of types that enable the definition of reactions and species. The reactions are applied to the biochemical entities in the modelled system that match the reactions' reactants and construct new entities as defined by the reactions' products. Thus a generalized reaction is a template for manipulating graphs of biochemical entities and contains structures which enable a reactant to match with species from a range of species types.

In a generalized reaction the reactant, product and modifier structures are *generic graphs* of a form similar to those graphs contained by species type structures. Whereas a species type can contain concrete bonds, which refer to pairs of binding sites, a generic graph can additionally contain generic bond structures, which refer to only one binding site. A generic bond simply represents a portion of a matching species graph which the reaction does not directly transform. A generic bond is identified by an associated symbol. The same symbol typically will occur in both the reactants and products indicating that a matching component is transferred from reactant to product.

3 Conclusions

SBML is a defacto standard for the exchange of biochemical reaction networks. The development of SBML is ongoing to ensure that it meets the full requirements of the Systems Biology community. The proposals described for model composition and multi-component species will ensure that SBML can represent highly complex biological systems such as signal transduction systems.

Acknowledgements

The development of SBML was originally funded by the JST (Japan). Support for the continued development of SBML today comes from the following sources: NHGRI/NIH (USA), NIGMS/NIH (USA), NEDO (Japan), BBSRC (UK), DARPA (USA), and the Air Force OSR (USA). The SBML development community includes members of the BioSPICE Model Definition Language, sysbio and sbml-discuss mailing lists. I thank Michael Blinov, Roger Brent, Fabian Campagne, Michael Hucka, Sarah Keating, Larry Lok, Nicolas Le Novère, Robert Phair and Jeremy Zucker for helpful comments.

References

1. Hucka, M., Finney, A., Sauro, H.M., Bolouri, H., Doyle, J.C., Kitano, H., Arkin, A.P., Bornstein, B.J., Bray, D., Cornish-Bowden, A., Cuellar, A.A., Dronov, S., Gilles, E.D., Ginkel, M., Gor, V., Goryanin, I.I., Hedley, W.J., Hodgman, T.C., Hofmeyr, J.H., Hunter, P.J., Juty, N.S., Kasberger, J.L., Kremling, A., Kummer, U., Le Novère, N., Loew, L.M., Lucio, D., Mendes, P., Minch, E., Mjolsness, E.D., Nakayama, Y., Nelson, M.R., Nielsen, P.F., Sakurada, T., Schaff, J.C., Shapiro, B.E., Shimizu, T.S., Spence, H.D., Stelling, J., Takahashi, K., Tomita, M., Wagner, J., Wang, J.: The Systems Biology Markup Language (SBML): A medium for representation and exchange of biochemical network models. Bioinformatics **19** (2003) 524–531
2. Hucka, M., Kovitz, B., Matthews, J., Schilstra, M., Finney, A., Bornstein, B., Shapiro, B., Keating, S., Funahashi, A.: The SBML website. Available via the World Wide Web at http://www.sbml.org/. (2003)
3. Finney, A., Hucka, M.: Systems Biology Markup Language: Level 2 and Beyond. Biochemical Society Transactions **31** (2003) 1472–1473
4. Ginkel, M., Kremling, A., Tränkle, F., Gilles, E.D., Zeitz, M.: Application of the process modeling tool ProMot to the modeling of metabolic networks. In Troch, I., Breitenecker, F., eds.: Proceedings of the 3rd MATHMOD. (2000) 525–528
5. Duncan, J., Arnstein, L., Li, Z.: Teranode corporation launches first industrial-strength research design tools for the life sciences at demo 2004. Available via the World Wide Web at http://www.teranode.com/about/pr_2004021601.php (2004)
6. Finney, A.: Systems Biology Markup Language (SBML) Level 3 Proposal: Model composition features. Available via the World Wide Web at http://www.cds.caltech.edu/~afinney/model-composition.pdf (2003)
7. Ginkel, M.: Modular SBML Proposal for an Extension of SBML towards level 2. In: Proceedings of 5th Forum on Software Platforms for Systems Biology (SBML Forum). (2002) Available via the World Wide Web at http://www.sbml.org/workshops/fifth/sbml-modular.pdf.
8. Webb, J.: BioSpice MDL Model Composition and Libraries. Available via the World Wide Web at http://bio.bbn.com/biospice/mdl/design/compose.html (2003)
9. DeRose, S., Maler, E., Orchard, D.: XML Linking Language (XLink) Version 1.0 W3C Recommendation. Available via the World Wide Web at http://www.w3.org/TR/2000/REC-xlink-20010627/ (2001)
10. Hodge, R.: Linking the levels of life from genes to cellular processes with the Genome Knowledge Base. Available via the World Wide Web at http://www.ebi.ac.uk/Information/News/ensembl_040203.pdf. (2003)
11. Goldstein, B., Faeder, J.R., Hlavacek, W.S., Blinov, M.L., Redondoc, A., Wofsy, C.: Modeling the early signaling events mediated by FcεRI. Molecular Immunology **38** (2001) 1213–1219
12. Le Novère, N., Shimizu, T.S.: StochSim: Modelling of stochastic biomolecular processes. Bioinformatics **17** (2001) 575–576
13. Le Novère, N., Shimizu, T.S., Finney, A.: Systems Biology Markup Language (SBML) Level 3 Proposal: Multistate Features. Available via the World Wide Web at http://sbml.org/multistates.pdf (2003)
14. Finney, A.: Internal discussion document possible extension to the Systems Biology Markup Language Complex Species and Species Graphs. Available via the World Wide Web at http://www.cds.caltech.edu/~afinney/CplxSpecies.pdf (2001)

General Stochastic Hybrid Method for the Simulation of Chemical Reaction Processes in Cells

Martin Bentele and Roland Eils

Div. Theoretical Bioinformatics, German Cancer Research Center (DKFZ),
Im Neuenheimer Feld 580, D-69120 Heidelberg, Germany
r.eils@dkfz-heidelberg.de
http://www.dkfz.de/ibios/index.jsp

Abstract. Stochastic approaches are required for the simulation of biochemical systems like signal transduction networks, since high fluctuations and extremely low particle numbers of some species are ubiquitous. Computational problems arise from the huge differences among the timescales on which the reactions occur, causing high cost for stochastically exact simulations. Here, we demonstrate a general hybrid method combining the exact Gillespie algorithm with a system of stochastic differential equations (SDEs). This technique provides a smooth and correct transition between subsets of 'slow' and 'fast' reactions instead of abruptly cutting the stochastic effects above a certain particle number. The method was successfully applied to mitochondrial Cytochrome C release in the CD95-induced apoptosis pathway. Moreover, this approach can also be used for other kinds of Markov processes.

1 Introduction

Biochemical networks are often simulated as well-stirred chemical reaction systems, for which a variety of deterministic and stochastic simulation methods exist. The deterministic approach is widely used and translates the reactions into a system of ordinary differential equations (ODEs). Let's assume a system with m molecule species and n reactions R_j with the reaction rates v_j and the stoichiometric coefficients ν_{ij}. The state is defined by the concentrations $(x_1, \ldots, x_m)$ and the time evolution of the system is simulated using an ODE solver for the equation system

$$\frac{dx_i}{dt} = \sum_{j=1}^{n} \nu_{ij} v_j(\mathbf{x}(t)), \qquad i = (1, \ldots, m). \tag{1}$$

The assumption of a continuous and predictive system is not always given, especially in signal transduction systems, where some molecule species might occur in very low numbers. A stochastically exact simulation method was provided by Gillespie [1]. Here, the system is defined by the discrete particle numbers

V. Danos and V. Schachter (Eds.): CMSB 2004, LNBI 3082, pp. 248–251, 2005.

$(X_1, \ldots, X_m)$. The time evolution is simulated by mapping random numbers to the time point and the type of the next molecular reaction events according to the *propensity* functions $a_j(\mathbf{X})$, the temporal probability density for R_j.

An approximate solution of the stochastic system is given by the chemical Langevin equations [2], a set of stochastic differential equations:

$$dx_i(t) = \sum_{j=1}^{n} \nu_{ij} a_j(\mathbf{x}(t))dt + \sum_{j=1}^{n} \nu_{ij} a_j^{\frac{1}{2}}(\mathbf{x}(t))dW_j, \tag{2}$$

with a deterministic and a stochastic term (W_j: Wiener process). For high reaction rates, the second term becomes negligible.

2 General Hybrid Solver for Multi-scale Problems

Obviously, each of the three methods is advantageous for a certain range of particle numbers and/or reaction rates, but none of them is an adequate method if there are huge differences among them, which is a typical situation in signal transduction systems, for example. In case of low particle numbers, stochastic effects can influence the system's behaviour even qualitatively, but the exact stochastic approach, which triggers each single molecular event, is not feasible for (sub-)systems with high particle numbers and reaction rates.

Subsets of Reactions and Threshold Criteria. In this work, the timescale problem was approached by splitting the reaction system into three reaction subsets, each of them corresponding to one of the upper simulation methods. Since in general, the reaction subsets are not independent of each other, the interactions between them have to be considered carefully. In addition, the assignment of the reactions to the subsets has to be managed dynamically.

In order to separate stochastically highly relevant reactions from those ones, which are well-described by Langevin equations or even by deterministic methods, rigorous quantitative criteria have to be introduced. Note, that these criteria often depend on the specific problem. Two examples are given here:

$$S = \left\{ R_j \;\middle|\; \exists i \mid \frac{1}{\sqrt{dt}} \cdot \frac{\breve{\sigma}_{ij}(dt)}{x_i} > \epsilon_1 \right\} \quad \text{or} \quad S = \left\{ R_j \;\middle|\; \exists i \mid \frac{|\,\nu_{ij}\,|}{X_i} > \epsilon_2 \right\}, \tag{3}$$

where $R = \{R_1, \ldots, R_n\}$ is the set of all reactions, $S \subseteq R$ the subset of all reactions to be treated stochastically exact using the Gillespie algorithm and $\breve{\sigma}_{ij}(dt)$ the standard deviation of the molecule concentration x_i under the assumption that it is influenced by R_j only ($\breve{\sigma}_{ij}(dt)$ scales with $\sqrt{dt}$). Thus, a reaction is treated stochastically exact as soon as it causes a *critical relative fluctuation* of the concentration of at least one species. The second criteria addresses extremely low particle numbers, where the discrete nature causes increments, which might be much higher than the respective standard deviation. The subset $L \subseteq R$ covers those reactions, which influence molecule species with particle numbers in a higher range and which cause fluctuations, that are not negligible. The threshold between L and the subset $D = R \setminus \{S \cup L\}$ described by ODEs can also be defined by (3a) with an appropriate ϵ_1. The assignment of the reactions to S and L is denoted by the flags $(s_1, \ldots, s_n)$ and $(l_1, \ldots, l_n)$.

Generalized System State. The system of m molecule species and n reactions between them is now described by a 'generalized' state vector:

$$(x_1, \ldots, x_m, p_1, \ldots, p_n),$$
$$(r_1, \ldots, r_n), \quad \textit{random numbers}, \ 0 \leq r_j < 1.$$

Here, $\{p_j\}$ denotes the probability, that a reaction event of type R_j, $R_j \in S$, has not occurred yet. This can be determined by using the propensity functions:

$$p_j(t + dt) = p_j(t) - p_j(t)a_j(\mathbf{x}(t))dt \quad \Rightarrow \quad dp_j(t)/dt = -p_j(t)a_j(\mathbf{x}(t)). \tag{4}$$

Starting with $p_j(t_0) = 1$, the random numbers r_j are mapped to the putative time point $t_0 + \tau_j$ of the next reaction of type R_j:

$$p_j(t_0 + \tau_j) = r_j. \tag{5}$$

Interaction Between the Reaction Subsets. Between the stochastic reaction events, many reactions of the other two subsets occur. These changes are determined by propagation of the system of ODEs and Langevin equations. The correct handling of the interactions between the subsets is critical and approached as follows: Both parts of the generalized system state - the concentrations x_i and the probability functions p_j of the 'exact' stochastic reaction subset S - are propagated using the same solver. Thus the influence of the deterministic and the Langevin reaction subset on the reaction probabilities p_j is entirely considered. Once the time point of a stochastic reaction R_j is reached according to (5), the reaction is executed by updating the concentrations $\{x_i\}$.

Solving the Langevin Equations. In order to numerically solve the stochastic part of the Langevin equation (2), a Wiener process is generated with finite increments $\Delta W = \mathcal{N}(0, \Delta t)$ [3] ($\mathcal{N}$: normal random variable with variance Δt). This numerical process will be combined with the propagation of the deterministic term of (2) and the deterministic part of the stochastic term. The step size of the ODE solver is limited to the step size Δt introduced for generation of the Wiener process, but smaller step sizes due to adaptive step size control of the solver are possible. The deterministic part of the stochastic term is propagated in the following way:

$$d\tilde{x}_i(t) = \sum_{j=1}^{n} l_j \cdot (1 - s_j) \cdot \nu_{ij} \sqrt{a_j(\mathbf{x}(t)) \cdot dt} \, . \tag{6}$$

Multiplied with a unit normal random number $n_i = \mathcal{N}(0, 1)$, it corresponds to an increment of the Wiener process. After each completed time step, this term is added to the deterministic part and new normal random numbers $\{n_i\}$ are generated.

Algorithm for the Hybrid Solver:

1. Set the initial concentrations $(x_1, \dots, x_m)$ and set $t_0 = t_{start}$.
   Set the probabilities of no reaction $(p_1, \dots, p_n) = (1, \dots, 1)$.
   Generate the random numbers $(r_1, \dots, r_n)$.
2. Determine the stochastic subset $S \subset R$ and $L \subset R$ using the upper criteria.
3. Start ODE solver at time $t = t_0$ with the following task:
   Propagate the generalized system $(x_1, \dots, x_m, p_1, \dots, p_n)$ according to
   $$\frac{dx_i(t)}{dt} = \sum_{j=1}^{n}(1 - s_j) \cdot \nu_{ij} v_j(\mathbf{x}(t)) \quad \textit{and} \quad \frac{dp_j(t)}{dt} = -s_j \cdot p_j(t) a_j(\mathbf{x}(t)).$$
   Stop at the putative time point of the next stochastic reaction event $t_{stop} = t_0 + \tau_{j^*}$ with $\tau_{j^*} = \min_{j \in S}\{\tau_j\}$, to be determined implicitly by (5).
   Within the solver: add the stochastic Langevin increments after each solver step as described above (see (6)).
4. Execute next stochastic reaction R_{j^*} by updating $x_i = x_i + \nu_{ij^*} \forall i$.
5. Generate new random number r_{j^*} and set $p_{j^*} = 1$.
6. Set $t_0 = t$ and go to step 3, or go to step 2 for updating $\{s_j\}$ or $\{l_j\}$.

3 Conclusion

In summary, a method is provided, which is stochastically exact in the sensitive low-particle number range, but still efficient for the fast reaction subset, where the Gillespie approach would no longer be feasible. Moreover, the stochastic effects are well approximated in the higher range by use of SDEs. Note, that in general, it doesn't make sense to treat the stochastic behaviour in an exact way below a certain threshold and to abruptly ignore the even higher stochastic fluctuations above, as it would be realized in a pure stochastic-deterministic approach. In addition, all interactions between the reaction subsets are correctly considered. Therefore the exactness of the algorithm is only limited by the exactness of the numerical solution of the ODEs and SDEs. The method was applied to the mitochondrial Cytochrome C release, whose duration [4] could be explained by the stochastic effects of an ensemble of many individual mitochondria. Here, the hybrid method ! was about 20 times faster than the exact algorithm. Since the applied stochastic methods are directly derived from the master equation, this hybrid approach can also be used for any other kind of Markov processes.

References

1. Gillespie, D. T.: Exact stochastic simulation of coupled chemical reactions. J. Phys. Chem. **81** (1977) 2340–2361
2. Gillespie, D. T.: The chemical Langevin equation. J. Chem. Physics **113(1)** (2001) 297–306
3. Kloeden, P. E. and Platen, E.: Numerical Solution of Stochastic Differential Equations. Springer, Berlin, 1992
4. Goldstein, J. C. et al.: The coordinate release of cytochrome c during apoptosis is rapid, complete and kinetically invariant. Nature Cell Biol **2** (2000), 156–162

The Biodegradation Network, a New Scenario for Computational Systems Biology Research

Florencio Pazos[1,*], David Guijas[1], Manuel J. Gomez[2], Almudena Trigo[2], Victor de Lorenzo[3], and Alfonso Valencia[2,#]

[1] bioALMA
www.bioalma.com/
[2] Protein Design Group, National Center for Biotechnology, CNB-CSIC
www.pdg.cnb.uam.es
[3] Bacterial Biotechnology Department. National Center for Biotechnology, CNB-CSIC
[*] Present address: Structural Bioinformatics Group, Department of Biological Sciences, Imperial College, London.
[#] Corresponding author: Alfonso Valencia,
Protein Design Group, National Center for Biotechnology CNB-CSIC
Cantoblanco, Madrid 28049
Valencia@cnb.uam.es

Abstract. The study of the global properties of complex metabolic networks in organisms such as E. coli and Yeast has become a priority in the new area of Computational Systems Biology, leading to the first models about the organization and evolution of these networks. Here we propose the analysis of biodegradation networks as an alternative to this, now classical, studies of metabolic networks. The biodegradation reactions carried out by communities of bacteria in contaminated environments offer interesting computational and biological advantages and challenges. Here we first describe our preliminary results on the analysis of the systems properties of the biodegradation network and the comparison with the other metabolic networks, and second describe the computational developments necessary for the analysis of the information. In particular we introduce the MetaRouter system, as an integrated information system for Biodegradation. Furthermore we describe the current work for improving the system with an alternative enzyme-centric view and new rule-based systems for the prediction of the capacity of biological systems to degrade new chemical compounds.

Some microorganisms have acquired the ability to catabolise external chemical compounds that do not form part of their Standard Metabolism, such as chemical pollutants (xenobiotics) result of industrial activity. The natural microbial communities found in these environments are composed of a complex mixture of species and strains working coordinately. The final chain of reactions, leading to chemical degradation is a puzzle of reactions carried out by enzymes from the various microorganisms. The xenobiotic compounds are modified until the point in which they can enter into de central metabolism or they are reduced to non-toxic forms. During the process the bacterial communities may obtain two advantages, first they eliminate compounds dangerous for them, and second they obtain energy from new sources.

V. Danos and V. Schachter (Eds.): CMSB 2004, LNBI 3082, pp. 252–256, 2005.

From a biological point of view the biodegradation network presents very interesting properties that differentiate it from the standard metabolic pathways, i.e. it has an inter-species composition, and it is the result of a fast adaptation to new environmental conditions. For Biotechnology the study of biodegradation offers interesting possibilities for the cleanup of contaminated soils and waters by degrading the hazardous compounds (Bioremediation). Interestingly they can also be seen as a new scenario for analysis with a Systems Biology viewpoint, offering possibilities different and complementary to the study of the organization and evolution of biological networks and biochemical pathways.

Hundreds of reactions has been experimentally characterized up to now in different conditions, accounting for an interesting collection of data for analysis, even if they are probably only a small proportion of the larger biological reality. The available information about the environmental conditions in which these reactions are taking place is still one of the limiting factors specially when compared with the immense diversity of possible ecological niches. Indeed all this area of research is embedded in the larger framework of the description of all the species (genomes) present in a given ecosystems (a field now known as "metagonomics"), which first spectacular results have been very recently published. (Venter et al., 2004). At the Bioinformatics level there is still quite a work ahead of us, including linking the various disperse sources of information, which include: the chemical structure and reactivity of the organic compounds, the sequence, structure and function of the enzymes participating in this processes, the information related with the different genomes, their organization, structure and gene expression regulation. All this area of research can

We have recently carried out a first study of the general properties of the known biodegradation network (Pazos et al., 2003). The main observation of this study is the scale-free structure of this network, including the input / output reactions, with characteristics similar to the ones observed for the standard metabolic networks. A similarity that fits well with the biological model that describes the collective behavior of the biodegradation networks as similar to the one of single organisms, even if its nature and evolution are necessarily very different. This first analysis of the biodegradation network also allowed us to propose a first model for its evolution (see Fig. 1).

To continue with the study of the biodegradation network it is now absolutely necessary to organize all the available information in a coherent database, with substantial capacity for the interaction with experimental biologist working in this domain (a number of initiatives in this area are in progress, see Wackett et al., 2004).

In this context we have developed MetaRouter, as a Bioinformatics system for maintaining heterogeneous information related with Biodegradation including facilities for updating, query, modification and data mining (Pazos et al., submitted). The core of the system is a relational database where the information on chemical compounds, reactions, enzymes and organisms is stored. The current data set includes 740 organic compounds (2,167 synonyms), 820 reactions, 502 enzymes and 253 organisms. For the chemical compounds, the following information is included: name, synonyms, SMILES code, molecular weight, physical-chemical properties (density, evaporation rate, melting point, boiling point and water solubility), chemical formula, image of the chemical structure, canonical three-dimensional structure in PDB format, and links to other databases. The reactions are described in terms of substrates and products, catalyzing enzyme and links to other databases. The enzyme information includes the name, Enzyme Commission (EC) code, organisms where the gene is present, database sequence identifiers, and links to other databases. The main public sources of information for obtaining the initial set of data were UMBBD, ChemFinder (http://chemfinder.cambridgesoft.com), and the ENZYME and SwissProt databases.

The research capacity implemented in the system includes the localization of biodegradative pathways for one (or a set of) compounds, by finding pathways between those complex(es) and the entry points into the central metabolism, or between pairs of compounds. These searches can be restricted by a number of criteria, such as length of the pathway, required enzymes or organism(s), properties of the compounds involved (for example, highly soluble) (See Fig. 2 for a representation of the typical output of the system).

The current work in the MetaRouter includes two key extensions. First we have developed an alternative representation of the network using the proteins (enzymes) as nodes, instead of the compounds (Trigo et al., in preparation). The enzyme-centric view of standard metabolism has been essential for the analysis of the biological and evolutionary properties of the networks in other studies (Pereira-Leal et al., 2004). Second, we have developed a machine learning approach for the prediction of the "biodegradability" of new compounds, based on their chemical descriptors (Gomez et al., submitted 2004). These descriptors include a definition of chemical triads of atoms in the structures (taking into account the type of chemical bonds) and the solubility of the compounds (from ChemFinder). The analysis of the network revealed the following categories of compounds: Non-biodegradable compounds, 353 compounds that do not participate as substrates in any reaction, or are connected to pathways that do not lead to the standard metabolism. 533 biodegradable compounds including the 38 compounds that can also be found in the common metabolism, and other compounds placed in the path to central metabolism. The final set of 329 compounds belong to paths that finally produce Carbon Dioxide, and as such are not directly connected to common metabolism compounds. For the combination of the information, and to generate the predictions, we have used the C4.5 package (Quinlan, 1993) to build decision trees and to generate propositional rules that can be used as predictors of “biodegradability”. The average sensitivity of the cross-validated predictions in between 62 and 69% for the three classes, with Average Specificity ranging from 69 to 84% for the predictions in a two combined classes schema. The examination of the rules generated by C4.5 allows the interpretation of the chemical characteristics of the compounds that are better related with the capacity of the biological systems to metabolize them, opening the possibilities for their study in the laboratory, and for predicting the biodegradative fate of a new chemical compound before releasing it in the environment.

Acknowledgements

We want to acknowledge the support of our colleges in bioALMA, Protein Design Group and Victor De Lorenzo´s laboratory. This work was supported in part by EU contracts QLK3-CT-2002-01933, QLK3-CT-2002-01923, and INCO-CT-2002-1001, by grant BIO2001-2274 of the MCyT and by the Strategic Research Groups Program of the Autonomous Community of Madrid.

References

Ellis, L.B.M., Hou, B.K., Kang, W., and Wackett, L.P. (2003) "The University of Minnesota Biocatalysis/Biodegradation Database: Post-Genomic Datamining" Nucleic Acids Research 31: 262-265.

Pazos F, Valencia A, De Lorenzo V. (2003) The organization of the microbial biodegradation network from a systems-biology perspective. EMBO Rep. 4:994-999.

Pereira-Leal JB, Enright AJ, Ouzounis CA. (2004) Detection of functional modules from protein interaction networks. Proteins. 54: 49-57

Quinlan, J.R. (1993) C4.5: Programms for Machine Learning. Morgan Kaufmann, San Francisco, CA.

Venter JC, Remington K, Heidelberg JF, Halpern AL, Rusch D, Eisen JA, Wu D, Paulsen I, Nelson KE, Nelson W, Fouts DE, Levy S, Knap AH, Lomas MW, Nealson K, White O, Peterson J, Hoffman J, Parsons R, Baden-Tillson H, Pfannkoch C, Rogers YH, Smith HO. (2004) Environmental genome shotgun sequencing of the Sargasso Sea. Science. 304: 66-74.

Wackett, L.P., Dodge, A.G., and Ellis, L.B.M. (2004) "Microbial Genomics and the Periodic Table" (minireview) Applied and Environmental Microbiology, 70: 647-665.

Figures

The Chemical compounds are represented with circles proportional to their molecular weight. The width of the reaction arrows is proportional to the evolutionary conservation of the corresponding catalyzing enzyme. The degree of connectivity of the chemical compounds is represented with a color scale (from black to light blue). For the entry of new compound (red circle) the most likely possibilities proposed in the

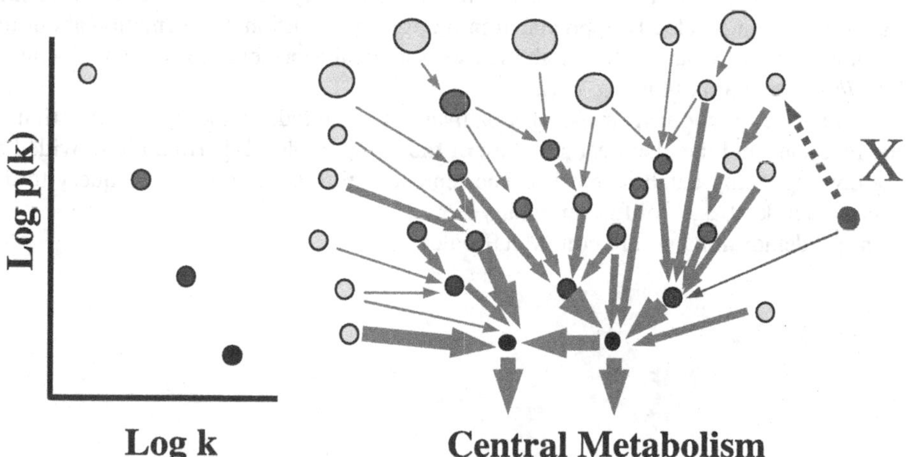

Fig. 1. Schematic representation of the properties and hypothetical evolution of the biodegradation network

model to account for the observed scale-free structure is the connection with a preferentially connected node (hub), that in turn is closed to the central metabolism. The graphic represents the typical log-log plot of connectivity (k) against number of elements with that connectivity (p(k)). The position of the different types of compounds in this plot is shown. (Adapted from Pazos et al., 2003).

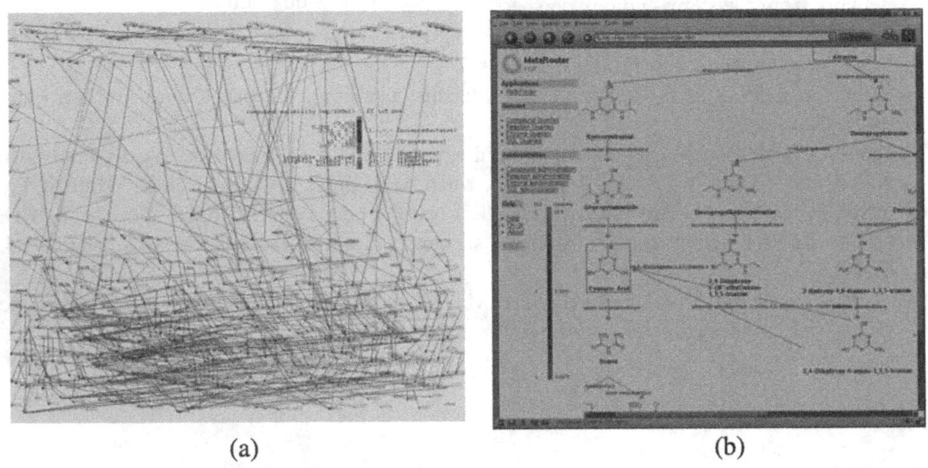

(a) (b)

Fig. 2. Example of the use of the Metaruter database for the analysis of Biodegradation

The full biodegradation network is represented in 2a, the various colors and lines correspond to different type of chemical compounds and reactions. The upper part represents entries into the system and the lower one compound that connect directly with the standard central metabolism. The larger density of hubs close to the central metabolism can be clearly appreciated in the figure (Additional information about the organization of the network and alternative graphical representations are available at: http://pdg.cnb.uam.es/biodeg_net).

In panel 2b a zoomed image of a pathway is provided, including the direction of the reactions and the chemical structure of the compounds. Hyperlink are provided to the corresponding database entries. The representation is the result of a query to the Metarouter database. For further details see:

http://demos.almabioinfo.com/MetaRouter

Brane Calculi

Interactions of Biological Membranes

Luca Cardelli

Microsoft Research

Abstract. We introduce a family of process calculi with dynamic nested membranes. In contrast to related calculi, including some developed for biological applications, active entities here are tightly coupled to membranes, and can perform interactions on both sides of a membrane. That is, computation happens *on* the membrane, not inside of it.

1 Introduction

A biological cellular membrane is an oriented closed surface that can perform various molecular functions. Membranes are not just containers: they are coordinators and active sites of major activity[1]. Large functional molecules (proteins) are embedded in membranes, with consistent orientation, and can act on both sides of the membrane simultaneously. The consistent orientation of such proteins induces an orientation on the membrane. Freely floating molecules interact with membrane proteins, and can be sensed, manipulated, and pushed across by active molecular channels. Membranes come in different kinds, distinguished mostly by the proteins embedded in them, and typically consume energy to perform their functions.

One of the most remarkable properties of biological membranes is that they form a two-dimensional fluid (a lipid bilayer) embedded in a three-dimensional fluid (water). That is, both the structural components and the embedded proteins freely diffuse on the two-dimensional plane of the membrane (unless they are held together by specific mechanisms). Moreover, membranes float in water, which may contain other molecules that freely diffuse in that three-dimensional fluid. Membrane themselves are impermeable to most substances, such as water and protons, so that they partition the three-dimensional fluid.

Many membranes are highly dynamic: they constantly shift, merge, break apart, and are replenished. But the transformations that they can support are rather limited, partially because orientation must be preserved, and partially because membrane transformations need to be fairly continuous. For example, it is possible for a membrane to gradually buckle and create a bubble that then detaches, or for such a bubble to merge back with a membrane, but it is not possible for a bubble to "jump across" a membrane (only small molecules can do that).

---

[1] "For a cell to function properly, each of its numerous proteins must be localized to the correct cellular membrane or aqueous compartment." [9] p.675.

V. Danos and V. Schachter (Eds.): CMSB 2004, LNBI 3082, pp. 257 – 278, 2005.

The fluid-within-fluid structure inspires the basic organization of our Brane Calculi[2], which is characterized by two commutative monoids, each representing a kind of fluid. The specific transformations that we have selected are further inspired by (some of) the biological constraints. However, within the general structure of Brane Calculi there is scope for refining or ignoring such constraints.

One of the constraints one may adopt is the preservation of orientation (e.g., membranes of different orientation should not merge). A related constraint is *bitonality*, which requires nested membranes to have opposite orientations, so that the orientations can be coded by coloring systems in two tones, as in Figure 1, where P and Q represent arbitrary subsystems. Preservation of bitonality means that reactions must preserve the even/odd parity with which components are nested inside membranes: note that P and Q remain on the same color background in each reaction. This means, in particular, that in a sequence of bitonal reactions there is never any actual mixing of fluids from inside and outside any given membrane, although external fluids can be brought inside if safely wrapped in another membrane. Bitonality is common in cellular-scale living systems. Although not universal, it inspires a collection of basic reactions that are biologically implementable, and that are different from those of calculi that are not biologically inspired.[3]

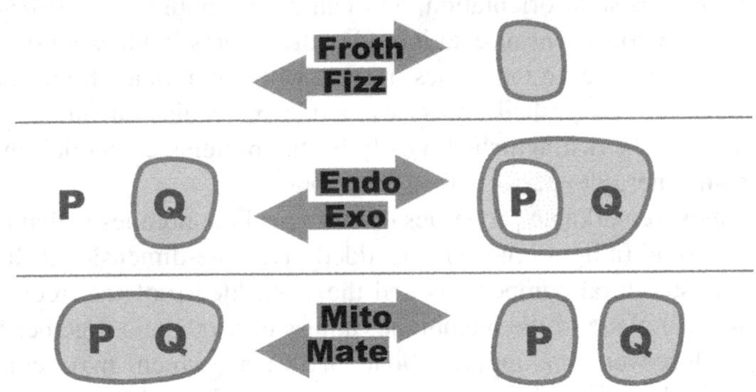

Fig. 1. Examples of Bitonal Reactions

The reactions illustrated in Figure 1 can be formalized and studied on their own [2]. However, in this paper we use them only as informal guides for more detailed calculi, where the reasons "why" those reactions happen are made more apparent.

[2] "Brane" is a common abbreviation for "membrane" in physics.

[3] The framework in which Brane Calculi are formalized originates in the study of calculi for mobile agents [3]. In that context, *sandboxing* an applet on its arrival at a site is, in fact, a bitonal operation: it maintains the separation between safe regions (of internal origin) and unsafe regions (of external origin). We are not aware of proposals to use sandboxing as a basic operations in that context; here, it corresponds to phagocytosis.

2 Basic Framework

2.1 Syntax and Reactions

The basic structure of Brane Calculi consists of two commutative monoids with replication: we use m for composition of *systems*, with unit k, and | for composition of *membranes*, with unit 0. Replication (!) is used to model the notion of a "multitude" of components of the same kind, which is in fact a standard situation in biology. Quantitative refinements are possible [12] and certainly desirable.

Systems consist of nested membranes, and membranes consists of collection of *actions*. Actions are left unspecified at the moment, and are detailed in the following sections. The familiar notion of structural congruence of processes [11] is applied to systems and membranes, characterizing their fluidity properties. Reactions happen only at the level of systems, and are caused only by actions on membranes.

Syntax

Systems	P,Q ::= k ¦ PmQ ¦ !P ¦ σhPi	nests of membranes
Branes	σ,τ ::= 0 ¦ σ|τ ¦ !σ ¦ a.σ	combinations of actions
Actions	a,b ::= ...	(detailed later)

We abbreviate a.0 as a, and 0hPi as hPi, and σhki as σhi.

Structural Congruence

PmQ 7 QmP	σ|τ 7 τ|σ
Pm(QmR) 7 (PmQ)mR	σ|(τ|ρ) 7 (σ|τ)|ρ
Pmk 7 P	σ|0 7 σ
!k 7 k	!0 7 0
!(PmQ) 7 !Pm!Q	!(σ|τ) 7 !σ|!τ
!!P 7 !P	!!σ 7 !σ
!P 7 Pm!P	!σ 7 σ|!σ
0hki 7 k	
P7Q ⇒ PmR 7 QmR	σ7τ ⇒ σ|ρ 7 τ|ρ
P7Q ⇒ !P 7 !Q	σ7τ ⇒ !σ 7 !τ
P7Q ∧ σ7τ ⇒ σhPi 7 τhQi	σ7τ ⇒ a.σ 7 a.τ

Basic Reactions

$$P \} Q \;\Rightarrow\; P m R \} Q m R$$
$$P \} Q \;\Rightarrow\; \sigma h P i \} \sigma h Q i$$
$$P 7 P' \wedge P' \} Q' \wedge Q' 7 Q \;\Rightarrow\; P \} Q$$

We write $\}$ * for the reflexive and transitive closure of $\}$.

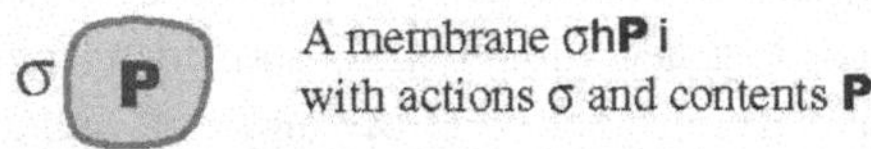

Fig. 2. Brane Graphical Notation

Within this framework, our Basic Brane Calculus is the one gradually introduced in Sections 3 and 4. Possible extensions are discussed in Section 5. Orthogonally, one could add restriction operators to both systems and membranes, in the style of π-calculus [11], with extrusion rules such as $((\nu n)\sigma)hPi\ 7\ (\nu n)(\sigma hPi)$ if $n C fn(P)$. The bound names n would be the ones used in the following sections to identify pairs of related actions and co-actions.

3 Bitonal Interactions

Bitonal interactions [2] are inspired by endocytosis/exocytosis (the second reversible reaction in Figure 1). Endocytosis is the process of incorporating external material into a cell by "engulfing" it with the cell membrane (without breaking the membrane or letting the material cross it). Exocytosis is the reverse process.

3.1 Definitions

Endocytosis, thus described, is an uncontrollable process that can engulf an arbitrary amount of material. We are interested in more controllable interactions, therefore we specialize endocytosis into two basic operations: *phagocytosis*, engulfing just one external membrane, and *pinocytosis*, engulfing zero external membranes. In addition we have *exocytosis*, which is itself sufficiently controllable. Each action usually comes with a co-action that it is intended to interact with, indicated by the symbol $^{|}$ (pinocytosis does not have a co-action).

Bitonal Actions

Actions $a ::= \ldots \mid J_n \mid J^{|}_n(\sigma) \mid K_n \mid K^{|}_n \mid G(\sigma)$ phago J, exo K, pino G

Precedence: $a.\sigma|\tau$ stands for $(a.\sigma)|\tau$, and $!\sigma|\tau$ stands for $(!\sigma)|\tau$. The subscripted names n are used to pair-up related actions an co-actions; we omit them when there is no ambiguity. Co-phago is indexed by a membrane σ; this σ becomes the new membrane that engulfs the outside material: conceptually it is related to a piece of the old membrane. Exo causes irreversible mixing of membranes: since membranes are fluids, there is in general no way to untangle two membranes once they have merged. Incidentally, this implies that merging is often not a desirable operation.

Bitonal Reactions

Phago	$J_n.\sigma|\sigma_0 h P i \; m \; J'_n(\rho).\tau|\tau_0 h Q i \;\; \} \;\; \tau|\tau_0 h \rho h \sigma|\sigma_0 h P i \; i m Q i$
Exo	$K'_n.\tau|\tau_0 h K_n.\sigma|\sigma_0 h P i m Q i \;\; \} \;\; P \; m \; \sigma|\sigma_0|\tau|\tau_0 h Q i$
Pino	$G(\rho).\sigma|\sigma_0 h P i \;\; \} \;\; \sigma|\sigma_0 h \rho h k \; i m P i$

One can see that the parity of nesting of P and Q is preserved in all these reactions, hence they preserve the bitonal coloring of those subsystems.

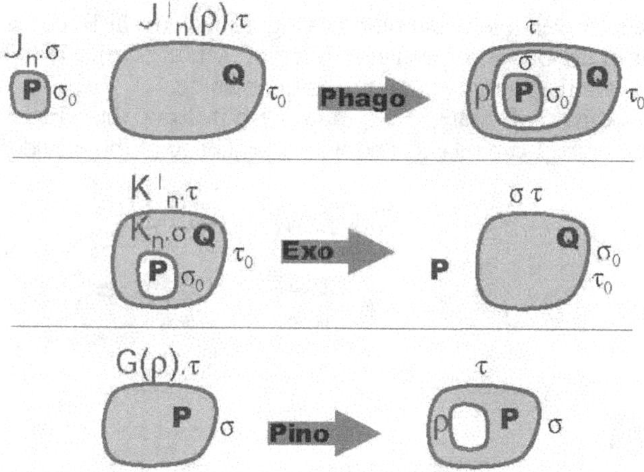

Fig. 3. Phago, Exo, Pino (shaded for emphasis)

3.2 Derived Bitonal Interactions

The Mito reaction, as illustrated in Figure 1 is another uncontrollable process that can split a membrane at an arbitrary place. To make it more controllable, we specialize it into two basic operations: *budding*, splitting off one internal membrane, and *dripping*, splitting off zero internal membranes. In addition we have *mating* (a.k.a. merging or fusion), the obvious merging of membranes, which is itself sufficiently controllable.

These three bitonal operations, mating, budding, and dripping, can be derived from the previous three. The derivations are not meant to be biologically significant: they are just a test of expressive power. In practice one would want to consider these as primitives at the same level as Phago, Exo, and Pino, since they all have direct implementations in cellular mechanisms.

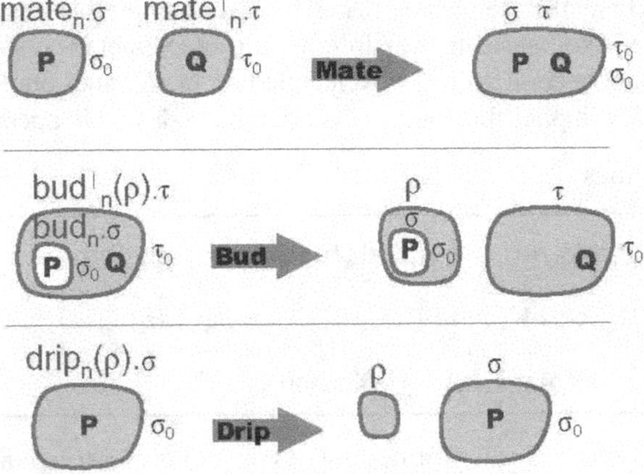

Fig. 4. Mate, Bud, Drip (shaded for emphasis)

Mate causes irreversible membrane mixing, as in Exo. In Bud, the fresh membrane ρ that surrounds the bud is a parameter of the co-action, similarly to the situation with Phago. Drip is similar to Pino, but towards the outside.

The encodings of Mate, Bud, and Drip follow the single basic idea that Mito/Mate in Figure 1 can be encoded with a sequence of three Endo/Exo operations.

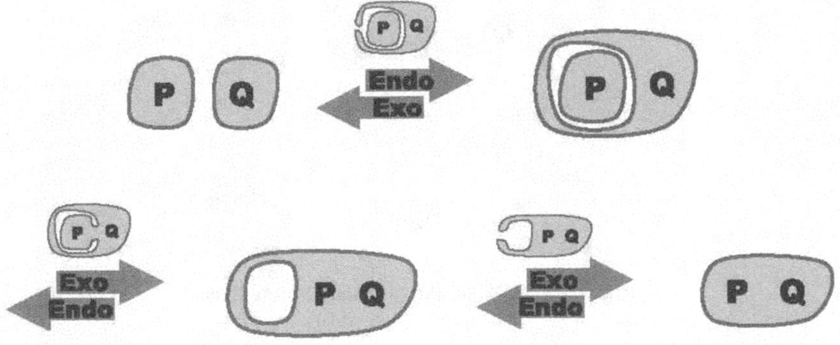

Fig. 5. Mito/Mate by 3 Endo/Exo (basic technique)

Mate

$\text{mate}_{n}.\sigma$ @ $J_{n}.K_{n'}.\sigma$

$\text{mate}^{|}_{n}.\tau$ @ $J^{|}_{n}(K^{|}_{n'}.K_{n''}).K^{|}_{n''}.\tau$

$\text{mate}_{n}.\sigma|\sigma_0 hP\,i$ m $\text{mate}^{|}_{n}.\tau|\tau_0 hQ\,i$ } * $\sigma|\sigma_0|\tau|\tau_0 hP m Q\,i$

$$\mathrm{mate}_n.\sigma|\sigma_0 hP\,i \;m\; \mathrm{mate}^{\perp}_n.\tau|\tau_0 hQ\,i \;=$$
$$J_n.K_{n'}.\sigma|\sigma_0 hP\,i \;m\; J^{\perp}_n(K^{\perp}_{n'}.K_{n''}).K^{\perp}_{n''}.\tau|\tau_0 hQ\,i \quad \}_{\text{Phago } n}$$
$$K^{\perp}_{n''}.\tau|\tau_0 hK^{\perp}_{n'}.K_{n''}hK_{n'}.\sigma|\sigma_0 hP\,i\,i \;m\; Q\,i \quad \}_{\text{Exo } n'}$$
$$K^{\perp}_{n''}.\tau|\tau_0 hK_{n''}|\sigma|\sigma_0 hk\,i\,i\; P \;m\; Q\,i \quad \}_{\text{Exo } n''}$$
$$\sigma|\sigma_0|\tau|\tau_0 hP \;m\; Q\,i$$

Bud

$$\mathrm{bud}_n.\sigma \;@\; J_n.\sigma$$
$$\mathrm{bud}^{\perp}_n(\rho).\tau \;@\; G(J^{\perp}_n(\rho).K_{n'}).K^{\perp}_{n'}.\tau$$
$$\mathrm{bud}^{\perp}_n(\rho).\tau|\tau_0 h\mathrm{bud}_n.\sigma|\sigma_0 hP\,i \;m\; Q\,i \quad \} * \rho h\sigma|\sigma_0 hP\,i\,i \;m\; \tau|\tau_0 hQ\,i$$

$$\mathrm{bud}^{\perp}_n(\rho).\tau|\tau_0 h\mathrm{bud}_n.\sigma|\sigma_0 hP\,i \;m\; Q\,i \;=$$
$$G(J^{\perp}_n(\rho).K_{n'}).K^{\perp}_{n'}.\tau|\tau_0 hJ_n.\sigma|\sigma_0 hP\,i \;m\; Q\,i \quad \}_{\text{Pino}}$$
$$K^{\perp}_{n'}.\tau|\tau_0 hJ^{\perp}_n(\rho).K_{n'}hk\,i \;m\; J_n.\sigma|\sigma_0 hP\,i \;m\; Q\,i \quad \}_{\text{Phago } n}$$
$$K^{\perp}_{n'}.\tau|\tau_0 hK_{n'}h\rho h\sigma|\sigma_0 hP\,i\,i\,i \;m\; Q\,i \quad \}_{\text{Exo } n'}$$
$$\rho h\sigma|\sigma_0 hP\,i \;m\; \tau|\tau_0 hQ\,i$$

Drip

$$\mathrm{drip}_n(\rho).\sigma \;@\; G(G(\rho).K_n)).K^{\perp}_n.\sigma$$
$$\mathrm{drip}_n(\rho).\sigma|\sigma_0 hP\,i \quad \} * \rho h\,i \;m\; \sigma|\sigma_0 hP\,i$$

$$\mathrm{drip}_n(\rho).\sigma|\sigma_0 hP\,i \;=$$
$$G(G(\rho).K_n)).K^{\perp}_n.\sigma|\sigma_0 hP\,i \quad \}_{\text{Pino}}$$
$$K^{\perp}_n.\sigma|\sigma_0 hG(\rho).K_n hk\,i\,i \;m\; P\,i \quad \}_{\text{Pino}}$$
$$K^{\perp}_n.\sigma|\sigma_0 hK_n h\rho hk\,i\,i\,i \;m\; P\,i \quad \}_{\text{Exo } n}$$
$$\rho h\,i \;m\; \sigma|\sigma_0 hP\,i$$

3.3 Example: Viral Infection, Part 1

Certain kinds of viral infection mechanisms represent an ideal example of bitonality in action. A virus is too big to just cross a cellular membrane. It can either punch its DNA or RNA through the membrane, essentially performing a Mate, or it can enter by utilizing standard cellular endocytosis pathways, as shown in Figure 6.

The Semliki Forest virus consists of a capsid containing the viral RNA (the nucleocapsid). The nucleocapsid is surrounded by a membrane that is similar to the cellular membrane (in fact, it is obtained from it "on the way out"). This membrane is however enriched with a special protein that plays a crucial trick on the cellular machinery, as we shall see shortly. The virus is brought into the cell by phagocytosis, thus wrapped by an additional membrane layer; this is part of a standard transport pathway into the cell. As part of that pathway, an endosome compartment merges with the wrapped-up virus. At this point, usually, the endosome causes some reaction to happen in the material brought into the cell. In this case, though, the virus uses its special membrane protein to trigger an exocytosis step that deposits the naked nucleocapsid into the cytosol. The careful separation of internal and external substances that the cell usually maintains has now been subverted. The nucleocapsid is in direct contact with the inner workings of the cell, and can begin doing damage. First, the nucleocapsid disassembles itself, depositing the viral RNA into the cytosol. This vRNA then follows three distinct paths. First it is replicated (either by cellular proteins, or by proteins that came with the capsid), to provide the vRNA for more copies of the virus. The vRNA is also translated into proteins, again by standard cellular machinery. Some proteins are synthesized in the cytosol, and form the building blocks of the capsid: these self-assemble and incorporate a copy of the vRNA to form a nucleocapsid. The virus envelope protein is instead synthesized in the Endoplasmic Reticulum, and through various steps (through the Golgi apparatus) ends up lining transport vesicles that merge with the cellular membrane, along another standard transport pathway. Finally, the newly assembled nucleocapsid makes contact with sections of the cellular membrane that are now lined with the viral envelope protein, and buds out to recreate the initial virus structure outside the cell.

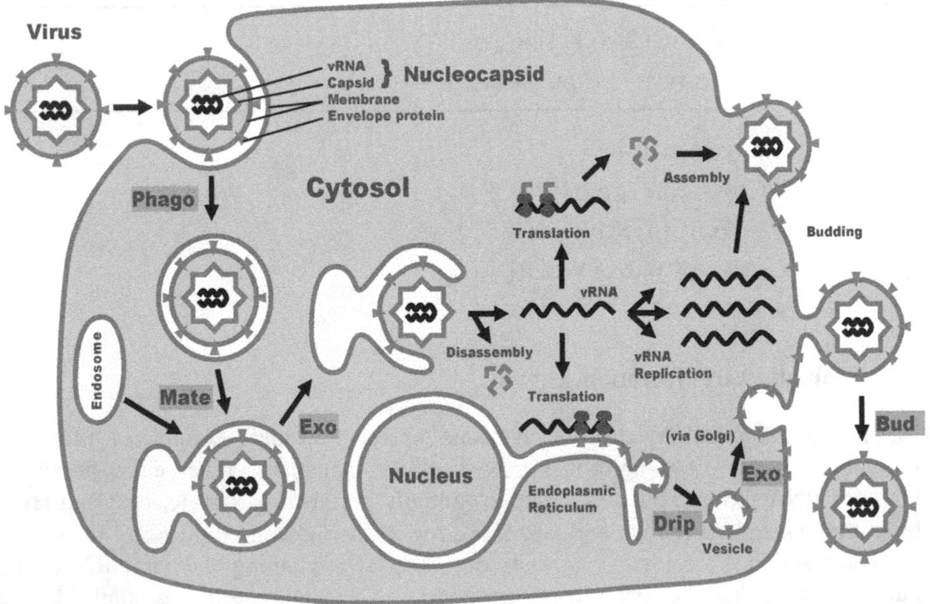

Fig. 6. Viral Infection and Reproduction ([1] p.279)

The initial and final stages of the virus lifecycle can be coded up as follows.

virus	@	J.Khnucap i
nucap	@	!budlXhvRNA i
cell	@	membranehcytosol i
membrane	@	!J'(mate)l!K'
cytosol	@	endosome m Z
endosome	@	!mate' l!K'h i
viral-envelope	@	bud'(J.K)
envelope-vesicle	@	K.viral-envelopeh i

In the first phase (infection, Figure 7), the nucleocapsid (i.e., the capsid with the viral RNA inside, abbreviated "nucap") places itself in the cytosol:

virus m cell } * membranehnucapmcytosol i

We next assume that, by interaction with the available cellular machinery in the cytosol, the nucap causes the production of some number of copies n and m of envelope-vesicles and nucaps, leaving some modified cytosol'. (In section 4.6 we detail the mechanisms involved, including the unspecified cytosol', X, Z, Z'.)

nucap m cytosol } * nucap^{n} m envelope-vesicle^{m} m cytosol'

In the final phase (reproduction, Figure 8), the virus reassembles itself outside the cell:

membranehnucap m envelope-vesicle m Z' i } * membranehZ' i m virus

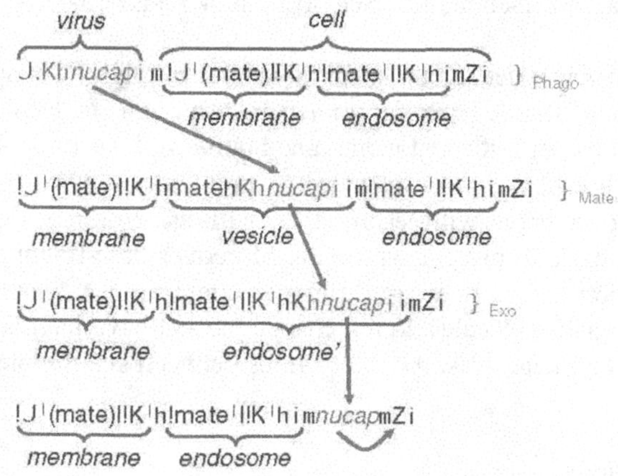

Fig. 7. Viral Infection

The level of abstraction in this code has been chosen to be as close as possible to the one in the picture. This is important, because we rarely understand the finest details of biological processes, and even if we did, we still would not want to model every molecule individually. The reality of virus infection is of course much more complex, and the modeling could be correspondingly refined. But one has to be able to choose an appropriate level of abstraction: Brane Calculi aim to provide such a level of abstraction for dynamic membrane transformations.

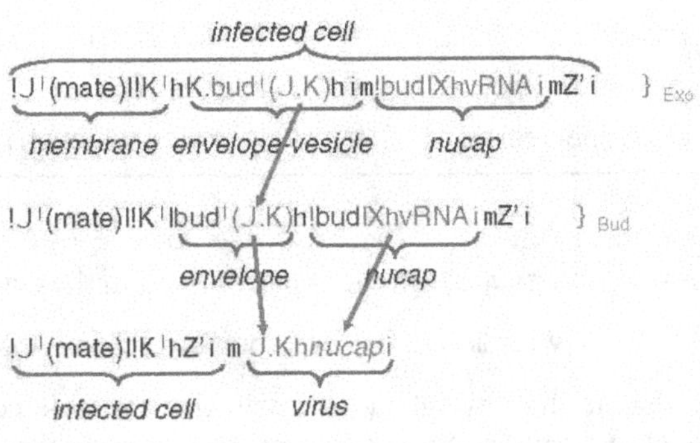

Fig. 8. Viral Reproduction

4 Molecules

We have not discussed free-floating molecules so far, to emphasize membrane interactions. Still, a primary function of membranes and of their embedded proteins is to shuttle molecules across, and it is important to include this ability in our models. In this section we discuss only *small* molecules, the ones that can easily cross or be transported across membranes. See sections 4.7 and 5.4 for a discussion of *large* molecules.

Membranes may let certain small molecules through by simple diffusion. Usually, however, they shuttle specific molecules through molecular channels that are implemented by sophisticated membrane-bound proteins (represented by our *actions*). Membranes are also a favorite mooring point of catalysts that cause free-floating molecules to interact with each other without crossing the membrane (e.g. in processes as basic as protein synthesis). Moreover, free-floating molecules can act as communication tokens between different membranes. A simplifying assumption for now is that small molecules do not change, do not have internal structure, and do not interact among themselves. All interactions between small molecules are mediated by membranes.

4.1 Definitions

Membranes can bind molecules on either sides of their surface, and can release molecules on either sides of their surface. Usually, coordinated bindings and releases happen completely or not at all, as in the antiporter in Figure 10. Because of this, we integrate in a single new action the ability to bind and release multiple molecules simultaneously.

Molecules and Molecular Actions

Systems P,Q ::= ... ¦ m	systems extended with molecules mBM
p,q ::= $m_1 m \dots m m_k$	multisets of molecules
Actions a,b ::= ... ¦ $p_1(p_2) \rightrightarrows q_1(q_2)$	bind&release of molecules
B&R p_1 m $p_1(p_2) \rightrightarrows q_1(q_2).\sigma | \sigma_0$ h p_2 m P i } q_1 m $\sigma | \sigma_0$ h q_2 m P i	

A set of molecules M is added to the syntax of systems. A bind&release action is added to the set of actions. This action (Figure 9) binds, in general, a multiset of molecules outside the membrane (p_1) and a multiset of molecules inside the membrane (p_2); if that is possible, it instantly releases a multiset of molecules outside the membrane (q_1) and a multiset of molecules inside the membrane (q_2). (Conservation of mass or energy is not enforced, and must be designed in.)

Fig. 9. Bind and Release

Obvious special cases are the separate binding and release on a single side; we omit **k**(**k**):

$p_1(\mathbf{k})\rightrightarrows$	bind outside	$\rightrightarrows q_1(\mathbf{k})$	release outside
$\mathbf{k}(p_2)\rightrightarrows$	bind inside	$\rightrightarrows \mathbf{k}(q_2)$	release inside

4.2 Example: Chemical Reactions

A chemical reaction between molecules can be represented as a *catalyst*: an always empty membrane that enables a reaction via an appropriate bind-outside&release-outside action. Therefore, an explicit catalyst has to be present for a certain reaction to happen. This may be a bit artificial for simple chemistry, but most biological reactions are actively controlled or enhanced by catalysts.

p 56667 q	@	! p(**k**) ⇉ q(**k**) h i	Chemical reaction
p 6<<9 q	@	p 56667 q m q 56667 p	Reversible reaction

For example, the reaction forming a peptide bond between two amino acids (with residues R^1 and R^2) can be written:

R^1R^2PeptideBonding @

R^1-COOH m H_2N-R^2 56667 R^1-CO-HN-R^2 m H_2O

4.3 Example: Compartment Conditions

We can use an appropriate bind-inside&release-inside action to model chemical reactions that are specific to a given compartment; we call these *conditions* of the compartment. For example, certain chemical reactions happen only at a certain acidity, which is a compartment-wide property. An appropriate condition on the membrane of a compartment can represent acidity, and the evolution of conditions and compartments can represent changes of acidity. For example, the merging of a vesicle carrying some reagents with an endosome having a certain acidity condition, can cause the reagents to react after the merge because they find themselves in a compartment with the right acidity condition.

p5J7q @ ! k(p) ⇉ k(q) Condition causing p to change into q

p6L9q @ p5J7q m q5J7p Reversible condition

p5J7qlσhPi Compartment-wide condition affecting P

p5J7qlσhpi } p5J7qlσhqi A condition-driven reaction

4.4 Example: Molecular Pumps and Channels

A plant vacuole is a specialized membrane that stores nutrients, e.g. salt. The breakdown of ATP on the external surface of the vacuole, via a *proton pump*, is used to charge the interior of the vacuole with protons (H^+). In general, several other specialized pumps and channels can be powered by such a charge. In a plant vacuole, a passive (but selective) *ion channel* can let chlorine ions (Cl^-) in, attracted by the

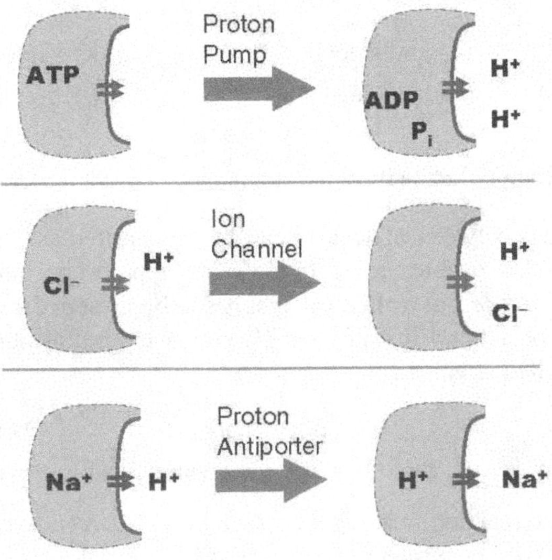

Fig. 10. Molecular Channels

excess electric charge of H^+. Transporting sodium ions (Na^+) inside is more difficult, because those are naturally repelled by the excess charge of H^+. A *proton antiporter*, however, can swap an Na^+ outside with an H^+ inside.

Each pump and channel is represented by a replicated bind&release action. These actions are then assembled as the membrane of an initially empty vacuole.

Plant Vacuole

ProtonPump	@	! ATP(k) ⇉ ADPmP$_i$(H^+mH^+)
IonChannel	@	! Cl^-(H^+) ⇉ k(H^+mCl^-)
ProtonAntiporter	@	! Na^+(H^+) ⇉ H^+(Na^+)
PlantVacuole	@	ProtonPump I IonChannel I ProtonAntiporter h i

This is of course a qualitative representation of the process. Attaching reaction rates to the actions, as in Stochastic π-calculus [12] should yield quantitative modeling. Accurately modeling this situation should be quite interesting, because the reaction rates depend on the concentrations on both sides of the membrane.

4.5 Examples: Molecularly-Triggered Membrane Interactions

Molecular interactions can trigger membrane interactions, simply by sequencing the two kinds of actions on a membrane. In the following example, membrane A produces a molecule that stimulates membrane B to eat A:

Eat Me

A	@	⇉n(k). JhPi
B	@	n(k)⇉. J$^\perp$(ρ)hQi
A m B	}	JhPi m n m n(k)⇉. J$^\perp$(ρ)hQi
	}	JhPi m J$^\perp$(ρ)hQi
	}	hρhhPi i m Qi

Pinocytosis, in reality, may incorporate molecular nutrients into the cell. Our basic pinocytosis operation does not do that, but it can be used as follows to recognize and incorporate external nutrients. Here n is a nutrient molecule, and C is a cell that recognizes it, transports it, and stores it in an internal vesicle.

Seek and Store

seek$_n$	@	!n(k)⇉. G(⇉k(n).mate$_{store}$)
store	@	!mate$^\perp_{store}$
C	@	seek$_n$hstoreh i i
n m C	} }	seek$_n$h ⇉k(n).mate$_{store}$h i m storeh i i
	}	seek$_n$hmate$_{store}$hn i m storeh i i
	}	seek$_n$hstorehn i i

4.6 Example: Viral Infection, Part 2

We can now complete the central part of the virus reproduction cycle, as shown in Figure 11.

In section 3.3, we still had to provide a mechanism for the following reaction:

nucap m cytosol } * nucap^{n} m $\text{envelope-vesicle}^{m}$ m cytosol’

This can be obtained by the following definitions.

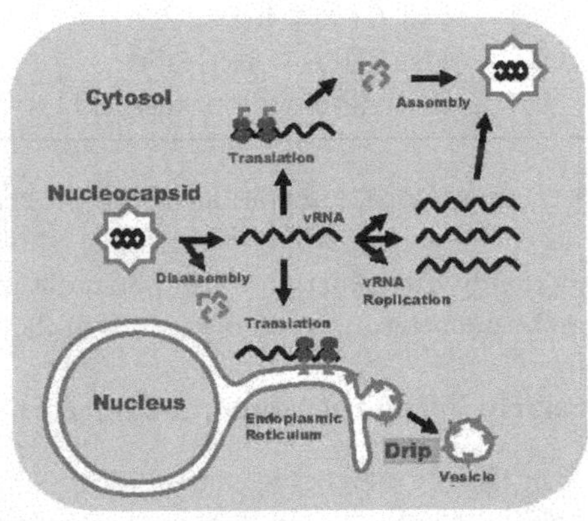

Fig. 11. Nucleocapsid Replication (detail of 0Figure 11)

Nucleocapsid structure

nucap @ capsidhvRNA i

capsid @ !bud I disasm

disasm @ disasm-trigger(vRNA)⇉vRNA(k)

a) vRNA replication **(Figure 11** ***middle-right)***

vRNA-repl @ vRNA 56667 vRNA m vRNA

b) Capsomer translation and nucleocapsid assembly **(Figure 11** ***top-right)***

capsomer-tran @ !vRNA(k)⇉vRNA(k).drip(capsomers)h i

capsomers @ vRNA(k)⇉k(vRNA).capsid

c) Envelope protein translation and transport **(Figure 11** ***bottom-right)***

ER @ !vRNA(k)⇉vRNA(k). drip(K.viral-envelope)hNucleus i

Cytosol contents

cytosol @ endosome m !disasm-trigger
m vRNA-repl m capsomer-tran m ER

A nucap particle is defined as a capsid containing vRNA (we do not model any other content of the capsid, for simplicity). The capsid surface is capable of either budding from the cell (as in section 3.3), or of disassembling the nucap by pushing the vRNA outside the capsid in response to some trigger molecule found in the cytosol (we do not model the fate of the disassembled capsid). There are then three paths that the newly freed vRNA follows:

(a) vRNA is replicated by the standard cellular machinery found in the cytosol:

vRNA-repl m vRNA } vRNA-repl m vRNA m vRNA

(b) The cellular machinery (modeled here by a fictitious empty membrane "capsomer-tran" with an active surface) translates vRNA into capsomer proteins that self-assemble (by dripping) into an entity that inserts vRNA from the cytosol into an empty capsid, hence producing a nucap:

capsomer-tran m vRNA } * capsomer-tran m nucap

(c) The E.R. translates vRNA into viral-envelope proteins that are collected (by dripping) into envelope-vesicles that are ready to merge (K) with the cellular membrane as shown in section 3.3:

ER m vRNA } * ER m vRNA m envelope-vesicle

Finally, the cytosol is defined as containing all the ingredients needed for this process.

The whole reaction then works as follows. By the disassembly of the nucap, we first obtain (where !budh i is the capsid residue):

nucap m cytosol } * cytosol m vRNA m !budh i

Then, the vRNA gets replicated *(a)*, and the cytosol can interact to assemble nucaps *(b)* and produce envelope vesicles *(c)*, obtaining any number of copies n,m,p of the respective components, and some residue:

nucap m cytosol } * nucap^{n} m $\text{envelope-vesicle}^{m}$ m cytosol m vRNA^{p} m !budh i

4.7 Protein Complexes

The handling of protein complexes requires more sophistication in the structure of molecules. See for example the κ-calculus [5], for an expressive notation for molecular complexes that includes state parameters and binding constructs, and that can realistically model protein interaction networks. Our bind&release mechanism and the rewrites of κ-calculus should mutually generalize; we think this is a promising direction for combining complexation with membrane operations. Here we just describe a simple extension of our framework, by adding complex formation, m_1:m_2, between simple molecules:

Molecular Complexes

Systems	P,Q ::= ... ¦ c	systems extended with complexes c
Complexes	c,d ::= m ¦ c:d	basic molecules mBM, or complexation
	p,q ::= c_1m...mc_k	multisets of complexes
Actions	a,b ::= ... ¦ $p_1(p_2)$⇉$q_1(q_2)$	bind&release of complexes
B&R	p_1 m $p_1(p_2)$⇉$q_1(q_2)$.σ|$σ_0$hp_2mP i } q_1 m σ|$σ_0$hq_2mP i	

Then, we can use the bind&release operator to express, e.g. complexation on the inside surface of a membrane:

k(m_1mm_2)⇉k(m_1:m_2)

Protein synthesis in the E.R. has the following structure: membrane bound ribosomes take amino acids (bound to tRNA) from one side of the membrane, and produce complexes (polypeptides) on the other side of the membrane. Hence, decomplexation, membrane-crossing, and complexation are combined in a single process. A completely satisfactory description of this process, though, probably requires either restriction [5], to model the identity of the polypeptide being assembled, or some further notions of complexation with membrane-bound proteins.

5 Extensions

In this section we discuss possible extensions that fit well into the Brane Calculi framework.

5.1 Communication

Although much can be done with purely combinatorial operators, as in the Basic Brane Calculus considered so far, it is possible to add communication operations in the style of CCS or BioAmbients, assuming a substitution τ{p←m} of name m for name p in τ.

On-Membrane Communication (CCS style)

Actions	a,b ::= ... ¦ p2p_n(m) ¦ p2p'_n(m)
peer to peer	: p2p_n(m).σ | p2p'_n(p).τ | ρhPi } σ | τ{p←m} | ρhPi

Cross-Membrane Communication (BioAmbients style)

Actionsa,b ::= ... ¦ s2s_n(m) ¦ s2s'_n(m) ¦ p2c_n(m) ¦ p2c'_n(m) ¦ c2p_n(m) ¦ c2p'_n(m)

to sibling: s2s_n(m).σ\|$σ_0$hPi m s2s'_n(p).τ\|$τ_0$hQi } σ\|$σ_0$hPi m τ{p←m}\|$τ_0$hQi

to child: p2c_n(m).σ\|$σ_0$hp2c'_n(p).τ\|$τ_0$hQi m Pi } σ\|$σ_0$hτ{p←m}\|$τ_0$hQi m Pi

to parent: c2p'_n(p).τ\|$τ_0$hc2p_n(m).σ\|$σ_0$hQi m Pi } τ{p←m}\|$τ_0$hσ\|$σ_0$hQi m Pi

5.2 Choice

A choice operation can be added to membranes:

Choice

Branes	σ,τ ::= ... ¦ σ+τ

Its main impact is that all reaction rules must then consider more complex normal forms for membranes, of the form (a.σ+σ_1)|σ_0hPi instead of a.σ|σ_0hPi. There may be ways to hide this complexity behind appropriate notation, particularly in absence of binding operators.

A good use for choice is to express a shuffle operator a.b.σ+b.a.σ, which is natural when considering individual proteins triggered by two independent binding sites. On the other hand, common forms of choice can be embedded directly in the notation for molecules [5]. Choice at the system level, instead of the membrane level, does not seem very realistic.

Exercise: define (without using choice) a pair of *isolation* actions isl$_n$, isl$'_n$, such that:

$$\text{isl}'_n.\sigma|\sigma_0\text{hPi m isl}_n.\tau|\tau_0\text{hQi } \} * \sigma\text{hki m } \tau|\tau_0\text{h0h}\sigma_0\text{hPi i m Qi}$$

that is, isl$'_n$.σ, when triggered by its co-action, isolates σhki as the only residual, and makes the rest of its membrane and its contents inaccessible. Then, use a pair of isolation actions in parallel to implement a limited form of choice.

5.3 Atonal Transport

Although we have emphasized bitonal operators, there are situations in which simple in-out transport operators, as in BioAmbients [14], may be preferable. One example is when representing a protein with multiple interaction domains as a (fictitious) membrane (see [14] for a detailed discussion). When a protein is represented that way, protein transport in/out of a (real) membrane takes the form of *atonal operations* (ones that do not preserve bitonality). Atonal situations may also arise at higher levels of organization, as when a cell enters the bloodstream through a vessel wall.

The following transport operations are similar to the ones in BioAmbients:

Actions	a,b ::= ... ¦ in$_n$ ¦ in$'_n$ ¦ out$_n$ ¦ out$'_n$
In	in$_n$.σ|σ_0hPi m in$'_n$.τ|τ_0hQi } τ|τ_0hσ|σ_0hPi m Qi
Out	out$'_n$.τ|$\tau_0$hout$_n$.σ|σ_0hPi m Qi } σ|σ_0hPi m τ|τ_0hQi

Alternatively, one can think of adding a single atonal primitive to Phago/Exo/Pino in order to encode In/Out. A simple solution is:

$$\text{wrap}(\sigma).\tau|\tau_0\text{hPi } \} \ \sigma\text{h}\tau|\tau_0\text{hPi i}$$

so that wrap + exo = out, and wrap + phago + exo = in.

It is conceivable that a simple type system may keep the bitonal and atonal parts of a system separate. It is also conceivable that empty membranes (representing molecules) may harmlessly assume a double tonality, violating bitonality only in a weak sense. This could be achieved by restricting In/Out to the empty membrane case:

SmallIn	$in_n.\sigma l\sigma_0 hki\ m\ in^{\mid}_n.\tau l\tau_0 hPi\ \}\ \tau l\tau_0 h\sigma l\sigma_0 hki\ m\ Pi$
SmallOut	$out^{\mid}_n.\tau l\tau_0 hout_n.\sigma l\sigma_0 hki\ m\ Pi\ \}\ \sigma l\sigma_0 hki\ m\ \tau l\tau_0 hPi$

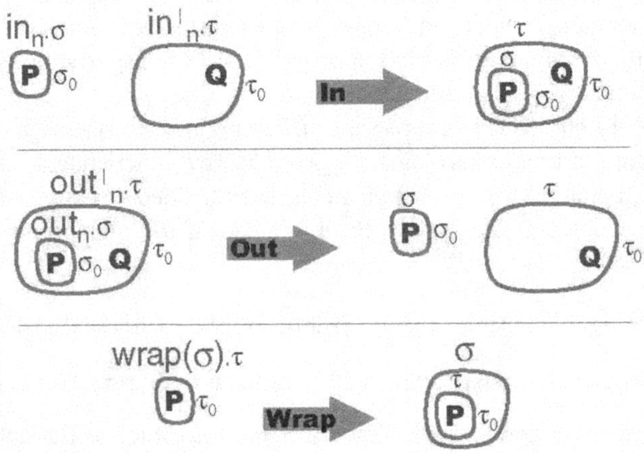

Fig. 12 Atonal Reactions

This way, although k really changes tone in reactions, the systems is consistently bitonal both before and after reactions. Again, a minimal atonal extension could consist of:

SmallWrap	$wrap(\sigma).\tau l\tau_0 hki\ \}\ \sigma h\tau l\tau_0 hkii$

Exercise: show that it is possible to represent small molecules m as empty membranes $mol_m hki$, for an appropriate definition of mol_m, in such a way that an operation similar to bind&release of Section 4.1 is definable. Hints: choice is useful; limit the exercise to sequential bind&release of individual molecules, rather than atomic bind&release of multiple molecules.

5.4 Free-Floating Proteins as Membranes

Free-floating proteins are large molecules with complex dynamic behavior and multiple independent domains of interaction: they can interact with membranes and with each other, and can act as catalysts for smaller molecules. In section 4.7 we have discussed how to model protein complexes directly. It may also seem reasonable to model such large molecules as “small membranes”, that is, as membranes σhi with multiple surface actions but (normally) empty contents.

In this view, a free floating protein inside a membrane is just a membrane inside a larger membrane. This idea and the issues it raises are discussed in [14]. (A different proposal is to assume multi-domain molecules as primitive [4][5].)

One problem with representing molecules as membranes, in general, is that molecules can "squeeze through" membranes or through their channels, while membranes cannot. Situations where large molecules cross membranes are however, limited, and can sometimes be modeled by other mechanisms. One common case is when proteins and RNA cross the nuclear membrane through its pores. The nuclear membrane is a double membrane, so crossing it can be modeled bitonally by Phago and Exo through the lumen. (This is slightly artificial, but an accurate geometrical modeling of the nuclear double membrane and its toroidal pores would in any case require a 3D calculus.)

The problem of complex formation and breaking ([14], Section 3.2) also has a bitonal solution. Assuming proteins are represented as empty membranes σhi, τhi with all their domains on σ and τ, then complexation is simply merging of two such membranes, σlτhi (modulo some interaction). Breakup can be achieved by Pino, to recreate internally the protein fragments, ρhσ$_1$hi mτ$_1$hii followed by Bud to separate them, ρ$_1$hσ$_2$hii mρ$_2$hτ$_2$hii, and finally by two Exo, σhi m τhi.

Enzyme interactions ([14], Section 3.3) also have a bitonal solution for enzymes reacting with proteins (as opposed to small molecules). Two proteins σhi,τhi can bind to an enzyme ρhi by Phago, ρhρ$_1$hσhii mρ$_2$hτhiii, followed by Mate to bring them in contact, ρhρ$_3$hσhi mτhiii, followed by their interaction, e.g. again Mate, ρhρ$_3$hσlτhiii, followed by Exo to release the catalyzed product, σlτhi m ρhi. However, the production of enzymes has to be modeled as the production of membranes, not of molecules, and this might be awkward.

5.5 Bitonal Brane Calculi

While the operations of the Basic Brane Calculus are bitonal in nature (i.e. they preserve the nesting parity of subsystems, with the exception of molecules in bind&release), the calculus framework does not build-in bitonality.

A proper Bitonal Brane Calculus would, instead, adopt a syntax of alternating colored brackets σ$_1$hσ$_2$fσ$_3$hσ$_2$f...gigi, with an assumption that the tone-dual of a reaction is also a reaction. (This could also be achieved by type distinction, instead of syntactic distinctions.) All the figures resulting from such a calculus could be consistently shaded in two alternating tones, and atonal operations like In, Out, or Wrap could not be directly supported because they would violate the alternation.

Exercise: show that a bitonal calculus (with Phago+Exo+Pino and alternating brackets) can emulate the atonal calculus (with Phago+Exo+Pino+Wrap). Hint: double walling.

6 Encoding Brane Calculi

Are Brane Calculi really novel, of can they be easily encoded in other calculi? The obvious comparison is with the closely related BioAmbients Calculus. Let us consider the simplest possible idea for a translation $P^\dagger$ into BioAmbients, namely "in brane" actions (Figure 13).

$$\sigma\langle P\rangle^\dagger \triangleq [\sigma^\dagger \mid P^\dagger]$$

Where the membrane σ is converted into a process inside a membrane [...], at the same level as the translation of P. Consider now the induced translation of Exo:

$$\text{Exo} \qquad K^{\perp}_{n}.\tau|\tau_0\langle K_n.\sigma|\sigma_0\langle P\rangle \circ Q\rangle \rightarrow P \circ \sigma|\sigma_0|\tau|\tau_0\langle Q\rangle$$
$$\text{Exo}^\dagger \qquad [K^{\perp}.\tau^\dagger \mid \tau_0^\dagger \mid [K.\sigma^\dagger \mid \sigma_0^\dagger \mid P^\dagger] \mid Q^\dagger] \rightarrow P^\dagger \mid [\sigma^\dagger \mid \sigma_0^\dagger \mid \tau^\dagger \mid \tau_0^\dagger \mid Q^\dagger]$$

Where we would have to devise an appropriate definition for $K^{\perp}.\tau^\dagger$ and $K.\sigma^\dagger$ so that $\text{Exo}^\dagger$ had the prescribed behavior. A problem, though, is already apparent. The Exo rule separates P from $\sigma|\sigma_0$ on the r.h.s., and it can do so because the separation between $\sigma|\sigma_0$ and P is built into the term $K_n.\sigma|\sigma_0\langle P\rangle$ on the l.h.s.. In $\text{Exo}^\dagger$, though, the process $K.\sigma^\dagger \mid \sigma_0^\dagger \mid P^\dagger$ on the l.h.s. is a featureless composition; how does the rule "know" to split off $P^\dagger$ precisely at that position?

To avoid this loss of structure, it is necessary to put more structure in the translation:

$$\sigma\langle P\rangle^\dagger \triangleq [[\sigma^\dagger] \mid P^\dagger] \qquad \text{or:}$$
$$\sigma\langle P\rangle^\dagger \triangleq [\sigma^\dagger \mid [P^\dagger]] \qquad \text{"Ball bearing" encoding}$$

This requires more complicated encodings of operations, which need to cross multiple level of brackets and therefore have atomicity problems.

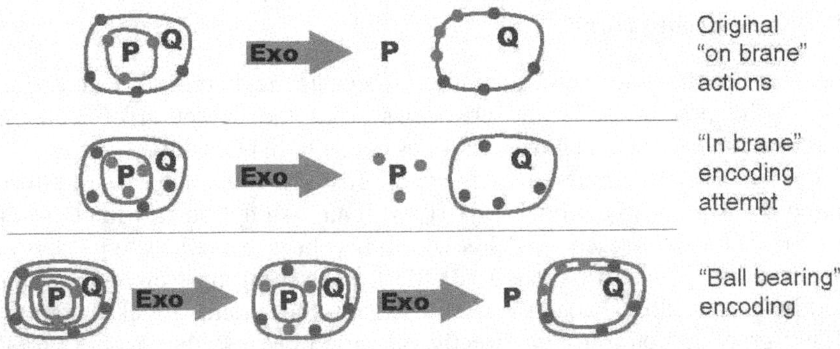

Fig 13. Exo Encodings

Our suspicion is that an encoding of Brane Calculi in Ambients-like calculi may be possible, but it is not easy and almost certainly not practically usable.

7 Conclusions

How are "bio"-calculi different from other process calculi? Both in Brane Calculi and in BioAmbients, (and in BioSPI [13], before that), we have used standard concepts and techniques developed for calculi of concurrency and mobility. We believe that Brane Calculi are beginning to confront some of the pragmatic issues discovered with BioAmbients, by emulating more closely biological processes, in the same way that BioAmbients removed the need for some artificial encodings in BioSPI.

The issue of choosing "realistic" primitives is a tricky one. At one extreme, only the precise mechanisms that have an existing biological implementation are realistic, and those usually have extremely sophisticated and still only partially understood molecular-level implementations. However, even without understanding the molecular details, it is possible to distinguish operations that work via dedicated molecular machinery from those that do not. At the other extreme, biological systems have general constraints and invariants that determine which operations are at least in principle realistic (and which are not). Membrane orientation is one such invariant: it is actively maintained by living cells by consistently orienting proteins on the membrane surface. Bitonality is another invariant, at least in some regimes of operation; it derives from certain transformations of oriented membranes that produce deeper nestings: the basic bitonal structure of a cell and its organs is due to such transformations that happened during evolution ([1] p. 556). These biological invariants suggest a different set of "potentially realistic" basic operations for concurrent calculi than ones that had been considered before.

Another basic aspect of biological membranes is their nature as a two-dimensional fluid embedded in a three-dimensional fluid; this is in fact more fundamental than any orientability or bitonality considerations. This means that there are at least two commutative monoids involved, and not the single one usually seen in process calculi. The formalization of these two monoids adds complexity, but supports the notion of computation *on* the membrane, that is, of computation that is directly aware of conditions on both sides of the membrane. Trying to emulate this fluid-in-fluid structure by other encodings is awkward (see Section 6), although the issue has been valiantly confronted in BioAmbients.

References

[1] B.Alberts, D.Bray, J.Lewis, M.Raff, K.Roberts, J.D.Watson. Molecular Biology of the Cell. Third Edition, Garland.

[2] L.Cardelli. Bitonal Membrane Systems – Interactions of Biological Membranes. To appear.

[3] L.Cardelli and A.D.Gordon. Mobile Ambients. Theoretical Computer Science, Special Issue on Coordination, D. Le Métayer Editor. Vol 240/1, June 2000. pp 177-213.

[4] V.Danos, M.Chiaverini. A Core Modeling Language for the Working Molecular Biologist. 2002.

[5] V.Danos and C.Laneve. Formal Molecular Biology. Theoretical Computer Science, to Appear.

[6] L.H.Hartwell, J.J.Hopfield , S.Leibler , A.W.Murray: From molecular to modular cell biology. Nature. 1999 Dec 2;402(6761 Suppl):C47-52.

[7] H. Kitano: A graphical notation for biochemical networks. BIOSILICO 1:169-176, 2003.

[8] K.W. Kohn: Molecular Interaction Map of the Mammalian Cell Cycle Control and DNA Repair Systems. Molecular Biology of the Cell, 10(8):2703-34, Aug 1999.

[9] H.Lodish, A.Berk, S.L.Zipursky, P.Matsudaira, D.Baltimore, J.Darnell. Molecular Cell Biology. Fourth Edition, Freeman.

[10] H.H.McAdams, A.Arkin : It's a noisy business! Genetic regulation at the nanomolar scale. Trends Genet. 1999 Feb;15(2):65-9.

[11] R.Milner. Communicating and Mobile Systems: The π-Calculus. Cambridge University Press, 1999.

[12] C.Priami. The Stochastic pi-calculus. The Computer Journal 38: 578-589, 1995.

[13] C.Priami, A.Regev, E.Shapiro, and W.Silverman. Application of a stochastic name-passing calculus to representation and simulation of molecular processes. Information Processing Letters, 80:25-31, 2001.

[14] Regev, E.M.Panina, W.Silverman, L.Cardelli, E.Shapiro. BioAmbients: An Abstraction for Biological Compartments. Theoretical Computer Science, to Appear.

[15] Systems Biology Markup Language. <www.sbml.org>

Author Index

Lecture Notes in Bioinformatics

Vol. 3388: J. Lagergren (Ed.), Comparative Genomics. VII, 133 pages. 2005.

Vol. 3370: A. Konagaya, K. Satou (Eds.), Grid Computing in Life Science. X, 188 pages. 2005.

Vol. 3318: E. Eskin, C. Workman (Eds.), Regulatory Genomics. VII, 115 pages. 2005.

Vol. 3240: I. Jonassen, J. Kim (Eds.), Algorithms in Bioinformatics. IX, 476 pages. 2004.

Vol. 3082: V. Danos, V. Schachter (Eds.), Computational Methods in Systems Biology. IX, 280 pages. 2005.

Vol. 2994: E. Rahm (Ed.), Data Integration in the Life Sciences. X, 221 pages. 2004.

Vol. 2983: S. Istrail, M.S. Waterman, A. Clark (Eds.), Computational Methods for SNPs and Haplotype Inference. IX, 153 pages. 2004.

Vol. 2812: G. Benson, R.D.M. Page (Eds.), Algorithms in Bioinformatics. X, 528 pages. 2003.

Vol. 2666: C. Guerra, S. Istrail (Eds.), Mathematical Methods for Protein Structure Analysis and Design. XI, 157 pages. 2003.